Nature and Nurture During Middle Childhood

To the memory of Steven G. Vandenberg

Nature and Nurture During Middle Childhood

John C. DeFries

Robert Plomin

David W. Fulker

BLACKWELL
Oxford UK & Cambridge USA

First published 1994

Blackwell Publishers
238 Main Street,
Cambridge, Massachusetts 02142
USA

108 Cowley Road
Oxford OX4 1JF
UK

Library of Congress Cataloging-in-Publication Data
Nature and nurture during middle childhood / John C. DeFries, Robert Plomin, David W. Fulker
   p. cm.
  Includes bibliographical references and indexes.
  ISBN 1–55786–393–8 (acid-free paper)
  1. Nature and nurture—Longitudinal studies. 2. Individual differences in children—Longitudinal
studies. 3. Children, Adopted—Colorado—Psychology—Longitudinal studies. 4. Colorado
Adoption Project. I. DeFries, J. C., 1934– . II. Plomin, Robert, 1948– . III. Fulker,
David W., 1937– .
  BF341.N374 1994 155.42′48234—dc20 93–22617
                            CIP

British Library Cataloguing in Publication Data
A CIP catalogue record for this book is available from the British Library.

Typeset in 10 on 12 pt Ehrhardt by Pure Tech Corporation, India
Printed in Great Britain by T. J. Press Ltd, Padstow, Cornwall

This book is printed on acid-free paper

Contents

List of Figures

List of Tables

List of Collaborators

Laura A. Baker   University of Southern California

Julia M. Braungart   University of Notre Dame

Lon R. Cardon   Stanford University

Stacey S. Cherny   University of Colorado, Boulder

Robin P. Corley   University of Colorado, Boulder

Judy Dunn   The Pennsylvania State University

Susan Felsenfeld   University of Pittsburgh

Scott L. Hershberger   United States International University, San Diego

Hsiu-zu Ho   University of California, Santa Barbara

Shirley McGuire   University of California, San Diego

Jenae M. Neiderhiser   The Pennsylvania State University

Stephen A. Petrill   Case Western Reserve University

Richard Rende   Columbia University

Chandra Reynolds   University of Southern California

Sally-Ann Rhea   University of Colorado, Boulder

Stephanie Schmitz   University of Colorado, Boulder

Clare Stocker   University of Denver

Lee Anne Thompson   Case Western Reserve University

Sally J. Wadsworth   University of Colorado, Boulder

Keith E. Whitfield   The Pennsylvania State University

Preface

Although dramatic life transitions occur during middle childhood, surprisingly little is known about the origins of individual differences in behavioral development at this age. The goal of this book is to contribute toward a better understanding of individual differences during this important developmental epoch. First, we review what little is known about the genetic and environmental provenances of development in middle childhood. Then, we attempt to fill in some of the gaping holes in our knowledge with new behavioral genetic results obtained from a large-scale, longitudinal study, the Colorado Adoption Project (CAP).

Beginning in 1975, CAP investigators studied the children, parents (including the biological parents of the adopted-away children), and home environments of 245 adoptive families and 245 matched nonadoptive families when the children (and their younger siblings) were 1, 2, 3, and 4 years of age. The children were studied in the laboratory at 7 years of age, and were administered telephone tests and interviews at 9 and 10 years. Questionnaire data were collected at 5, 6, and 8 years. The CAP design facilitates analyses of genetic and environmental influences in development via both parent–offspring and sibling comparisons. The parent–offspring comparisons include "genetic" parents (biological parents and their adopted-away offspring), "environmental" parents (adoptive parents and their adopted children), and "genetic-plus-environmental" parents (nonadoptive parents and their children). The sibling design includes nonadoptive siblings (biological siblings reared together in nonadoptive families) and adoptive siblings (genetically unrelated children reared in the same adoptive homes). This dual parent–offspring and sibling adoption design, combined with its longitudinal and multivariate assessments, makes the CAP uniquely suited to broach issues of nature and nurture in development.

This is the third in a projected five-book series describing the CAP results. The first two books presented results in infancy (Plomin & DeFries, 1985) and early childhood (Plomin, DeFries, & Fulker, 1988). A fourth book is planned for early adolescence, when complete CAP results at 9 and 10 years and new results at 11 and 12 years will be presented. The fifth book will focus on adolescence, including 13, 14, and 15 years, and a laboratory test session at 16 in which the children are administered the same battery of tests that their parents completed over a decade and half earlier.

As we approach the 20th anniversary of the CAP, a particularly pleasing feature of this book is that it was written in collaboration with colleagues, staff, and present and former students who have been involved in the CAP over the years. This strategy has greatly broadened the perspective of this book and added to the expertise brought to bear on middle childhood. However, this is not just another edited volume. We are all part of a team committed to the CAP and, in this sense, all of the contributors could be co-authors of all of the chapters. All have contributed to the conceptual, methodological, and substantive developments that made this work possible – not to mention conducting a total of 9,784 test sessions during the past 17 years! We gratefully acknowledge the contributions of the talented people with whom we have had the privilege to work and look forward to our continued collaboration with them. We also wish to thank the CAP testers (Bo Bishop, Leza Clymer, Kim Corley, Lara Cunning, Beth Landt, and Diane Perry), our data manager (Annie Johnson), the administrative and office staff at the Institute for Behavioral Genetics (Agnes Conley, Dianne Johnson, Martha Norton, Lee Nickerson, and Jeri Titchenal), and the scores of other individuals who have contributed to the success of this long-term longitudinal study. We are especially grateful to staff members of the Lutheran Family Services and the Denver Catholic Community Services, whose cooperation made the CAP possible, and to the hundreds of families who have endured our yearly intrusions into their lives with such good will and grace.

The book begins with chapters that review behavioral genetic research in middle childhood (Chapter 1), provide an overview of the CAP (Chapter 2), and describe adoption design methodology (Chapter 3). The other chapters are more empirically oriented and present new findings on the following topics: longitudinal analyses of change and continuity of general cognitive ability (Chapter 4), multivariate and developmental analyses of specific cognitive abilities (Chapter 5), predictions from infant cognition (Chapter 6), school achievement (Chapter 7), speech and language disorders (Chapter 8), personality and temperament (Chapter 9), perceptions of self-competence (Chapter 10), behavioral problems and stress of entering school (Chapter 11), body size and obesity (Chapter 12), motor development (Chapter 13), sex differences (Chapter 14), nonshared environment and sibling differences (Chapter 15), family relationships (Chapter 16), genetic influences on "environmental" measures (Chapter 17), early family environment and outcomes in middle childhood (Chapter 18), home environment and cognitive development (Chapter 19), genotype–environment interaction and correlation (Chapter 20), and applied issues relevant to adoption (Chapter 21). A final chapter summarizes and integrates the findings presented in the previous chapters and outlines future research directions that may provide a better understanding of nature and nurture during middle childhood.

We are grateful for the continuous support of the National Institute of Child Health and Human Development since 1977 for the collection of CAP data at 1–4,

7, and 13–16 years of age (HD-10333 and HD-18426). Beginning in 1988, the testing of CAP children during early adolescence has been supported by a grant from the National Institute of Mental Health (MH-43899). Since 1978, the National Science Foundation has awarded grants (BNS-7826204, BNS-8200310, BNS-8505692, BNS-8643938, BNS-8806589, and BNS-9108744) that enabled us to assess mother–child and sibling interactions. The CAP was launched in 1976 with the aid of funds from the University of Colorado's Biomedical Research Support Grant and a small grant from the National Institute of Mental Health (MH-28076). The William T. Grant Foundation supported the project during 1976–1979, and launched the testing of CAP children at 7 years of age in 1983. The Spencer Foundation provided support from 1982 to 1984 for the purpose of testing younger adopted and nonadopted siblings of the probands at 5 and 7 months of age, and from 1985 to 1988 for the extension of CAP testing into early adolescence. Finally, our research and thinking about nature and nurture during childhood has profited immensely from our participation in the Early Childhood Transitions Research Network of the John D. and Catherine T. MacArthur Foundation. The research reported in Chapter 6 was also supported by grants from NSF (BNS-7826202) and NICHD (HD-19802). Lee Anne Thompson was supported by NICHD training grant HD-07289 during data collection and by HD-21947 and MH-46512 while the chapter was written. Stephen A. Petrill and Sally J. Wadsworth were supported by NICHD and NIMH training grants HD-07176 and MH-16880, respectively.

We dedicate this book to the memory of Steven G. Vandenberg, who died on August 27, 1992. Professor Vandenberg, our beloved colleague at the Institute for Behavioral Genetics, contributed generously to the planning of CAP. Before coming to the University of Colorado in 1967, he served as director of the other major longitudinal behavioral genetic study of development, the Louisville Twin Study. In 1970, he and the senior author of this book co-founded the journal, *Behavior Genetics*. His prolific writings and seminal twin studies, dating back to the 1950s, support our designation of Professor Vandenberg as the father of the modern era of human behavioral genetics.

1 Nature and Nurture in Middle Childhood

Middle childhood is the stage during which children begin to adapt to life outside the family, especially to the multifaceted demands of school and relationships with peers. Major developmental shifts occur. Most notably, cognitive reorganization is marked by the emergence of concrete operations. Dramatic changes also occur in children's social cognitions and interpersonal behavior. Parents, teachers, and peers have increased expectations regarding social behavior, affect regulation, and self-control. Moreover, behavioral problems at this age begin to be predictive of later psychopathology.

In recent testimony before the US House and Senate Appropriations Committees, Alan Kraut, Executive Director of the American Psychological Society, noted the paucity of research pertaining to middle childhood as follows:

... many problems of adolescence and young adulthood – problems of school dropouts, unwanted pregnancies, gangs, alcohol and drug abuse, and AIDS, among others – have their roots in the middle childhood years. We need to know about the development of a whole series of middle childhood skills dealing with decision making, resolving conflicts, fighting off peer pressure, building self-confidence, and many others, including traditional academic functioning, if we are to legitimately address these problems. The middle childhood years – 5 to 11 – are just those least understood by our nation's developmental researchers. (1992, p. 7)

Because middle childhood is such an interesting and important developmental period, it is surprising that so little research has focused on it. The purpose of this chapter is to provide an overview of previous behavioral genetics research during middle childhood. We begin with the domains that have been studied most throughout the life-span: cognitive abilities, personality, and psychopathology. Next we mention other domains of middle childhood that are explored for the first time in the Colorado Adoption Project (CAP). Finally, in the last part of the chapter we discuss developmental, multivariate, and environmental analyses which are the foci of several chapters in this volume.

General Cognitive Ability

The classic adoption study by Skodak and Skeels (1949) suggested an increase in genetic influence on cognitive ability from 4 to 7 years of age. Longitudinal testing of IQ in the Louisville Twin Study also provided evidence that heritability increases from early to middle childhood (Wilson, 1983). It is especially noteworthy that the results of both studies suggest that genetic influence on IQ in the early school years is nearly as strong as genetic influence in adolescence and adulthood.

Results obtained from other adoption studies in middle childhood (Burks, 1928; Fisch, Bilek, Deinard, & Change, 1976; Leahy, 1935) and two more recent cross-sectional adoption studies (Horn, Loehlin, & Willerman, 1979; Scarr & Weinberg, 1977) are also consistent with this conclusion. Especially strong evidence for genetic influence comes from the two studies that provide direct tests of genetic influence in the form of resemblance between biological mothers and their adopted-away offspring (Horn et al., 1979; Skodak & Skeels, 1949). Adoption studies that rely on the indirect comparison between familial correlations in nonadoptive and adoptive families are somewhat less consistent, especially the few comparisons using the sibling adoption design rather than parent–offspring comparisons (Plomin & Loehlin, 1989).

Other smaller, cross-sectional twin studies of IQ in middle childhood tend to confirm the results of the longitudinal Louisville Twin Study (Koch, 1966; Segal, 1986). However, two twin studies in middle childhood using the Ravens Coloured Progressive Matrices for Children as an index of IQ found little evidence for heritability (Garfinkle & Vandenberg, 1981), probably due to low reliability of the measure at this age (Foch & Plomin, 1980; Knaack, 1978).

In summary, it seems fairly well established that genetic influences contribute substantially to individual differences in IQ by the early school years. Previous CAP analyses have focused on the development of general cognitive ability during infancy and early childhood. These results, and new analyses of middle childhood, are summarized in Chapter 4.

Specific Cognitive Abilities

There is certainly more to cognition than IQ. Behavioral geneticists have assessed broad factors of specific cognitive abilities such as verbal ability, spatial ability, perceptual speed, and memory (DeFries, Vandenberg, & McClearn, 1976; Plomin, 1988). Specific cognitive abilities are especially interesting during middle childhood. Although the antecedents of such factors emerge during early childhood, factors similar to those observed in adulthood can be assessed by the early school years. The two twin studies in middle childhood that reported results for IQ subtests (Segal, 1986; Wilson, 1975, 1986) both provide evidence for substantial genetic influence on verbal ability. The results for the performance subtests are more discrepant, although performance IQ showed substantial genetic influence in both studies. In these two twin studies, subtest profiles were analyzed using trend correlations to assess patterns of cognitive strengths and weaknesses. Genetic influence was found in both studies for these trend correlations.

Studies that focus on factors of specific cognitive abilities, rather than on subtests from tests designed to assess IQ, are more illuminating. Although the twin studies in middle childhood involve small sample sizes, they suggest genetic influence on most specific cognitive abilities. For example, the results of a reanalysis of Primary Mental Abilities scores obtained by Koch (1966) for 5- to 7-year-old twins (Plomin & Vandenberg, 1980) suggest substantial genetic influence on verbal and spatial abilities, but less on perceptual speed. Two other twin studies in middle childhood, designed to assess specific cognitive abilities, obtained evidence for strong genetic influence on verbal ability, but little genetic influence on tests of memory (Foch & Plomin, 1980; Garfinkle & Vandenberg, 1981). A recent twin study, the Western Reserve Twin Project (WRTP), investigated specific cognitive abilities during the early school years for a sample of 146 pairs of monozygotic (MZ) and 132 pairs of dizygotic (DZ) twins from 6 to 12 years of age (Thompson, Detterman, & Plomin, 1991). Eight tests from the CAP battery of specific cognitive abilities were employed to assess the group factors of verbal ability, spatial ability, perceptual speed, and memory. This study yielded heritabilities of about 70% for verbal and spatial abilities and perceptual speed, and a lower heritability (45%) for memory. From computer-based testing of information-processing tasks, WRTP investigators have also collected extensive data which are currently being analyzed.

In summary, although only tentative conclusions can be drawn, evidence for genetic influence is strongest for measures of verbal ability, somewhat less for spatial ability, much less for perceptual speed, and weakest for memory. It is interesting that research in adolescence and adulthood shows a similar pattern of heritability (DeFries et al., 1976; Plomin, 1988). CAP analyses of specific cognitive abilities in early and middle childhood are discussed in Chapter 5.

Personality

Personality includes diverse domains of behavior that are not primarily cognitive in nature, for example, overt behaviors such as activity level, feelings such as emotionality, preferences such as sociability, and attitudes such as traditionalism. Parent and tester ratings have been widely used to assess temperament in infancy and early childhood, whereas self-report questionnaires have been extensively employed in twin studies of personality during adolescence and adulthood (Goldsmith, 1993). In contrast, very few genetic studies of personality have been conducted on middle childhood using either ratings or self-reports (Loehlin, 1992; Plomin, 1986).

There have been a few small twin studies using parental ratings in middle childhood that generally suggest genetic influence (Matheny & Dolan, 1980; Scarr, 1969; Willerman, 1973). Two studies in middle childhood employed personality ratings by testers and also obtained some evidence for genetic influence (Goldsmith & Gottesman, 1981; Scarr, 1969). However, two studies that used observational measures reported mixed results concerning the heritable nature of personality traits (Plomin & Foch, 1980; Scarr, 1966).

Two adoption studies obtained personality data using self-report questionnaires for children whose average age was middle childhood, but who ranged in age from 4 to 16 years in one study (Scarr, Webber, Weinberg, & Wittig, 1981) and from 3 to 26 in the other (Loehlin, Horn, & Willerman, 1981; Loehlin, Willerman, & Horn, 1982, 1985). Parent–offspring comparisons in these studies yielded little evidence for genetic influence. Both studies reported correlations for adoptive siblings near zero, suggesting that shared environmental influence is also negligible. Both studies also included small samples of nonadoptive siblings (40 pairs and 24 pairs, respectively); however, these data yielded mixed results, generally suggesting little genetic influence.

Chapter 9 presents CAP analyses that assess genetic influence on personality in middle childhood using both parent–offspring and sibling adoption designs. Personality measures included ratings of sociability, emotionality, activity, and attention span/persistence by parents, testers, and teachers, and for older children, self-report.

Psychopathology

In recent years, there has been a marked increase in interest in the genetics of childhood psychiatric disorders. This work has been reviewed by Rutter and colleagues (1990) and will not be discussed here because the focus of the CAP is normal behavioral development. However, in contrast to diagnosed psychiatric disorders, behavioral problems are so common in middle childhood that they can be considered part of normal development. A twin study of behavioral problems in middle childhood found substantial genetic influence on commonly occurring externalizing problems such as aggressiveness and hyperactivity, as well as internalizing problems such as anxiety, based on parental ratings (O'Connor, Foch, Sherry, & Plomin, 1980). A recent twin study replicates this finding, especially for externalizing problems, using parental ratings on the Child Behavior Checklist (Edelbrock, Rende, Plomin, & Thompson, in press). A British twin study also obtained evidence for substantial genetic influence for hyperactivity and attentional syndromes based on both parent and teacher ratings (Goodman & Stevenson, 1989a, b). These and other ongoing twin studies suggest that the strong familial resemblance found for hyperactivity-related syndromes in family studies is largely genetic in origin (reviewed by Deutsch & Kinsbourne, 1990). They also support the conclusion of genetic influence on hyperactivity reached in two earlier adoption studies using retrospective reports by parents of their own childhood hyperactivity (Cantwell, 1975; Morrison & Stewart, 1973) and in a study of half-siblings (Safer, 1973).

Chapter 11 in this volume reports the results of the first adoption study of behavior problems and their relation to stress.

Other Domains

Next to nothing was known prior to the CAP about other domains of development in middle childhood such as school achievement, speech and language disorders, interests, change in body size and risk of obesity, motor development, self-competence and confidence, and family relationships. Each of these areas in middle childhood is addressed for the first time by a chapter in this volume.

Other Analyses

The truism that both genes and environment are required for development does little to describe or explain behavior. In contrast, assessing the etiology of individual differences provides an important first step toward the understanding of behavioral development. For example, genetic differences among children may not contribute importantly to individual differences for some observed or measured characters. Although examples of this are difficult to find, individual differences in neonatal temperament appear to be due primarily to nongenetic influences (Riese, 1990). Alternatively, it is possible that genetic differences can be largely responsible for observed differences. This is the case for height, for which heritability is very high; at least 80%. However, individual differences in the vast majority of behavioral phenotypes are due substantially to both genetic and environmental influences. As indicated in the foregoing review, this conclusion applies to the three major domains that have been investigated most thoroughly in adulthood – cognitive abilities, personality, and psychopathology – and it is likely that this conclusion will also apply to middle childhood.

Although assessing the genetic and environmental etiologies of individual differences is an important first step in the analysis of behavioral development, it is *only* a first step. Developmental, multivariate, and more detailed environmental analyses are important further steps.

Developmental analysis

The earliest studies of twins were explicitly developmental. For example, the first study of twins, reported by Francis Galton in 1876, investigated whether twins became more or less similar during development. A renascence of this developmental perspective in behavioral genetic research has occurred in recent years.

Two major types of developmental questions can be asked. First, do genetic and environmental components of variance change in their relative magnitude? For example, does heritability – the portion of phenotypic variance that can be ascribed to genetic influences – change with development? It is generally assumed at least implicitly – and explicitly in the former Soviet Union (Mangan, 1982) – that environmental factors play an increasingly important role during development. However, results obtained to date indicate the reverse: If heritability changes during development, it increases (Plomin, 1986). The best example is cognitive development, which shows a nearly linear increase in genetic influence from infancy through early childhood (Fulker, DeFries, & Plomin, 1988), perhaps continuing throughout the life-span (McGue, Bouchard, Iacono, & Lykken, 1993).

Developmental changes in environmental components of variance are also interesting. Again, the best example involves cognitive development. Shared family environment – those environmental influences that cause differences between families and similarities of siblings who grow up in the same family – appears to account for as much as 25% of the variance in childhood, but diminishes to negligible levels after adolescence when children begin to leave the family emotionally and, eventually, physically (Plomin, 1988).

The second type of question addresses genetic and environmental contributions to developmental change and continuity. Regardless of the magnitude of heritability at two ages, to what extent do genetic effects at one age also affect the other age? Although much less is known about this issue, it appears that cognitive development shows a surprising degree of genetic continuity from infancy through adulthood (DeFries, Plomin, & LaBuda, 1987). For personality, genetic contributions to change as well as continuity are likely, especially earlier in life (Plomin et al., 1993; Plomin & Nesselroade, 1990).

Most of the chapters in this volume employ CAP data from infancy through middle childhood for the purpose of considering nature–nurture issues from a developmental perspective. Chapter 4 presents the most recent CAP attempts to assess genetic and environmental contributions to developmental change and continuity for general cognitive ability.

Multivariate analysis

Another important step beyond the basic nature–nurture question is multivariate genetic analysis, which extends the univariate genetic analysis of the variance of a single trait to the analysis of the covariance between traits. Multivariate genetics facilitates the analysis of the extent to which individual differences in different characters are due to the same genetic and environmental influences. For example, do the genes that influence verbal ability also influence spatial ability? From a genetic perspective, a multivariate approach is important because it is highly unlikely that completely different sets of genes affect all of the behaviors that we investigate. As indicated in the next chapter, the CAP has been multivariate from the start, risking approbation as a "fishing expedition" in order to fulfill its goal of broadly assessing behavioral development. The following chapters in this volume attest to the value of this decision by reporting analyses of these extensive CAP measures. Several chapters are explicitly multivariate in nature. Chapter 5 presents state-of-the-art multivariate genetic model-fitting analyses of specific cognitive abilities. Other multivariate genetic analyses described in this volume include analyses of associations between school achievement and cognitive abilities (Chapter

7), between stress and behavioral problems (Chapter 11), and between mental and motor development (Chapter 13).

Environmental analysis

A third new direction for genetic research is the analysis of specific environmental issues within the context of a behavioral genetics design. Environmentalists are beginning to realize that it is necessary to use genetic designs in order to assess environmental influences (e.g., Wachs, 1993). In a corresponding manner, behavioral geneticists are beginning to see that genetic research is enhanced by including specific measures of the environment in their genetic designs (e.g., Plomin & Rende, 1991).

Two major themes have emerged that have profound importance for understanding environmental influences on development. The first is nonshared environment. In addition to providing the best available evidence for environmental influences, genetic research has led to an important discovery about the processes by which environmental influences affect development. For decades, it had been widely and quite reasonably assumed that the family environment was highly efficacious during early development. This would explain why children in the same family resemble one another – they grew up in the same family. However, we now know that genetic covariation between relatives accounts for most of this similarity, at least for most personality measures. For example, first-degree relatives adopted apart are just as similar as first-degree relatives living together. Growing up in the same family does not add to their resemblance. Moreover, pairs of genetically unrelated individuals reared together (adoptive sibling pairs) correlate negligibly on both self-report personality questionnaires and cognitive tests after adolescence. Thus, for such measures, shared rearing environment does not produce similarity.

Results such as these indicate that the most important environmental influences are those which are not shared by children growing up in the same family. This leads to the question, why are children in the same family so different? In order to answer this question, developmentalists must study more than one child per family, using measures of the environment specific to each child (Dunn & Plomin, 1990; Plomin & Daniels, 1987). As discussed in Chapter 16, this strategy makes it possible to assess the ways in which siblings in the same family experience different environments, and to determine if differences in siblings' experiences are related to differences in developmental outcomes.

The second theme to emerge from environmental studies that employ behavioral genetic designs is that widely used measures of the environment show genetic influence when treated as dependent measures (Plomin & Bergeman, 1991). Finding genetic influence on environmental measures suggests that these measures assess genetically influenced characteristics of parents or children. For example, in developmental psychology, environmental measures typically assess parental behavior, such as time spent reading to children. Differences among parents in the amount they read to their children could in part be due to genetically influenced characteristics of the parents, such as their facility and interest in language.

Alternatively, parental behavior might reflect some characteristics of their children. About a dozen previous genetic analyses of environmental measures, including a recent study focused on this issue (Plomin, Reiss, Hetherington, & Howe, in press), document ubiquitous genetic influence on ostensible measures of family environment. Results of these studies suggest two general questions for future developmental research (Plomin & Neiderhiser, 1992). First, what are the mechanisms by which heredity affects measures of the environment? Second, if genetics contributes to environmental measures as well as to developmental outcomes, is it possible that associations between environmental measures and developmental outcomes are mediated genetically? This second question lies at the heart of the interface between nature and nurture: The ways in which people interact with their environment, their experiences, may be influenced by genetic differences.

Several chapters in this volume report the results of environmental analyses of CAP data during middle childhood. Chapter 15 attempts to identify specific sources of nonshared environment. Chapter 17 examines genetic influences on "environmental" measures used in the CAP from infancy to middle childhood and also investigates the etiology of correlations between these measures and antecedent and contemporaneous measures of temperament. Chapter 18 assesses genetic contributions to longitudinal relationships between early environment and outcomes in middle childhood. Genetic mediation of associations between home environment and general cognitive ability is the focus of Chapter 19. Finally, Chapter 20 addresses issues of genotype–environment interaction and correlation.

In the following chapter, the history of the CAP will be reviewed and its design and sample will be characterized.

2 The Colorado Adoption Project

Just 20 years ago, the social and behavioral sciences were beginning to awaken from decades of environmentalism; however, these stirrings did not lead to a welcoming embrace for the nascent field of human behavioral genetics. The intemperate response to Arthur Jensen's 1969 paper, which had broached the topic of genetic differences between ethnic groups, tarred genetic research on individual differences with the same brush and nearly led to its demise. Criticism of genetic research on individual differences then began to focus on alleged fakery in Sir Cyril Burt's twin research on intelligence (Kamin, 1974), leading to an apparently authoritative indictment by a respected historian (Hearnshaw, 1979). Although that indictment is now being questioned (Fletcher, 1990; Joynson, 1989), the attempts to denigrate earlier research in behavioral genetics paradoxically had a positive effect on the field. Instead of arguing that the case had already been made for the importance of heredity in behavioral development, behavioral geneticists designed and initiated new studies that could address issues of nature and nurture more definitively.

The Colorado Adoption Project (CAP) was conceived by Robert Plomin and John C. DeFries during this turmoil of the early 1970s. Our colleague, Steven G. Vandenberg, to whom this book is dedicated, participated in its planning. In 1980, David W. Fulker began a collaborative analysis of the early CAP data; then, after joining the faculty of the Institute for Behavioral Genetics, University of Colorado, in 1983, he became a CAP co-investigator.

The purpose of this chapter is to provide an overview of the CAP, including its design, sample, and measures. More details about these topics can be found in previous CAP books on infancy (Plomin & DeFries, 1985) and early childhood (Plomin, DeFries, & Fulker, 1988).

Design

By 1974, it had become apparent to us (see Plomin & DeFries, 1985) that a full adoption study of behavioral development was needed to complement the previous twin studies. The standard twin design compares the similarity of identical twins and fraternal twins. It tests for the effect of the twofold greater genetic similarity of identical as compared to fraternal twin pairs. If heritable influences cause individual differences for a particular aspect of behavioral development, pairs of identical twins should be more similar for the trait than fraternal twins.

The full adoption design provides a very different way to assess the relative impact of genetic and environmental influences. Parents who rear their own biological children share both heredity and family environment with them. Thus, resemblance between parents and their offspring in such nonadoptive families cannot be ascribed unambiguously either to hereditary or to family environmental influences.

The full adoption design is a quasi-experimental design that separates the two sources of resemblance between nonadoptive ("genetic-plus-environmental") parents and offspring. Biological parents who relinquish their offspring for adoption at birth are "genetic" parents, and adoptive parents who adopt children early in life are "environmental" parents. By comparing the resemblance between biological parents and adopted children to that between adoptive parents and adopted children, the full adoption design can disentangle the possible genetic and environmental sources of resemblance for nonadoptive parents and their children. Even the most vociferous critics of behavioral genetics acknowledge the power of the adoption design (e.g., Kamin, 1981; Lewontin, 1975). Although the full adoption design requires only the two groups of genetic parents and environmental parents, inclusion of a comparison group of nonadoptive parents and their children adds to the power of the design and facilitates the testing of causal models.

Our interest in behavioral development led us to plan a longitudinal, prospective adoption study. Although the longitudinal design is the cornerstone of developmental analysis of change and continuity, few longitudinal genetic studies had been conducted. Balancing the desirability of frequent assessments and the economics of a long-term, large-scale, longitudinal project, we decided to test the children yearly. Our long-range plan was to study this unique sample from infancy to 16 years of age, when the children would be administered the same tests that their parents had completed at the outset of the study more than a decade and a half earlier.

Developmental genetics, which was beginning to emerge in the 1970s, pointed to genetic sources of change as well as continuity during development. If genetic influences change from childhood to adulthood, as we suspected, genetic resemblance between offspring who are children and parents who are adults will be attenuated. That is, if genetic influences in childhood differ from those in adulthood, biological parents and their adopted-away offspring at a *young* age may not manifest hereditary resemblance even if the trait is heritable both in childhood and in adulthood. However, if evidence for genetic influence emerges from the parent–

offspring adoption design, such a finding would indicate not only that the trait is heritable in childhood and in adulthood, but also that the same genetic factors are expressed in both childhood and adulthood (Plomin, 1986; Plomin & DeFries, 1981).

Our expectation was that genetic factors would increasingly account for parent–offspring resemblance as the offspring developed from infancy to childhood, adolescence, and young adulthood. Because we did not expect to find much evidence for genetic influence early in life using the parent–offspring design, we have been surprised by the extent to which evidence for genetic influence has emerged from the CAP parent–offspring design, which implies that there is indeed some genetic continuity from even early childhood to adulthood.

In order to triangulate this issue of genetic change and continuity, we added to the CAP parent–offspring design by planning the first sibling adoption study in childhood in which the proband and the proband's younger sibling would be studied individually at the same age. The rationale for adding siblings was to compare the relationship of siblings tested at the same age to that of parents and offspring separated in age by decades. We knew of course that many of our nonadoptive parents would give birth to other children following the birth of the CAP proband. These siblings, like nonadoptive parents and their children, are first-degree relatives who are 50% related genetically. Siblings growing up together in the same family share environmental influences that are likely to differ from those shared by parents and the children. Because such siblings share both heredity and environment, their resemblance cannot be attributed unambiguously to either. In contrast, the sibling adoption design provides a means of testing the extent to which sibling resemblance is due to shared environment. About a third of adoptive families adopt a second child. These adoptive siblings are genetically unrelated, but are reared in the same family and share environmental influences as do nonadoptive siblings. Thus, the resemblance of adoptive siblings directly tests the importance of shared environment in producing sibling resemblance. To the extent that correlations for nonadoptive siblings exceed those for adoptive siblings, the sibling adoption design also indirectly tests for genetic influence.

The sibling adoption design is especially valuable for genetic analyses of childhood measures for which comparable adult measures are not available. For example, infant novelty preference, motor development, and the stress of the transition to school cannot be assessed in adults. Although the parent–offspring design makes it possible to ask whether these childhood measures are related genetically to adult characteristics – such as cognitive abilities, athletic interest and talent, and psychopathology, respectively – sibling or twin data are needed for genetic analysis of measures specific to childhood. For this reason, many analyses reported in this book for middle childhood rely on the sibling adoption design. Beginning at 4 years of age, over 300 pairs of twins are also currently being tested on the CAP battery; however, too few twins have as yet been tested to include their data in this volume.

The utility of the sibling design also became important in an unanticipated way. In 1985, Judy Dunn, an international authority on sibling relationships, began studying the CAP siblings as siblings, rather than as individuals assessed separately.

Initially, additional visits were made to CAP families with siblings in order to observe sibling and mother–sibling interactions when the younger siblings were 4 years old. A second visit occurred about three years later, and subsequent interviews are currently being obtained 6, 7, and 8 years after the first sibling visit. Chapters 15 and 16 present some of the results from this project.

As we had anticipated (DeFries, 1975), the CAP was begun just in time. The sexual revolution of the 1960s had resulted in unprecedented numbers of babies relinquished for adoption; however, during the early 1970s, the availability of contraceptives, legalized abortions, and an increase in the number of unmarried mothers who kept their children combined to lower drastically the number of infants available for adoption (Sklar & Berkov, 1974). The decrease was most dramatic for Caucasian neonates, the primary group of adoptees in the Rocky Mountain area, with yearly declines of about 25% from 1970 to 1973. Thus, it would be nearly impossible to begin such a study again, at least in the United States.

Our goal was to include 250 adoptive families and 250 nonadoptive families. Our design required a large sample for three reasons. Most importantly, quantitative genetic analyses are based on correlations or covariances of relatives (for example, parent–offspring correlations in nonadoptive and adoptive families), and such comparisons entail large standard errors. Adequate power to detect reasonable parameter estimates thus requires large samples. Second, our long-term goal was to continue to study the CAP sample until the children had reached 16 years of age, when they could be administered exactly the same battery of tests that their parents took before they were born. Thus, it was critical that the foundation sample be able to sustain two decades of inexorable attrition, and yet provide an adequate final sample. Third, it was necessary that the foundation sample be large enough to yield adequate numbers of sibling pairs in both the adoptive and the nonadoptive families.

Sample

In 1974, we began working with the two largest adoption agencies in the Rocky Mountain region, Lutheran Social Services of Colorado and Denver Catholic Community Services. Testing of biological parents began in 1975 and the first adopted children were studied at 1 year of age in 1976. Biological mothers and about 20% of the biological fathers were tested before they relinquished their children for adoption, typically during the last trimester of pregnancy. After their release from the hospital, the infants were taken to a foster home for about four weeks until the legal requirements concerning relinquishment were fulfilled. The children were then placed in their adoptive homes, at the average age of 29 days. The adoptive parents of these adopted-away children were recruited to participate in the study after the adoption was finalized, when the child was about 7 months old on average.

Hospital birth records and subsequent telephone interviews were employed to select nonadoptive parents who matched the adoptive families for five criteria: sex of proband, number of children in the family, age of father (±5 years), National Opinion Research Center (NORC) rating of occupational status of the father (±8 points), and total years of the father's education (±2 years). This matching procedure ensured comparability between the adoptive families and nonadoptive families (whom we often refer to as control families), but it was not intended for the purpose of statistical analyses based on paired observations.

Nearly 90% of the biological parents and more than 95% of the adoptive and nonadoptive parents report that they are Caucasian; the rest are primarily Hispanic and Oriental.

Because the adoption agencies eliminated very few prospective parents and did not use wealth as a criterion for placements, CAP adoptive and nonadoptive parents are representative of the Denver metropolitan area in terms of socioeconomic status. Biological parents are on average younger (20 years) than adoptive parents (33 years) and nonadoptive parents (31 years); thus, their years of education and occupational status are lower than those of the adoptive and nonadoptive parents. However, the average years of education and occupational status of the biological grandparents (i.e., the parents of the CAP biological parents) are comparable to those of the adoptive and nonadoptive grandparents.

Although these demographic data support the internal coherence of the CAP design, the issue of representativeness extends also to comparisons with other samples. Occupational status of the CAP families is similar both in means and variances to a stratified random sample from a Denver suburban area taken from the 1970 census data (Plomin & DeFries, 1985). On the basis of national norms for the US white labor force, the CAP sample was found to be somewhat above the national average in occupational status, but nearly representative in terms of variance.

The CAP also appears reasonably representative of the Caucasian middle-class population in relation to cognitive measures. On the Bayley scale at 1 and 2 years of age (Bayley, 1969), and the Stanford-Binet (Terman & Merrill, 1973) at 3 and 4, the CAP means are about half a standard deviation above the standardization samples for these tests. The variances are about 80% of those reported for the standardization samples. At the year 7 laboratory visit, on the WISC-R test of intelligence (Wechsler, 1974), the average IQ of the probands is 113 with a standard deviation of 10.7 (for details, see Chapter 4). During the year 7 visit, the parents of the CAP children are administered the WAIS-R (Wechsler, 1981). The average IQ scores of the adoptive and nonadoptive parents when tested on this occasion are 111 and 112, respectively, with corresponding standard deviations of 11.5 and 11.1.

The CAP families have been extremely cooperative: 85% of the families enrolled during the home testing at 1 year of age also completed the laboratory testing session at 7 years. The present analyses at 7 years include data from 204 adopted probands and 216 nonadopted probands. At 7 years, 73 younger siblings of the adopted probands have also been tested. These include 19 younger siblings who are themselves adopted probands and whose biological parents were also tested. The number of younger siblings in the nonadoptive families who have been tested at 7 years is 76. Because testing is still in progress at 9 and 10 years, fewer children have been tested at these ages.

Measures

The adoptive and nonadoptive probands and their younger siblings are tested in their homes at 1, 2, 3, and 4 years of age. The children are tested in the laboratory at 7, 12, and 16 years. Telephone testing and interviews are conducted at 9, 10, 11, 13, 14, and 15 years, and some questionnaire information is collected during the other years (5, 6, and 8). Procedures and measures are described in previous CAP books for testing of parents, and for assessing home environments and offspring at 1 and 2 years (Plomin & DeFries, 1985) and during early childhood at 3 and 4 years (Plomin et al., 1988). These measures will be reviewed very briefly. The present analyses focus on middle childhood, from 7 to 10 years. Measures employed in the laboratory visit at 7 years and in the telephone interviews and tests at 9 and 10 years will be described in greater detail.

The breadth of measures in the CAP was the result of a feeling of responsibility for creating a unique resource of longitudinal data on behavioral development. Because it is highly unlikely that such a study could again be accomplished in the United States, we have attempted to collect as much valid and reliable data as possible during relatively short testing periods.

Parental measures

The CAP biological, adoptive, and nonadoptive parents completed a 3-hour test battery that assesses general and specific cognitive abilities, personality, interests and talents, common behavioral problems, commonly used drugs, and miscellaneous information.

Cognitive abilities   Thirteen tests were employed from the Hawaii Family Study of Cognition (DeFries et al., 1974). The tests yield a median internal reliability of .86 and a median test-retest reliability of .80. The scores were adjusted for age, age squared, and gender separately for each group of parents because age is significantly related to cognitive scores and is confounded with parental type; i.e., the biological parents are on average 12 years younger than the adoptive and nonadoptive parents. A first unrotated principal component is used as an index of general cognitive ability. Varimax rotation following a principal-components analysis yields four factors accounting for 61% of the total variance. The four specific cognitive abilities factors represent Verbal ability, Spatial ability, Perceptual Speed, and Memory. The battery yields a highly congruent factor structure across the three parental types.

Personality   Two self-report measures were used to assess personality. Form A of Cattell's Sixteen Personality Factor Questionnaire (16PF: Cattell, Eber, & Tatsuoka, 1970) yields 16 primary scales, and second-order factors that include Extraversion and Neuroticism. An adaptation of a temperament survey developed by

Buss and Plomin (1975) is used to obtain self-report and "mate ratings" of personality.

Interests and talents Artistic, group sports, individual sports, mechanical and domestic interests and talents were assessed. Other interest scores obtained in the CAP include the amount of television viewing, reading, and religiosity.

Behavioral problems Frequently occurring behavioral problems were assessed: depression, hysteria, sociopathy, phobias, compulsive behavior, sleep problems, motion sickness, menstrual problems, headaches, and speech problems. Detailed information concerning the parents' smoking history was elicited as well as information about alcohol consumption and coffee drinking.

Miscellaneous measures These measures included demographic information such as education, occupation, and ethnicity, as well as questions about handedness, food preferences, hours of sleep, frequency of colds, height, and weight.

Child measures in infancy and early childhood

The information obtained during the 2- to 3-hour home visits at 1, 2, 3, and 4 years of age included measures of mental development, communication, personality, behavioral problems, motor development, and miscellaneous measures, such as the child's health, interests in toys and other objects, and handedness. These data were collected via standardized tests, tester ratings, parental ratings, interviews, and videotaped observations of mother–child interaction. Because test-retest reliability is generally not known for these measures, 2-week test-retest data were obtained from 26 1-year-olds for all measures and from 32 3-year-olds.

Cognitive development The CAP administered the Bayley Mental Scale (Bayley, 1969) at 1 and 2 years of age and the Stanford-Binet Intelligence Scale (Form L-M: Terman & Merrill, 1973) at 3 and 4 years. Test-retest correlations in the CAP are .80 for the Bayley at 1 and .82 for the Stanford-Binet at 3 years.

A major CAP effort during early childhood was directed toward the development of test batteries of specific cognitive abilities for 3-year-olds (Singer, Corley, Guiffrida, & Plomin, 1984) and 4-year-olds (Rice, Corley, Fulker, & Plomin, 1986). Factor analyses indicated success in assessing Verbal, Spatial, Perceptual Speed, and Memory abilities. An unrotated first-principal-component score at 3 years correlates .61 with Stanford-Binet at 3 years and its test-retest reliability is .75. Test-retest reliabilities of the Verbal, Spatial, Perceptual Speed, and Memory abilities are .67, .40, .68, and .64 at 3 years and .81, .78, .77, and .56 at 4 years.

Language Production and comprehension were assessed at the 2- and 3-year home visits using the Sequenced Inventory of Communication Development (SICD: Hedrick, Prather, & Tobin, 1975). Test-retest reliabilities of the Expressive, Receptive, and Total scores at 3 years of age are .73, .70, and .79, respectively.

Personality   A multi-method strategy has been employed to assess personality using information from testers, parents, and videotaped observations. The Infant Behavior Record (IBR: Bayley, 1969) was used to rate children's behavior during administration of the mental tests at 1 and 2 years of age. At 3 and 4, a revised version of the IBR was employed. Factors are scored for Affect-Extraversion, Task Orientation, and Activity as suggested by Matheny (1980). Test-retest reliabilities for these three scales at 1 year were .76 for Affect-Extraversion and .60 for Task Orientation, but only .06 for Activity. IBR scales at 3 years yielded test-retest reliabilities from .50 to .75.

The Colorado Childhood Temperament Inventory (CCTI: Rowe & Plomin, 1977) was also administered at each age. The CCTI is a parental rating instrument that is an amalgamation of the EAS temperament dimensions (Buss & Plomin, 1975, 1984) and the nine dimensions of temperament postulated in the New York Longitudinal Study (Chess & Thomas, 1984). Average ratings of mothers and fathers are used at 1, 2, and 3 years of age to improve the reliability of the ratings; at 4, mothers' ratings are used because fathers were not asked to complete the CCTI at 4 years. The same six scales are constructed from each year's data: Emotionality, Activity, Sociability, Attention Span, Reactions to Foods, and Soothability. The median alpha reliability of the six scales is .80, and the median 1-week test-retest reliability is .66 for 1-year-olds in the CAP.

Interests   At 1, 2, and 3 years, mothers were interviewed concerning their children's liking of objects classified as gross motor, fine motor, and musical objects. At age 4, scales of interests and talents were used that were more similar to those obtained for the parents.

Behavioral problems   Common developmental problems were assessed by parental ratings on a dimension of difficult temperament (Daniels, Plomin, & Greenhalgh, 1984) as well as ratings of specific problems such as sleeping, eating, and elimination at 1, 2, and 3 years. At 4, the parental rating version of the Child Behavior Checklist (CBCL: Achenbach & Edelbrock, 1983) was added to the CAP battery in order to include a standard measure of behavioral problems. The CBCL consists of 118 3-point items that are scored on two second-order factors, Internalizing and Externalizing, and a total score. The second-order factors and total score yield 1-week test-retest reliabilities of .88, .95, and .92, respectively, for a sample of children 4 and 5 years of age.

Motor development   Motor development was assessed at 1 and 2 years by the Bayley Motor Scale (Bayley, 1969) and at 3 years using an extension of the Bayley Motor Scale developed by the CAP staff. At 3, factor analysis suggested separate factors of fine motor skills such as drawing and throwing, and gross motor skills such as walking on tiptoe and standing on one foot.

Miscellaneous measures   Perinatal information, health, height and weight, and handedness were also assessed. Adoptees' perinatal information was obtained from the adoption agency files and includes gestational age and birth weight; similar information was obtained from nonadoptive mothers. Major health-related aspects of development were assessed during a 10-minute interview with the parent and via a form that was completed by the child's pediatrician. A simple rating of general health yielded a test-retest reliability of .55 at 1 year and .73 at 3 years. Height and weight were also assessed each year; test-retest reliabilities for height and weight are .94 and .96 at 1 year, respectively, and .92 and .90 at 3 years.

Finally, videotapes at 1 and 2 years were used to rate relative and absolute strength of hand preference (Rice, Plomin, & DeFries, 1984). Also at 3 years, testers recorded hand preference during the testing of motor development, a scale that yields a .78 test-retest reliability.

Environmental measures

The inclusion of measures of specific environmental factors within a longitudinal adoption design facilitates novel approaches to the analysis of environmental influences. The CAP includes two measures of the home environment, the Home Observation for Measurement of the Environment (HOME) and the Family Environment Scale (FES) questionnaire.

Home Observation for Measurement of the Environment The HOME (Caldwell & Bradley, 1978) uses observation and interview to assess characteristics of the home environment related to cognitive development of infants. The instrument, designed primarily for use in lower-class families, was made more sensitive to variation in middle-class homes in the CAP by extending the HOME dichotomous scoring to a quantitative scoring system which had greater stability (Plomin & DeFries, 1985). Although the HOME is traditionally scored for a total scale and six subscales, most CAP analyses are based on somewhat different scales derived from factor analysis of the quantitatively scored items. These scales include a composite measure (the unrotated principal component) representing a general HOME factor, and four subscales derived from rotated factors: Maternal Involvement, Encouraging Developmental Advance, Restriction–Punishment, and Toys. The test-retest reliability at 1 year for the HOME general factor is .86; the reliabilities of the four subscales are .74, .79, .69, and .84, respectively. In the CAP, the HOME was extended for use at 3 and 4 years (Plomin et al., 1988). At 3, a general factor yielded a test-retest reliability of .79, although reliabilities of three subscales based on rotated factors were mixed: .63 for Maternal Involvement, .43 for Encouraging Developmental Advance, and .92 for Toys. At 4 years, only two rotated factors emerged, Maternal Involvement and Toys.

Family Environment Scale The FES (Moos & Moos, 1981) is a 90-item self-report questionnaire that was completed by both mothers and fathers when the children were 1 and 3 years old. In the CAP, we altered the FES from a true–false format to a 5-point rating scale and used an average rating for the two parents. The items were scored on 10 scales, generally confirmed in CAP factor analyses: Cohesion, Expressiveness, Conflict, Independence, Achievement Orientation,

Intellectual–Cultural Orientation, Active–Recreational Orientation, Moral–Religious Emphasis, Organization, and Control. Factor analysis of the 10 scales yielded two major second-order factors. One, which we labeled Traditional Organization, includes Organization, Control, and Moral-Religious Emphasis. The other, which we refer to as Personal Growth, is defined primarily by Cohesion and Expressiveness, although loadings above .30 also occurred for low Conflict, Independence, Intellectual–Cultural Orientation, and Active–Recreational Orientation. These two factors are reminiscent of the two major dimensions of childrearing typically seen in the literature: control and warmth. Test-retest reliabilities of factor scores for these two dimensions were .97 and .89 when the CAP children were 1 year old.

Physical environment   In addition to the HOME and FES measures, which emphasize the social environment of the family, aspects of the physical environment were also assessed. These heterogeneous items, such as rooms per person and noise, were organized into three categories as suggested by Gottfried (1984): variety of experience, provision for exploration, and physical home setting. The three scales yielded reliabilities of .69, .61, and .77, respectively, at 1 year, and .73, .59, and .47 at 3 years.

Demographic information   Parental education and occupational status were also assessed. Occupational status was rated using the Socioeconomic Index (SEI). The SEI, originally developed by Duncan (in Reiss, Duncan, Hatt, & North, 1961) using ratings of the 1950 census classification of occupations based on prestige, income, and education, was updated from the 1970 census classification (Hauser & Featherman, 1977).

Procedures in middle childhood

In the early school years, chronological age becomes less important than the school calendar. Test scores, especially school achievement-test scores, differ as a function of time in the school year. For this reason, CAP children are tested in the summer following their first year of school, typically when they are 7 years old. These tests are administered at the Institute for Behavioral Genetics, University of Colorado.

The session lasts for 5 hours, including a 10-minute break for refreshments and a 1-hour lunch break. The test session protocol is outlined in Table 2.1. Similar to the assessments in earlier years, the year 7 test session includes measures of cognitive abilities, personality, motor development, and interests and talents. In addition, school achievement, attitudes, and stressful life events are assessed. Several videotaped observations of parent–child interaction are obtained. During the year 7 session, parents are administered the Wechsler Adult Intelligence Scale–Revised (WAIS-R: Wechsler, 1981). Resulting WAIS-R Full-Scale IQ scores correlate .73 with the unrotated first principal component from the test battery administered to these same parents seven years earlier.

Table 2.1 Test session protocol at year 7

Time in test session	Child	Child/Parent	Parent
0:00		Introduction, height, weight (videotaped)	
0:10	Human figure drawing (audiotaped)		WAIS-R, PIAT
0:15	WISC-R		
1:15	Refreshment break ————————————————————		
1:25	Motor tests		Interview Questionnaires
2:00	Interviews Questionnaires		
2:20		Structured parent/child interaction (videotaped)	
2:30	Lunch break ———————————————————————————		
3:30	Tests of specific cognitive abilities		(Repeat above with other parent)
4:00	Interviews Questionnaires		
4:15	School achievement tests		
4:50		Problem solving parent/child interaction (videotaped)	
5:00	Dismissal —————————————————————————————		

Teacher ratings are obtained once in the spring preceding the summer test session and again each following spring. Teachers rate performance for academic subjects, temperament, behavioral problems, and peer interactions and social competence. Parents are asked to deliver the rating form and cover letter to the teacher, who mails it back directly to us and is paid $10 upon receipt of the packet. The teacher questionnaires are distributed in March and April in order to ensure the teacher's familiarity with the child and to avoid any extra imposition due to special time constraints that may occur near the end of the school year (e.g., additional class activities, gradings, etc.)

At 9 and 10 years, a telephone interview and test battery are employed. Our pilot work indicated that the telephone can be used at this age to obtain extensive information about children and their parents. Survey researchers have found that the quality of information obtained via the telephone is comparable to that obtained in face-to-face interviews. For example, telephone interviews are comparable to in-person interviews in terms of respondents' ability to answer complex questions, willingness to provide personal information, response validity, and consistency of information (e.g., Groves & Kahn, 1979). Importantly, studies also indicate that subjects generally prefer telephone interviews.

In preparation for the telephone interviews and tests, parents are contacted by letter to explain the project and are called to schedule a convenient testing time for their child, to explain the testing procedure, and to confirm that the test packet has been received. Two telephone calls are made each summer when the children are approximately 9 and 10 years old. During the first call, a short interview is conducted with a parent and a longer interview with the child. The cognitive test battery is administered during the second call. Children receive $10 each year for their participation. Parents are asked to complete questionnaires as well.

Child measures in middle childhood

Selection of measures in middle childhood was guided by three considerations. First, the measures were designed to be as consistent as possible with the measures obtained at earlier ages. Second, the measures were selected to be as consistent as possible with the CAP adult measures. Third, some new measures were added to assess important domains not previously assessed in the CAP, for example, self-esteem and major life events specific to middle childhood such as the stress of entering school.

7-year-old cognitive abilities   An individually administered intelligence test (WISC-R: Wechsler, 1974) is used as a measure of general cognitive ability. Subtests from the WISC-R also contribute to the assessment of specific cognitive abilities. As indicated in Table 2.2, tests of specific cognitive ability were again selected to load on the four specific cognitive abilities factors.

Table 2.2   Tests of specific cognitive abilities at year 7

Factor	Test
Verbal	WISC-R Vocabulary
	Verbal fluency measure similar to the Things test (used in the CAP adult battery)
	WISC-R Information
	WISC-R Similarities
	WISC-R Comprehension
Spatial	WISC-R Block Design
	WISC-R Object Assembly
	WISC-R Picture Completion
	WISC-R Picture Arrangement
	WISC-R Mazes
	PMA Spatial Relations
Perceptual Speed	Modification of ETS Identical Pictures (used in CAP adult battery)
	Modification of Colorado Perceptual Speed (used in CAP adult battery)
Memory	Immediate and delayed Picture Memory (used in CAP adult battery)
	Modification of immediate and delayed CAP Names and Faces (used in CAP adult battery)

Telephone testing of cognitive abilities at 9 and 10 years   Telephone interviews with 9- and 10-year-olds about their development and environment seem reasonable and preferable to mailed-out paper-and-pencil questionnaires. However, assessing cognitive abilities over the telephone is novel. A preliminary study successfully administered via telephone an hour-long battery of 12 cognitive measures to 208 children from 9 to 16 years (Kent & Plomin, 1987). This procedure is followed in testing CAP children at 9 and 10, but uses a reduced 8-test battery that requires only about 40 minutes to administer.

To discourage cheating, a variety of procedural precautions are taken, although pilot work indicated that cheating is not an important problem. Subjects report that they are too busy to think about ways to cheat, that they have no inclination to cheat, and that they could not think of ways to cheat even if they had wanted to. Although our procedures to discourage cheating are only partial solutions, they collectively serve as effective deterrents. First, the child is told that cheating will invalidate any findings and that analyses will involve group comparisons so that there is nothing to be gained by cheating. Secondly, the test packet is mailed to the parent with the understanding that the packet will not be given to the child until the time of the test session. Thirdly, the test booklet has been constructed to make cheating difficult. The tests are very brief and they are paced individually so that it would be difficult for a child to look at other tests. Furthermore, the pages of each test are sealed together until, at the start of each test, the child is asked to rip the seal, a distinctive sound audible to the examiner. Our pilot work indicated that subjects enjoy the telephone testing and are conscientious about their work.

As described in detail elsewhere (Kent & Plomin, 1987), the eight tests assess verbal ability (WISC-R Vocabulary and WISC-R Similarities), spatial ability (ETS Hidden Patterns and ETS Card Rotations), perceptual speed (ETS Finding A's and Colorado Perceptual Speed), and memory (immediate and delayed Picture Memory and immediate and delayed Names and Faces). All tests loaded moderately on an unrotated first principal component. The first-principal-component score at 9 years correlates .60 with WISC-R Full Scale IQ at 7 years. These results indicate that the first principal component represents a reasonably reliable and valid dimension of general cognitive ability (Cardon, Corley, DeFries, Plomin, & Fulker, 1992).

As discussed in Chapter 5, the factor structures at 7, 9, and 10 years are quite similar and confirm the anticipated structure of Verbal, Spatial, Perceptual Speed, and Memory abilities. Coefficients of congruence (Gorsuch, 1983) across the three years exceed .95 for all factors. Confirmatory factor analyses indicated that factor loadings did not differ significantly across age or between males and females. Test-retest reliability correlations were .79 for the first principal component, .81 for Verbal ability, .73 for Spatial ability, .82 for Perceptual Speed, and .57 for Memory.

One final issue is the possibility of practice effects, given that the same tests of specific cognitive abilities are being used at 9 and 10 years of age. Five points are relevant in response to this issue. First, a mean effect is of little importance to our covariance analyses unless a ceiling is reached which restricts variance. Second, parallel forms are difficult to create. If two different forms are not strictly parallel, age-to-age stability would be underestimated. Third, test-retest analyses of pilot data indicated that, even during a 3-week test-retest interval, practice effects are not observed for verbal and memory tests and only slight effects are found for spatial and perceptual speed tests. Fourth, even tests that offer the greatest opportunity for practice (or cheating), especially the vocabulary test, show little improvement from test to retest. Fifth, the children return the test booklet to us in a self-addressed stamped envelope before they receive their payment; thus, they do not have an opportunity to review the tests prior to their follow-up testing.

School achievement   Beginning at 9 years, we attempt to obtain national percentile scores from standardized achievement tests. Although different tests are used in different school districts, the tests yield comparable information concerning national percentile rankings of students in major subject areas such as reading, language, mathematics, science, and social studies. Achievement test results are obtained by means of a letter sent from the parent to the child's home-room teacher and returned directly to us. Teachers are also asked yearly to rate the child's effort and to comment on the child's academic performance.

In addition, school achievement tests of reading and mathematics are administered at the year 7 test session. Reading is assessed by the Peabody Individual Achievement Test (PIAT) Reading Recognition subtest. For mathematics, three subtests (Numeration, Addition, and Subtraction) from the KeyMath Diagnostic Arithmetic Test are administered (Connolly, Nachtman, & Pritchett, 1976).

Language   Tests of language use include two subtests of the Clinical Evaluation of Language Function (CELF) and the Token Test for Children. Measures of language production are also obtained from the videotaped records of children solving a puzzle which elicits communication.

Personality   The Colorado Childhood Temperament Inventory (Rowe & Plomin, 1977) is completed by parents, testers, and teachers at 7, by parents and teachers at 8, 9, and 10 years, and by the children at 9 and 10. Parents also complete the Behavioral Style Questionnaire (McDevitt & Carey, 1978) at age 7. At the year 7 laboratory visit, testers also rate the child on the CAP-modified version of the Bayley Infant Behavior Record. Also included are videotaped observations of children interacting with the tester and with their parents in situations differing in degree of structure. As indicated in Table 2.1, these situations include the preliminary session with the child, parent, and tester involving introductions, recording of physical measurements, a structured game-playing situation, and a problem-solving task.

Interests and talents   At the year 7 session, children are interviewed about their interests and abilities using items similar to those included in the adult test booklet. Parents also rate their children. As at earlier years, parents are interviewed concerning the children's television viewing.

Social competence   At 7, parent and teacher ratings of social competence are obtained using a CAP measure developed from several sources, but primarily from the Walker-McConnell Scale of Social Competence (Walker & McConnell, 1988). The measure includes scales of Leadership, Confidence, and Popularity. At 9, teacher reports are obtained for this measure.

Self-perceptions of competence   At 9 and 10, children complete Harter's Self-Perception Profile (Harter, 1982), which assesses five dimensions of perceived self-competence: Self-Worth, Behavior Conduct, Athletic Competence, Scholastic Competence, and Physical Appearance.

Behavioral problems   At 7, 9, and 10 years, parents and teachers are asked to rate behavioral problems using the Child Behavior Checklist (Achenbach & Edelbrock, 1983) for which extensive norms are available. Because internalizing problems are more subjective than externalizing problems, self-report measures of Depression and Loneliness at 9 and 10 – Kandel's Depressive Mood Inventory (Kandel & Davies, 1982) and Ascher's Loneliness Questionnaire (Ascher, 1985) – are also included.

Motor development   At 7, the short version of the Bruininks-Oseretsky Test of Motor Proficiency (Bruininks, 1978) is administered. This measure has 14 items concerning agility, balance, bilateral coordination, strength, visual-motor control, and dexterity. Norms are available for ages 4 to 14. The 1-week test-retest reliability for the 14-item short version for 7 year olds is .87. This test also assesses behavioral laterality, and a laterality questionnaire is also administered to the children and their parents.

Height, weight, health, and diet   At year 7, information pertaining to height, weight, and diet is recorded. At each year, parents are interviewed concerning the child's health.

Environmental measures in middle childhood

At 7, environmental assessments include tester ratings, videotaped observations of parent–child interaction, interviews with parents and children, and parental questionnaires. A total of 15 minutes of videotape are collected for each parent–child pair for ratings of interactions as well as for personality ratings. In addition, as described in Chapters 15 and 16, the CAP sibling study collects extensive observational and interview data concerning parent–child and sibling interactions.

Family Environment Scale   Two questionnaires are administered to parents. As was done at 1, 3, and 5 years, parents complete the Family Environment Scale (FES: Moos & Moos, 1981) at 7, 9, and 10 years. For middle childhood, we modified the FES for use as a self-report instrument in order to assess the family environment as perceived by the children. Four items were chosen to assess each of five FES dimensions: Cohesion, Expressiveness, Conflict, Achievement, and Control. Children indicate whether each statement is "like my family" or "not like my family."

Dibble and Cohen's Parent Report   The second questionnaire used each year is the Parent Report (PR) of childrearing behavior developed by Dibble and Cohen (1974) from the work of Schaefer and Bell (1958). The 48-item PR consists of 16 categories of parental behavior, eight socially desirable and eight socially undesirable, with three behaviorally specific items written for each category. Three factors have been derived in the CAP: Acceptance (Acceptance of Child as a Person, Child-Centeredness, Sensitivity to Feelings, Positive Involvement, and Shared Decision Making), Inconsistency (Inconsistent Enforcement of Discipline, and Detachment), and Negative Control (Control through Guilt, Hostility, Anxiety, and Withdrawal of Relationship).

Self-Image Questionnaire for Young Adolescents   At 9 and 10 years, children also report on their perceptions of parenting and their family environment using relevant items from the Self-Image Questionnaire for Young Adolescents (Petersen, Schulenberg, Abramowitz, Offer, & Jarcho, 1984).

School stress   During the year 7 test session, children and parents are administered an inventory developed in the CAP to assess stress related to the first year of school. Open-ended interviews with children, parents, teachers, and clinicians were used to generate items. Eighteen items were chosen to represent academic concerns, peer relations, teacher relations, parent relations, and general hassles, as described in Chapter 11. Children are interviewed and parents complete a questionnaire that assesses whether each event occurred during the first year at school and how upsetting the event was. Two composite scores were constructed for children and parents: the total number of events and a sum of the upsettingness ratings, scores consistent with current approaches to the study of stress in childhood (e.g., Compas, 1987).

Life events   At 7, 9, and 10 years, the frequency and degree of upsettingness of major life events – such as death of a relative, hospitalization or health problems, and school and family problems – are assessed using Coddington's (1972) 33-item Social Readjustment Scale. Mothers are asked to indicate whether each event occurred during the past year. Mothers and children are then asked to rate the degree to which the child was upset by the event using a 4-point rating scale. At ages 9 and 10, children are also asked to rate the severity of major life events using the Life Events Scale for Adolescents developed by Brooks-Gunn based on her work concerning life stress in early adolescence (Brooks-Gunn & Petersen, 1984). As with the Coddington measure, children are asked to indicate whether an event occurred and to rate the extent to which they were upset by it.

Selective Placement

The major assumption of the adoption design is that adoptees are not matched to their adoptive parents on the basis of their biological parents' characteristics. To the extent that such selective placement occurs, estimates of both heredity and shared environmental influences will be inflated. Fortunately, selective placement is an empirical issue that can be tested by correlating characteristics of the biological and adoptive parents.

The two adoption agencies participating in the CAP were particularly appealing because, like other progressive agencies, they do not attempt to match adoptees to adoptive parents. They believe that this raises false expectations for adoptive parents and that the agency's resources are better spent counseling adoptive parents about the differences that are likely to occur. This policy and the long waiting-list to adopt a child combine to avoid selective placement in the CAP. Selective placement correlations for the demographic variables accessible to social workers are negligible for the biological and adoptive parents, as well as for the parents' parents. The median selective placement correlation is .04 for education and .02 for socio-economic status (Plomin et al., 1988). Moreover, selective placement is negligible for all CAP measures, including cognitive abilities and personality.

This overview of the CAP design, sample, and measures sets the stage for the following chapters. The next chapter describes CAP methodology, with a focus on model-fitting analyses. Subsequent chapters are more empirical, presenting some of the most interesting CAP results from infancy to middle childhood for cognitive abilities, language, school achievement, personality, stress, body size and obesity, motor development, sex differences, competence, relationships, nonshared environment, the nature–nurture interface, genotype–environment interaction and correlation, and applied issues relevant to adoption.

3 Adoption Design Methodology

Lon R. Cardon
Stacey S. Cherny

The full adoption design is one of the most powerful methods available for disentangling hereditary and environmental influences on continuous characters (Cavalli-Sforza, 1975). The strength of the method lies in its ability to yield direct estimates of *both* genetic and environmental effects, in contrast to other human behavioral genetic methods which infer such effects through comparisons of individuals having different degrees of genetic resemblance. Because the CAP uses a full adoption design with measures from adopted offspring and both their biological and adoptive parents, as well as matched sample of nonadoptive families, it takes complete advantage of the powerful adoption methodology for elucidating the sources of variation in behavioral traits.

Since the inception of the CAP in 1975, a variety of methodological tools has been advanced to extract the most information possible from adoptive and nonadoptive family data, with the aim of providing interpretable and replicable results. Several of these tools are extensions of the biometrical approach developed by Jinks and Fulker (1970) for genetic analysis of family data. These statistical approaches are used to fit models to observed data, estimate various genetic and environmental parameters, and test alternative hypotheses. More recent advancements combine these methods with the structural modelling procedure first set forth by Jöreskog (1973) to address more complex issues such as the shared etiology of multiple variables, patterns of genetic and environmental influences over time, and possible correlations between genotypes and environments. The structural modeling approach is emerging as a powerful and popular method for biometrical analysis of twin and family data (Neale and Cardon, 1992), and is proving to be an invaluable tool for multivariate and longitudinal analyses of the CAP data.

In this chapter, we describe the methods most often employed in analyses of CAP data. In the first section we present a brief overview of the quantitative genetic principles underlying the methods and describe the method of path analysis for illustrating relationships among family members. In the second section we develop the structural equation models for adoptive and nonadoptive sibling pairs, including univariate, multivariate, and longitudinal path models. The final section presents a description of the parent–offspring adoption models used in the CAP.

Quantitative Genetics and the Polygenic Model

We present here a brief overview of quantitative genetic theory as it relates to the adoption design. More detailed expositions are given in several introductory textbooks, notably those by Falconer (1989) and Mather and Jinks (1982).

The principles of quantitative genetics underlie all of the methods used for genetic inference in the CAP. Quantitative genetics is concerned with the sources of individual differences: the genetic and environmental factors which create differences among individuals as manifest in the variability of observed measures. The general theory of quantitative genetics is based upon Mendel's two laws: (1) the law of segregation, which states that individuals carry alternate forms of each gene that have equal probabilities of being transmitted to an offspring; and (2) the law of independent assortment, which states that genes which lie on different chromosomes or are otherwise unrelated (so-called "unlinked" genes) are transmitted from parents to offspring independent of one another. Fisher (1918) showed that under a Mendelian system with many genes contributing to a trait, the number of different gene assemblages (genotypes) becomes very large. The resulting genetic variation thereby accounts for the genetic etiology of individual differences.

The statistical methods used in the CAP assume this system of multiple genes contributing to individual differences in a phenotype, known as polygenic influence. Under the polygenic assumption, the different genotypes follow a normal distribution and have a genotypic mean and genotypic variance. If we had complete genetic information about the trait of interest and we could identify actual genotypes, it would be possible to assign values to genotypes and determine their variance directly. This type of molecular information is increasing with the recent advancements in recombinant DNA methodology, but it is not yet a practical reality for polygenic behavioral characters and we must infer the variance instead.

It is important to note that quantitative genetic theory does not require or assume information about the precise chromosomal location of genes or about the pathways linking DNA to observable behaviors. The gene–behavior pathway follows directly from the central dogma of molecular genetics, that sequences of DNA are converted into proteins (via RNA). It is the proteins that actually take part in cellular structure, function, and regulation that ultimately contribute to observable characters or phenotypes. The end result of this pathway – differences among individuals via gene expression through cells, tissues, and organs – is what we refer to as genetic influence.

Quantitative genetic theory treats environmental influences in a manner analogous to its treatment genetic effects. In human behavioral genetics, the "environment" is typically thought of as everything that does not originate in segregating genes, thus including such factors as prenatal trauma and environmentally induced changes in DNA, as well as traditional factors such as childrearing, school environment, and peer influence. Such environmental influences also contribute to individual differences in the phenotype. If we could measure these environmental influences we could assess their variance directly. Assessment of at least some environmental influences is more plausible than genotypic measurement at least at the present, and some research in the CAP has been directed toward this type of assessment (see Chapters 18 through 20). However, psychological characters are too complex for measurement of all trait-related environmental factors; thus, we typically infer this variance as well.

The preceding discussion of genetic and environmental influences and their contributions to a particular phenotype may be expressed more formally using the basic linear model of quantitative genetics (Falconer, 1989):

$$P = G + E, \tag{1}$$

where P represents an individual's phenotype, G is the genotypic value, and E is the environmental value for the individual. It is important to recognize in Eq. 1 that an individual's phenotype is assumed to be determined entirely by genetic and environmental factors. Since environmental factors are defined as all nongenetic influences, the assumption actually is that phenotypes are determined by genetic and nongenetic factors. Since each of these factors is expected to have a variance and to contribute to the variance of the phenotype, it follows that

$$V_P = V_G + V_E + 2\text{Cov}_{G,E} \tag{2}$$

where V_P, V_G, and V_E are the phenotypic, genotypic, and environmental variances, respectively, and $\text{Cov}_{G,E}$ is the genotype–environment (G–E) covariance. It is often convenient to rearrange this equation by dividing both sides by V_P (thus not changing the equation at all) to yield

$$\frac{V_P}{V_P} = \frac{V_G}{V_P} + \frac{V_E}{V_P} + \frac{2\text{Cov}_{G,E}}{V_P}$$

$$1 = h^2 + e^2 + 2hse, \tag{3}$$

where h^2 is the proportion of observed variance explained by genetic factors (the "heritability"), e^2 is the proportion of observed variance accounted for by environmental influences (the "environmentality"), and s is the genotype–environment correlation (G–E). When the G–E correlation is assumed to be zero, Eqs 2 and 3 reduce to

$$V_P = V_G + V_E \tag{4}$$

and

$$1 = h^2 + e^2. \tag{5}$$

As described later in this chapter, this assumption may be tested by analyzing parent and offspring data from the adoption design.

The basic quantitative genetic model may be expressed conveniently using the method of path analysis developed by the geneticist Sewall Wright (1921). Path analysis, which is an illustrative method of depicting familial relationships, has the advantage of specifying model assumptions explicitly and facilitating the derivation of expected correlations based on genetic and environmental parameters. A standardized path model of the linear phenotypic expectation of Eq. 1 is given in Figure 3.1. The one-way arrows shown from the unobserved, or latent, G and E variables toward the observed variable P reflect the causal genotypic and environmental influences on the phenotype. The curved arrow between G and E represents the genotype–environment correlation. Following the tracing rules of path analysis, the expected variance of P may be derived from the diagram:

$$V_P = h \cdot h + e \cdot e + hse + esh$$

$$= h^2 + e^2 + 2hse, \tag{6}$$

which is the same result developed algebraically in Eq. 3. Detailed descriptions of path analysis and tracing rules may be found in texts by Li (1975) and Loehlin (1987).

It is possible to partition further the genotypic and environmental factors and variances. For example, in many CAP models the environmental value (E) is divided into environmental factors which are shared by individuals living together, sometimes referred to as "common-environment" factors (C), and environmental factors which are not shared, called unique or specific environment factors (SE). A similar

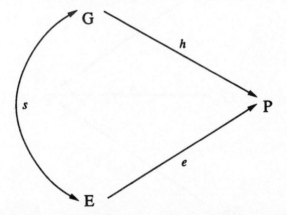

Figure 3.1 Standardized path diagram showing genetic (G) and environmental (E) influences on a phenotype (P). Parameters h and e represent the sq. roots of the heritability and environmentality of the phenotype, and s represents the genotype–environment correlation.

distinction may be made for genetic factors between additive genetic values (A), which are the sum of the average effects of individual alleles (alternate forms of a gene), and dominance deviations (D), which are due to nonlinear interactions between alleles at the same loci. Thus, the elements on the right side of Eq. 1 may be expanded as $G = A + D$ and $E = C + SE$, allowing the equation to be expressed as

$$P = A + D + C + SE \tag{7}$$

In analyses of the CAP data, the C and SE environmental distinction is often observed, but typically only the variance due to additive (A) genetic effects is estimated. Dominance deviations do not contribute to parent–offspring resemblance. Also, CAP analyses ignore nonlinear interactions between alleles at different loci (epistasis) [see Crow (1986) for a discussion of this assumption with respect to quantitative characters]. Thus, the typical CAP phenotypic expectation is

$$P = G + C + SE. \tag{8}$$

In the absence of G–E correlation, the resulting phenotypic variance is as follows:

$$V_P = V_G + V_C + V_{SE}, \tag{9}$$

or, in the standardized case:

$$1 = h^2 + c^2 + e^2, \tag{10}$$

where c^2 is the proportion of phenotypic variance explained by environmental effects shared by individuals in the same home and e^2 is due to nonshared environmental influences. A path diagram of this model is presented in Figure 3.2.

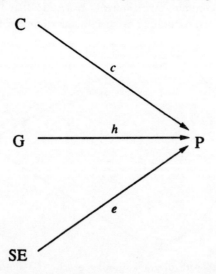

Figure 3.2   Path diagram of genetic (G), shared or common environmental (C), and specific environmental (SE) influences on an individual.

Path diagrams also facilitate derivation of expected variances and covariances among different traits. Multivariate path analysis is used to assess genetic and environmental contributions to covariances between variables, e.g., genetic and environmental effects that influence both height and weight or different cognitive abilities. A multivariate extension of the basic phenotypic model in Figure 3.2 is given in Figure 3.3. In this figure, **P**, **G**, **C**, and **SE** represent vectors of observed genetic, shared environmental, and unique environmental values, respectively, and the parameters h, c, and e are replaced by parameter matrices Λ_G, Λ_C, and Λ_{SE}. In this phenotypic model, path tracings yield the following expected covariance matrix of phenotypes:

$$\mathbf{Cov}_{P,P} = \Lambda_G \Lambda_G' + \Lambda_C \Lambda_C' + \Lambda_{SE} \Lambda_{SE}' \tag{11}$$

in which matrix transposes are shown with primes, e.g., Λ_G'. A comprehensive description of multivariate path analysis is given by Vogler (1985), who originally developed the method.

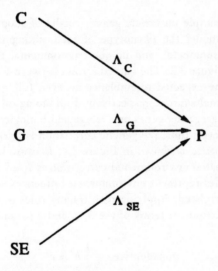

Figure 3.3 Multivariate path diagram of genetic and environmental influences on multiple phenotypes. **P**, **G**, **C**, and **SE** represent vectors of observed, genetic, shared environmental, and unique environmental values, respectively. Matrices Λ_G, Λ_C, and Λ_{SE} contain parameters representing the respective influences on the phenotypes.

Models of Sibling Resemblance

Univariate model

In the univariate and multivariate path diagrams just described, the parameters of genetic and environmental influence (h, c, and e parameters in the univariate model and the Λ_G, Λ_C, and Λ_{SE} in the multivariate model) are unknown quantities which are estimated from observed data. In nuclear families these influences are always confounded and there is no direct way to evaluate their relative importance. However, the adoption design disentangles these effects by comparing resemblances between individuals sharing different proportions of genes. For example, the phenotypic correlation between biological parents and their adopted-away offspring provides direct evidence for the importance of genetic factors, since, in the absence of selective placement, these individuals share only genetic influences. Similarly, the phenotypic correlation between adoptive parents and their adopted children provides direct evidence for the importance of shared environmental influences, since these individuals resemble one another because of environmental factors. A similar comparison may be made between genetically unrelated siblings reared together, who share only environmental influences, and natural siblings, whose covariance is comprised of one-half the additive genetic variance for the trait and all the shared environmental variance. In this section, we discuss some models of sibling resemblance in the CAP, and return to the more complicated parent–offspring models in the next section.

Figure 3.4 shows a simple univariate genetic model of adoptive and nonadoptive sibling data. In this model the phenotype of each sibling is determined by the genetic, common environmental, and unique environmental latent variables G, C, and SE, as shown in Figure 3.2. The curved arrows between G_1 and G_2 and between C_1 and C_2 represent the expected resemblance between full siblings due to genetic and shared environmental sources, respectively. Full biological siblings are expected to share 0.5 of their segregating genes on average and biologically unrelated siblings share no genes whatsoever; thus, the genetic correlation is fixed at 0.5 for biological and 0.0 for adoptive pairs, as shown in Figure 3.4. Because both sibling types are reared together, the shared environmental correlation is fixed at 1.0 for both types. Because the SE variables represent environmental influences which are individually unique, they are uncorrelated. Applying path tracing rules to Figure 3.4 yields the expected sibling correlations in terms of the h, c, and e parameters:

$$\text{nonadoptive } r = \frac{1}{2} h^2 + c^2$$

$$\text{adoptive } r = c^2 \tag{12}$$

Thus, c^2 is estimated directly from the adoptive sibling correlation and h^2 may be estimated as twice the difference between the nonadoptive and adoptive correlations, that is, $h^2 = 2(\frac{1}{2} h^2 + c^2 - c^2)$. The e^2 parameter may be obtained by subtraction; rearrangement of Eq. 10 gives $e^2 = 1 - (h^2 + c^2)$. This univariate model has two free parameters, h^2 and c^2, and one derived parameter, e^2.

Although the model parameters h^2, c^2 and e^2 may be estimated using these subtractive methods, typically they are estimated using multiple regression models or structural equation model-fitting procedures. The regression approach, or "DF model," developed by DeFries and Fulker (1985, 1988) for identical and fraternal twin applications, involves two fundamental forms. The first form, or "basic

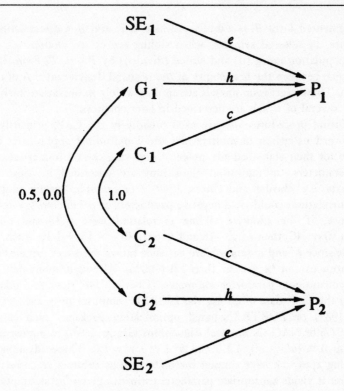

Figure 3.4 Path diagram of the relationship between genetic (G) and environmental (C, SE) influences for two siblings. The genotypes are expected to be correlated at values of 0.5 and 0.0 for biological and adopted siblings, respectively, and the shared environments are correlated 1.0 by definition for all individuals reared together.

model," is designed for use in selected samples and is based on the idea that if a trait is heritable, siblings with different degrees of genetic resemblance to selected probands should exhibit differential regression to the population mean. The dependent variable in this regression model is the observed variable of the co-sibling (cosib; C), which is regressed on the proband score (P) and the coefficient of relationship (R), which equals 0.5 for nonadoptive siblings and 0.0 for adoptive siblings. The complete form of this basic model with a regression constant, A, appears as

$$C = B_1 P + B_2 R + A, \tag{13}$$

in which B_1 adjusts the cosib score for average sibling resemblance and B_2 reflects the extent of differential cosib regression. Statistical significance of B_2 is taken as evidence of heritable influence on the trait. The second form of the regression model, the "augmented model," which is appropriate for both selected and unselected samples, includes an additional term for the interaction between proband score and coefficient of relationship (PR):

$$C = B_3 P + B_4 R + B_5 PR + A. \tag{14}$$

In this augmented form B_3 is a direct estimate of c^2 and B_5 a direct estimate of h^2. Furthermore, in selected samples, when sibling scores are modeled as deviations from the population mean (μ) and scaled (divided) by $\bar{P} - \mu$, B_4 estimates $h_g^2 - h^2$, the difference between the heritability of the proband deficit and that of the overall population. These regression models are applied widely in univariate analyses of the CAP data, several of which are described in later chapters.

Model-fitting procedures also are used broadly in the CAP, primarily for their straightforward extension to multivariate and longitudinal applications (discussed below) and for their statistical advantage of imposing certain numerical constraints on the parameters (minimization algorithms are described in most numerical analysis texts, e.g. Burden and Faires, 1989). For example, certain adoptive/non-adoptive correlations could yield negative parameters or parameters greater than 1.0. For instance, if the adoptive sibling correlation were $-.15$ and nonadoptive correlation were $.40$, then $\hat{c}^2 = -.15$ and $\hat{h}^2 = 2(.40 - -.15) = 1.10$. Such values are not feasible since h^2 and c^2 (and e^2) are variance ratios; variances cannot be negative and the ratios cannot be greater than 1.0 (100%). Although submodels of Eq. 14 facilitate constrained parameter estimates (Cherny, DeFries, & Fulker, 1992), analyses of the CAP data often use the LISREL computer program (Jöreskog and Sörbom, 1989) or FORTRAN-based optimization packages such as MINUIT (CERN, 1977) or NAG (Numerical Algorithms Group, 1990) to impose constraints of the form $0 \leq (h^2, c^2, e^2) \leq 1.0$ and $h^2 + c^2 + e^2 = 1.0$. Three advantages of this model-fitting approach over simpler methods such as subtractive correlation comparisons are: It yields appropriate parameter estimates given the assumption of each model; it provides standard errors for these parameter estimates; and it provides a goodness-of-fit test to aid in the evaluation of alternative models. The maximum-likelihood fitting functions used in model-fitting applications will be discussed in detail later in this chapter.

Multivariate and longitudinal models

Whereas genetic analysis of a single measure provides evidence for a genetic and/or environmental etiology of the measure, multivariate genetic analysis is concerned with the shared etiology of multiple measures. Any behavioral genetic design that can be used to decompose variation for an individual measure also can be used for multivariate genetic analysis. The adoption design employed in the CAP is no exception; indeed, the CAP is particularly well suited for multivariate analysis because the sample has been administered a very large number of measures relating to personality, cognition, scholastic achievement, language, health, and several other substantive areas. With the large number of measures available both within and across psychological domains, the CAP facilitates the examination of such issues as different etiologies for different traits, the existence of one or more genetic factors contributing to different variables, associations between the genes that influence different behaviors, and developmental change and continuity in the etiology of one or more characters. In this section, we describe several of the multivariate models employed in analyses of the CAP sibling data.

The essence of the multivariate models employed in the CAP is the decomposition of covariance between multiple measures obtained from the same individuals into genetic, shared environmental, and unique environmental components. In doing so, it is possible to separate each covariance component into portions which are common to some or all of the variables under consideration and portions which are specific to each variable. For example, the path diagram in Figure 3.5 shows three observed variables, P_1, P_2 and P_3, from one sibling which may be correlated because of common genetic, shared environmental, or unique environmental influences, G_C, C_C, and SE_C, for the three traits. In addition, each observed variable has specific, uncorrelated, underlying effects, G, C, and SE. This model, known as a "common factor" model, is a genetic extension of the single-factor model originally proposed by Charles Spearman (1904) in his theory of general intelligence ("g").

By tracing the paths in Figure 3.5 it may be seen that the covariance between P_1 and P_2 is comprised of a genetic component, $g_1 g_2$, a shared environment component, $c_1 c_2$, and a unique environment component, $e_1 e_2$; thus $\text{Cov}_{P_1,P_2} = g_1 g_2 + c_1 c_2 + e_1 e_2$. Similarly, the other covariances are $\text{Cov}_{P_1,P_3} = g_1 g_3 + c_1 c_3 + e_1 e_3$ and $\text{Cov}_{P_2,P_3} = g_2 g_3 + c_2 c_3 + e_2 e_3$. The variance of the first observed variable may be calculated as $V_{P_1} = (g_1^2 + g_{s1}^2) + (c_1^2 + c_{s1}^2) + (e_1^2 + e_{s1}^2)$, with similar expressions for V_{P_2} and V_{P_3}. The parentheses in this expression denote the variance components attributable to genetic, shared environmental, and unique environmental effects. Each component of variance/covariance may be conveniently represented as the matrix product $\Lambda_{(G,C,E)} \Lambda'_{(G,C,E)}$, where

$$\Lambda_G = \begin{pmatrix} g_1 & g_{s1} & 0 & 0 \\ g_2 & 0 & g_{s2} & 0 \\ g_3 & 0 & 0 & g_{s3} \end{pmatrix}$$

and

$$\Lambda_G \Lambda'_G = \begin{pmatrix} g_1^2 + g_{s1}^2 & g_1 g_2 & g_1 g_3 \\ g_1 g_2 & g_2^2 + g_{s2}^2 & g_2 g_3 \\ g_1 g_3 & g_2 g_3 & g_3^2 + g_{s3}^2 \end{pmatrix}$$

The matrices Λ_C and Λ_{SE} and their products are similarly expressed. Then, the observed covariance matrix may be calculated simply as $\text{Cov}_{P,P} = \Lambda_G \Lambda'_G + \Lambda_C \Lambda'_C + \Lambda_{SE} \Lambda'_{SE}$, which is the same expectation shown in multivariate form in Figure 3.3 and in Eq. 11.

As in the univariate model described above, comparisons of adoptive and nonadoptive sibling pairs may be used to estimate the genetic and environmental parameters. The expected cross-sibling covariances are

$$\text{nonadoptive Cov} = \frac{1}{2}\Lambda_G \Lambda'_G + \Lambda_C \Lambda'_C$$

$$\text{adoptive Cov} = \Lambda_C \Lambda'_C \tag{15}$$

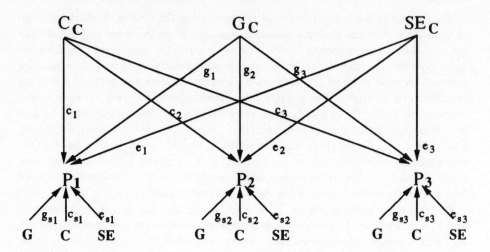

Figure 3.5 Common factor model of genetic and environmental influences on three phenotypes (P_1, P_2, and P_3). Genetic and environmental effects which influence all three phenotypes are subscripted c; effects which are unique to each phenotype have subscripts s.

where $\frac{1}{2}$ is a matrix with all diagonal elements fixed at $1/2$ and all off-diagonal elements fixed at 0.0. It should be recognized that these expressions are just multivariate extensions of those shown in Eq. 12, and, thus, elements of $\Lambda_C \Lambda_C'$ could be estimated directly from the adoptive sibling cross-covariances and $\Lambda_G \Lambda_G'$ could be estimated as twice the difference between the nonadoptive and adoptive sibling matrices, although in practice this subtractive procedure is seldom used.

This matrix representation permits calculation of the heritabilities, environmentalities, and genotype–environment correlations contributing to each variable. The heritabilities may be calculated as the diagonal of the "phenotypically standardized genetic covariance matrix" (Hegmann and DeFries, 1970):

$$\mathbf{Cov}_{G,G} = \mathrm{diag}\,(\mathbf{Cov}_{P,P})^{-.5}\,\Lambda_G \Lambda_G'\,\mathrm{diag}\,(\mathbf{Cov}_{P,P})^{-5}$$

$$\mathbf{h}^2 = \mathrm{diag}\,(\mathbf{Cov}_{G,G}) \qquad\qquad (16)$$

Shared and unique environmentalities (c^2 and e^2) are similarly calculated by substitution of Λ_C and Λ_{SE} for Λ_G. Genetic and environmental correlations are expressed by standardizing the genetic and environmental matrices:

$$\mathbf{R}_G = \mathbf{h}^{-1} \mathbf{Cov}_{G,G}\, \mathbf{h}^{-1}$$

$$\mathbf{R}_C = \mathbf{c}^{-1} \mathbf{Cov}_{C,C}\, \mathbf{c}^{-1}$$

$$\mathbf{R}_{SE} = \mathbf{e}^{-1} \mathbf{Cov}_{SE,SE}\, \mathbf{e}^{-1} \qquad\qquad (17)$$

where $\mathbf{h}$, $\mathbf{c}$, and $\mathbf{e}$ are diagonal vectors of the square roots of the corresponding elements in $\mathbf{h}^2$, $\mathbf{c}^2$, and $\mathbf{e}^2$.

In contrast to the common factor model in Figure 3.5, many multivariate analyses of the CAP data employ variants of the "tridiagonal" or "Cholesky decomposition" model (Gorsuch, 1983). For each covariance component, the Cholesky model specifies a latent factor for each variable, with all variables loading on the first factor, all variables except the first variable loading on the second factor, and so on. A path diagram of a three-variable Cholesky model is given in Figure 3.6, wherein this pattern of loadings may be observed. This type of model is useful for multivariate analysis because its statistical properties ensure a positive semi-definite covariance matrix and provide a "saturated" model having as many parameters as free variances/covariances to be used as a null model for chi-square tests of parameter significance; moreover, it is easily interpreted in terms of etiological components. From a genetic perspective, for example, the first factor, F_1, reflects genetic effects which impact all of the observed variables, similar to the common factor described above and shown in Figure 3.5. The second factor, F_2, represents genetic influences which contribute to the second and third variables after partialling out the common genetic effects of F_1. Similarly, the third factor, F_3, represents genetic influences on the third variable beyond those explained by the first two

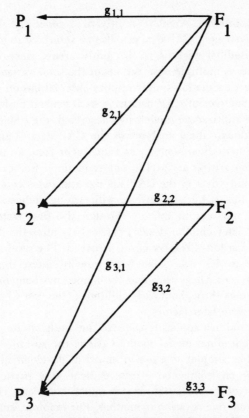

Figure 3.6 Tridiagonal or Cholesky model of three phenotypes. The latent factors F_1, F_2, and F_3 impact the respective phenotypes P_1, P_2, and P_3 and all subsequent ordered measures. In genetic applications the factors F are replaced by the covariance components G, C, and E.

factors. Thus, the Cholesky model can provide evidence for more than one set of genes influencing a group of variables. If all of the observed variables shared a common genetic background, the factor loadings from the second and third factors would be estimated as 0.0; however, if P_2 and P_3 manifest additional genetic influence, the factor loadings for F_2 and F_3 will be non-zero.

The Cholesky decomposition model may be formulated in a similar manner to that described for the common factor model, but the Λ matrices now take a triangular form. For the three-variable Cholesky shown in Figure 3.6, the genetic and environmental parameters may be estimated in Λ_G, Λ_C and Λ_{SE} matrices, where the form of these matrices is now

$$\Lambda_G = \begin{pmatrix} g_{1,1} & 0 & 0 \\ g_{2,1} & g_{2,2} & 0 \\ g_{3,1} & g_{3,2} & g_{3,3} \end{pmatrix}$$

The sibling expectations, heritabilities, environmentalities, and genetic and environmental correlations may be calculated using the exact procedures outlined in Eqs. 11 and 15–17.

In addition to facilitating extensive multivariate genetic analysis of data from children tested at given ages, CAP data can also be subjected to longitudinal genetic analysis. Any longitudinal analysis is also multivariate, since the 'multi' in the longitudinal case means multiple observations of the same variable. However, since the general multivariate situation simply implies observations on multiple measures, recent longitudinal analyses of CAP data have used general multivariate models.

Of the alternative multivariate models just described, the Cholesky model is most commonly used in longitudinal analyses of the CAP data. This model is readily interpretable in a longitudinal context, as subsequent factors may be indicative of new genetic variation arising at specific points in time. For example, analyses of general cognitive ability data in the CAP siblings and twins from ages 1 through 9 years have shown this pattern of effect, with strong indications of continuous influence of genes apparent in infancy through the first factor, and additional genetic influence at later childhood ages, particularly after the first year in school (Fulker, Cherny, & Cardon, 1993; see also Chapter 4). The results of these Cholesky model applications to IQ data have been equally interesting with regard to environmental influences, suggesting that the home environment exerts a continuous impact on IQ scores throughout early childhood (the first Cholesky factor), with little change apparent in later factors.

Aside from longitudinal applications based on multivariate models, the most frequently used longitudinal model in the CAP is the so-called "simplex" model (Guttman, 1954). The simplex is a useful model of development because it posits that genetic and/or environmental effects evident at a particular measurement occasion may be directly transmitted to subsequent occasions; that is, the components may persist from one occasion to another. For example, analyses of CAP data have suggested that genetic effects on IQ at ages 1 and 2 years continue to influence IQ later in childhood, although with diminishing effects over time. In addition, the simplex includes parameters representing time-specific variation which may be incorporated into the transmissible genetic and environmental influences. This portion of the model can highlight important ages at which new genetic or environmental factors impact a selected phenotype. This type of occasion-specific variation also has been observed in the CAP IQ data in that the persistent infant/toddler genetic influences appear to be augmented by new genetic variation shortly after completion of the first year in school (Cardon, Fulker, DeFries, & Plomin, 1992; see also Chapter 4).

In CAP applications, the simplex model is often combined with a common factor model to provide a general account of the developmental process underlying the phenotype of interest. Simplex and factor models differ in their treatment of repeated measures, and by fitting both models simultaneously it is possible to test which one better fits the observed data. The simplex model is designed for data in which the correlations are higher among adjoining occasions (adjacent to the diagonal of the correlation matrix) than among those more distantly related. Thus, the simplex model exhibits correlational dependence across occasions; the partial correlation between any two non-successive occasions, given the intermediary correlation, is zero. The common factor model, however, does not impose such rigorous longitudinal constraints. In this model, correlations which remain constant or even increase over time, or correlations between non-successive occasions which are unrelated to the intervening occasion, may be explained by alternative factor patterns. Thus, a common factor pattern may be observed when genetic and environmental effects are static and unchanging during development, or when they have delayed recurrent effects. This contrasts with the dynamic simplex pattern in which genetic and environmental effects may arise at any time(s), but must persist continuously to produce observable correlations across measurements.

A path diagram of the combined simplex and factor model is shown in Figure 3.7. The transmission process in the simplex model is characterized by the γ regression coefficients between successive occasions $(P_1, \ldots, P_n)$ and the common factor model by the factor loadings, c, of the general factor, G. Time-specific variation is estimated in the s parameters impacting each time point. This simplex/factor model was originally developed for twin applications by Eaves, Long, and Heath (1986) and Hewitt, Eaves, Neale, and Meyer (1988). Phillips and Fulker (1989) first applied the model to the CAP data and extended it to account for assortative mating and environmental transmission.

The general multivariate framework of covariance matrix expectations in Eq. 11 holds for the longitudinal simplex/factor model as well. The only difference is in the content of the Λ matrices. Phillips and Fulker (1989) and Cardon et al. (1992) have shown that all parameters in this model may be expressed in a single matrix of the type described here. An illustrative form of this matrix for four time points is

$$
\begin{array}{c}
\quad\quad F_G \quad\quad\quad\quad\quad\quad\quad\quad\quad F_{S_1} \quad\quad F_{S_2} \quad F_{S_3} \quad F_{S_4} \\
\begin{array}{c} P_1 \\ P_2 \\ P_3 \\ P_4 \end{array}
\left(
\begin{array}{llll}
c_1 & s_1 & & & \\
c_2 + \gamma_1 c_1 & s_1 \gamma_1 & s_2 & & \\
c_3 + \gamma_2(c_2 + \gamma_1 c_1) & s_1 \gamma_1 \gamma_2 & s_2 \gamma_2 & s_3 & \\
c_4 + \gamma_3[c_3 + \gamma_2(c_2 + \gamma_1 c_1)] & s_1 \gamma_1 \gamma_2 \gamma_3 & s_2 \gamma_2 \gamma_3 & s_3 \gamma_3 & s_4
\end{array}
\right)
\end{array}
$$

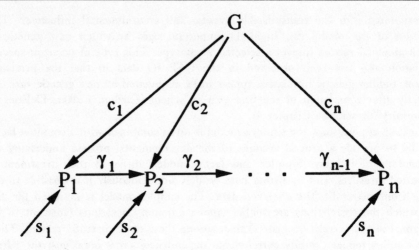

Figure 3.7 Path diagram of combined simplex and common factor model. The common factor is shown as G which influences the phenotypes, P_i, with factor loadings c_i. The simplex model is characterized by the age-to-age transmission parameters, γ. Specific effects originating at each age are shown as s parameters.

The form of this matrix shows that exclusion of selected parameters reduces the model to either a traditional simplex (omitting all c loadings) or a common factor model (omitting the γ coefficients) of the type described previously in this chapter. This matrix, with appropriate extensions for additional time points, is specified for each covariance component, Λ_G, Λ_C, and Λ_{SE}, thus providing a means to assess the genetic and environmental influences leading to continuity and change over the course of development.

Models of Parent–Offspring Resemblance

In addition to comparisons of adoptive and nonadoptive siblings, relationships between biological, adoptive, and nonadoptive (control) parents and their offspring may be used to infer genetic and environmental influences on a trait or traits. In principle, parent–offspring relationships represent the greatest asset of the adoption design: in the absence of selective placement effects, resemblances between biological parents who relinquish their offspring at a very young age and adopted-away children can only be attributable to genetic (or, for mothers, also prenatal) factors, whereas relationships between the parents who adopt those children and the children themselves can arise only from environmental similarities. Thus, each of these relationships yields a direct estimate of a causal component of individual differences.

In practice, however, the parent–offspring design has some limitations. It is intuitively obvious that if children are placed in adoptive homes which are similar to those they would have received in the presence of their biological parents, then the biological parent–offspring correlations will reflect some environmental effects as well as genetic influences. Similarly, if children reside with their biological parents for a sufficient duration that they receive lasting environmental influences, the biological parent–offspring correlations will be confounded with environmental variation. These two design aspects, selective placement and age of placement, can limit the utility of the parent–offspring adoption design. Fortunately, all of the CAP children were separated from their biological parents at very young ages (3 days on average). Moreover, the effects of selective placement, which are expected to be negligible by virtue of the CAP sample ascertainment process, can be examined in a model-fitting framework. An adoption study also may be criticized for being unrepresentative of the general population, as adoptive parents may tend to fall into the upper strata in education and socioeconomic status. The CAP parents are somewhat higher than the national average in terms of occupational status, but they are quite representative of the Denver region from which they were recruited (Plomin, DeFries, & Fulker, 1988).

One aspect of the CAP that requires special consideration in analyses and interpretations is the age of the children. The dramatic difference in ages of testing – parents as adults and offspring as infants and young children – complicates model expectations because the genes that influence childhood behavior may be quite different from those determining adult expression. For example, if the parents and offspring were tested at the same age, we would estimate the biological parent–offspring correlations as $1/2h^2$; however, if the offspring were tested as children and the parents were tested as adults, the expectation would be $1/2h_Ch_Ar_G$; i.e., the product of $1/2$ of the square root of the offspring (C: child) and parent (A: adult) heritabilities and the correlation between the two genotypes (r_G). There is insufficient information in just one correlation to estimate these three parameters. However, if additional information is available, for example, offspring and parent heritabilities (DeFries, Plomin, & LaBuda, 1987), the confound may be resolved and the adult–child genetic correlation may be estimated directly. In the absence of such information, parent–offspring resemblances still are not devoid of utility, as non-zero resemblance suggests heritability in childhood and adulthood, as well as substantial genetic continuity.

In several respects parent–offspring relationships permit much more searching analyses of individual differences and their etiologies than do the adoptive/non-adoptive sibling or classical twin designs. All of the sibling models discussed in this chapter are based on an assumption of random mating in the population for the trait of interest. If nonrandom, or assortative, mating occurs then the genetic parameters will be overestimated in sibling models in twin designs, genetic influences are underestimated, whereas the common environment is inflated (Eaves, Heath, & Martin, 1984). However, with data from spouse pairs, as in the CAP, the effects of assortative mating can be modeled explicitly and their bias examined. Inclusion of nonadoptive families in analyses also can elucidate correlations between genotypic and environmental factors, so-called G–E or G–C correlations. Since nonadoptive parents transmit both genetic and environmental effects to their offspring, G–E correlations may be estimated when data from both adoptive and nonadoptive families are available.

Several models of parent–offspring resemblance have been developed for application to the CAP data, nearly all of which include some convention for assessing assortative mating, environmental transmission, G–E correlation, and selective placement (Cardon, DiLalla, Plomin, DeFries, & Fulker, 1990; Coon, Fulker, DeFries & Plomin, 1990; DeFries et al., 1987; Fulker & DeFries, 1983; Phillips & Fulker, 1989; Rice, Carey, Fulker, & DeFries, 1989). Here we describe only the model of Fulker, DeFries, and Plomin (1988) which has been used most broadly in the CAP. This model is diagrammed for adoptive and biological families in Figure 3.8, in which P_{BM}, P_{BF}, P_{AM}, P_{AF}, and P_{AO} represent the phenotypes of biological mothers and fathers, adoptive mothers and fathers, and the adopted offspring, respectively, and P_M, P_F, and P_O represent the phenotypes of nonadoptive mothers, fathers, and offspring, respectively. Genetic transmission from the biological and nonadoptive parents to their offspring is shown in the paths labeled 1/2, and environmental transmission in the adoptive and nonadoptive families is shown as m (transmission from the mother) and f (transmission from the father). Heritabilities and nonshared environmentalities are shown as h and e. The new parameters in this

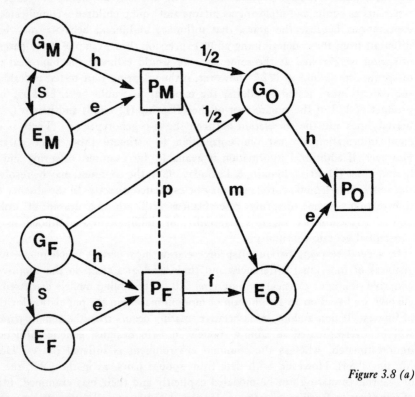

Figure 3.8 (a)

Figure 3.8 Parent–offspring model of nonadoptive (*a*) and adoptive (*b*) family relationships. Parameters representing assortative mating, genotype–environment correlation, and selective placement are shown as *p* and *q*, *s*, and x_{1-4}, respectively. Reprinted from "Genetic Influence on General Mental Ability Increases Between Infancy and Middle Childhood," by D. W. Fulker, J. C. DeFries, and R. Plomin, 1988, *Nature*, *336*: 767–769.

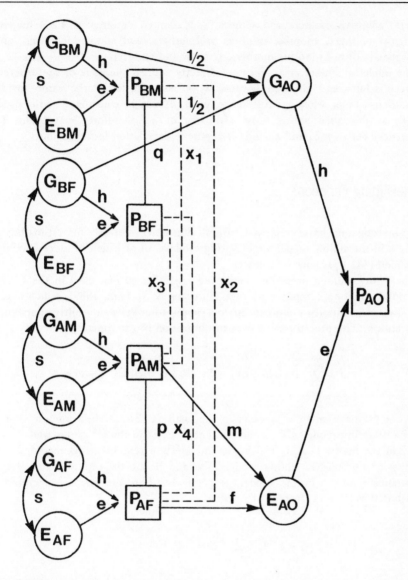

Figure 3.8(b)

parent–offspring model are those representing the G–E correlation (s), assortative mating (p and q for for biological and adoptive parents respectively), and selective placement (x_{1-4}). The assortment and placement parameters are drawn as conditional paths (Carey, 1986) which use slightly different path tracing rules than traditional path analysis, but greatly simplify diagrammatic model representation.

The model has a total of 11 free parameters: h, p, q, m, f, x_{1-4}, V_p, and V_c, where V_p and V_c symbolize variances of the parent and child scales. The s and e parameters are derived from the other parameters in the model. The model is fitted to three covariance matrices: a 5×5 matrix based upon data from biological mothers and fathers, adoptive mothers and fathers, and adopted children; a 4×4 matrix of biological mothers, adoptive mothers and fathers, and adopted children; and a 3×3 matrix of nonadoptive mothers, fathers, and children.

The model of Fulker et al. does not directly address the issue of age differences between adults and children because h is assumed to be the same for both generations. Thus, when age differences exist, h^2 may be considered as the product $h_{C}r_{G}h_{A}$ as described above. Low estimates of h^2, therefore, may result from differences between h_A and h_C and/or a genetic correlation less than unity.

Model-Fitting Functions

CAP investigators have employed several alternative methods for estimating parameters in the sibling or parent–offspring models, all of which produce maximum-likelihood (ML) parameter estimates.

The most common procedure employed is similar to that used in the LISREL structural equation computer package (Jöreskog & Sörbom, 1989), which assumes that the observations examined in any application have a multivariate normal distribution. The procedure involves minimization of the function

$$F = \sum_{k=1}^{m} N_k \{(\ln |\mathbf{\Sigma}_k| - \ln |\mathbf{S}_k| + tr(\mathbf{S}_k \mathbf{\Sigma}_k^{-1}) - p_k\}, \tag{18}$$

where m is the number of groups in the analysis (2 in sibling models, 3 in the parent–offspring model), N_k is the number of cases in the kth group, and p_k is the order of the matrix (e.g. 3, 4, 5 in the univariate parent–offspring model, $2 \times$ the number of variables in sibling models). $\mathbf{\Sigma}_k$ and $\mathbf{S}_k$ are the expected and observed covariance matrices in the model, respectively. This function is asymptotically distributed as χ^2 with degrees of freedom

$$df = \sum_{k=1}^{m} \frac{1}{2} p_k (p_k + 1) - t \tag{19}$$

where t is the number of parameters estimated in the model.

The LISREL package has internal routines for calculating the ML function, degrees of freedom, and standard errors of the estimates (although see Neale, Heath, Hewitt, Eaves, & Fulker, 1989). However, frequently the CAP models are too complex or too large for the LISREL program, and special-purpose programs must be written instead. These programs are then linked with available optimization packages such as MINUIT (CERN, 1977) or NAG (Numerical Algorithms Group, 1990) to minimize the function and calculate corollary quantities.

In recent applications of longitudinal sibling and parent–offspring models, CAP investigators have used an alternative fitting function. Because the CAP is ongoing and involves measurement of the same individuals over many years, longitudinal analyses are often encumbered with many different patterns of missing and non-missing data, particularly at later childhood ages where all individuals in the sample may not yet have reached the testing ages. Although in principle these types of analyses could be fitted using the ML equation in Eq. 18, these applications would be prohibitively costly and slow to execute on the computer. Consequently, these models have been fitted using the ML pedigree procedure of Lange, Westlake, and Spence (1976). In this technique, the following log-likelihood is calculated for each pedigree (sibling pair or family), and summed over all pedigrees:

$$L_k = \frac{1}{2}\{\ln |\Sigma_k| + (x_k - \mu)' \Sigma_k^{-1}(x_k - \mu)\} + c, \tag{20}$$

where Σ_k represents the matrix of expected covariances among sibling measures corresponding to the kth pattern of missing and non-missing data, x_k is the kth observed vector of family data, μ is a vector of estimated means, and c is a constant. Often the elements of μ are set to the observed means rather than estimated in order to reduce computation time, as done in analyses reported in Chapters 4, 5, 12, and 19. For model comparison purposes, twice the difference between two log-likelihoods is distributed asymptotically as χ^2 with degrees of freedom equal to the difference in the number of free parameters estimated in fitting the two models. The most important assumptions of this approach are that missing data are "missing at random" (Little & Rubin, 1987) and that non-missing data conform to a multivariate normal distribution. The latter assumption is also implicit in the more common ML function for covariance matrices. When imposing a listwise deletion procedure to allow use of the more common ML function, the former assumption is also implicit. In the case where there are no missing data, this pedigree function yields the same results as the ML covariance function.

Conclusions

The biometrical models used to analyze CAP data are formulated to extract the most information possible from the powerful adoption study design. These models combine elements of classical quantitative genetics and modern psychometrics to assess genetic and environmental influences on behavioral development. In addition to providing tests of genetic and/or environmental influence for single traits, they assess the etiology of correlations between different characters and of observed continuity and change, as well as analyze the effects of assortative mating, genotype – environment correlation, and selective placement on familial relationships. In this chapter, we have attempted to illustrate this general method of analysis and some of the approaches most frequently used to analyze CAP data. One of the great strengths of this general approach is that it allows rigorous testing of alternative models, thus providing a means to evaluate different hypotheses by their applicability in the substantive area of inquiry. This methodology has been employed in many of the analyses of CAP data reported in the following chapters.

4 General Cognitive Ability

Stacey S. Cherny
Lon R. Cardon

The etiology of individual differences in general cognitive ability has been a major focus of behavioral genetics since the very beginning of the subject in the last century, with Sir Francis Galton's (1869) classic family study *Hereditary Genius* providing the first empirical investigation of this issue. Although Galton's study is of historical significance, its design was necessarily limited. Because it was a family study, genetic and environmental influences could not be separately assessed. Prior to the 20th century, the nature of the hereditary mechanism was unknown, and the need for more appropriate designs, involving twins or adoptions, was not fully appreciated. Moreover, Galton's measure of cognitive ability was highly subjective, based on biographical information and not on psychometric measurement.

Galton laid the foundation for rigorous measurement through the development of linear regression, which others later built upon (Cattell, 1960, 1965). It was with the development of Binet's (1905, 1908) test of intelligence, its translation into English by Terman (1916), and a much clearer understanding of research designs for separating the effects of heredity and environment, that research into this question was finally placed on a firm scientific basis. In 1928, Barbara Burks, under the guidance of Terman, conducted the first adoption study of IQ. Later, Newman, Freeman, and Holzinger (1937) studied IQ in separated monozygotic twins. In addition, they studied a group of unseparated monozygotic and dizygotic pairs and established the statistical methodology of the twin study. Both studies were landmarks in the field of behavioral genetics, established the importance of hereditary influences on IQ, and stimulated a great deal of subsequent research. Thus, in 1981, Bouchard and McGue were able to report the results of more than 140 studies that had examined the importance of genetic and environmental influences on general intelligence employing a variety of research designs. The general picture obtained from their report is overwhelmingly one of a strong hereditary component, accounting for about 50% of variation, with a more modest contribution from the environment shared by children raised in the same family, this source of variation being about 10–20%.

Surprisingly, in spite of the large number of studies of IQ that have been carried out, very little beyond these rudimentary facts is present in the literature. For example, relatively little is known about the etiology of individual differences in cognitive development. Prior to the Colorado Adoption Project (CAP), only two long-term longitudinal studies attempted to address this issue. An early study of Skodak and Skeels (1949), in which adopted children were tested at least four times between infancy and adolescence, found increasing biological parent/child relationship with age, suggesting increasing heritability for IQ, or an increasing genetic correlation between child and adult measures of IQ. However, the study was limited by small numbers and a lack of clear goals at the outset. In contrast, the Louisville Twin Study (Wilson, 1972a, b, c, 1983) was fully prospective in character and studied cognitive development from infancy through adolescence in several hundred twin pairs. The focus of this study was IQ and it too suggested increasing heritability with age. However, in neither study was a thorough developmental analysis carried out. It was only comparatively recently that appropriate statistical methodology has been developed (Eaves, Long, & Heath, 1986; Hewitt, Eaves, Neale, & Meyer, 1988; Phillips & Fulker, 1989). In 1975, the Colorado Adoption Project was undertaken in order to investigate, in a fully prospective manner, the genetic and environmental determinants of child development. A major focus of the CAP has been on general and specific cognitive abilities and it is these measures that are most thoroughly investigated from a developmental perspective. In this chapter, we will describe some of this work relating to the development of general intelligence from ages 1, 2, 3, 4, 7, and 9. The following chapter discusses the development of specific cognitive abilities from ages 3, 4, 7, and 9.

Subjects and Measures

General cognitive ability data were obtained from 87 to 32 adoptive sibling pairs, 102 to 43 nonadoptive sibling pairs, and 300 to 278 singletons at 1, 2, 3, 4, 7, and 9 years of age, with the decreasing numbers arising from the ongoing nature of the study. The tests used were the Bayley Mental Development Index (MDI) (Bayley, 1969) at ages 1 and 2, the Stanford-Binet Intelligence Scale (Terman & Merrill, 1973) at ages 3 and 4, the Wechsler Intelligence Scale for Children – Revised (Wechsler, 1974) at age 7, and a first principal component score from the telephone-administered specific cognitive abilities test battery at age 9 (SCATPC) (see Kent and Plomin (1987) for a description of the year 9 test battery and its validation).

Table 4.1. Means and standard deviations for general cognitive ability from 1 to 9 years of age

Measure	Age	Adopted probands			Unrelated sibs			Nonadopted probands			Biological sibs		
		Mean	SD	N	Mean	SD	N	Mean	SD	N	Mean	SD	N
Males													
Bayley MDI	1	106.66	11.64	130	113.67	14.16	46	109.50	12.19	133	114.19	13.05	58
Bayley MDI	2	105.81	15.21	115	103.52	15.68	44	106.70	15.72	124	106.82	14.23	51
Stanford–Binet	3	102.70	13.31	113	101.89	16.83	37	105.91	15.35	113	106.14	16.43	49
Stanford–Binet	4	105.58	10.73	109	104.11	11.57	38	107.89	12.37	114	108.94	13.01	49
WISC–R IQ	7	113.48	11.02	107	113.44	13.41	25	115.38	11.41	116	118.00	8.80	42
SCATPC	9	-.13	1.00	99	-.14	1.33	12	.12	.99	92	.04	.91	25
Females													
Bayley MDI	1	106.40	11.73	113	110.64	11.39	44	107.88	12.57	113	114.23	11.11	44
Bayley MDI	2	109.05	14.06	101	109.91	15.85	43	110.70	16.10	105	113.98	15.85	42
Stanford–Binet	3	106.29	14.78	92	104.95	15.27	41	110.02	13.37	101	111.40	19.39	40
Stanford–Binet	4	108.53	12.38	86	109.78	14.29	40	111.17	10.88	100	111.79	17.89	39
WISC–R IQ	7	109.33	9.90	87	111.00	10.03	31	113.67	10.57	100	116.92	8.40	25
SCATPC	9	-.24	.92	81	.10	.81	23	.17	1.00	81	.37	.78	20

Mean Differences

In Table 4.1, means and standard deviations, by sex and family type, are presented for ages 1 through 9 years, along with sample sizes. A multivariate analysis of variance (MANOVA) with sex and adoptive status (adopted proband versus nonadopted proband) as between-subjects factors and age as a within-subjects factor, was performed. Of the between-subjects effects, only adoptive status was significant ($F_{1,286} = 7.71, p < .01$). However, adoptive status accounted for less than 3% of the total variance. The main effect of age was also significant, but should not be interpreted since different scales are used at different ages. More importantly, the year 9 test is based on a standardized principal component score which employs a completely different scale of measurement from the tests at the other ages. The sex-by-age interaction was also significant ($F_{5,282} = 6.02, p < .01$), but, again, the proportion of variance that this effect explains is very small ($< .1\%$). Since these significant effects only explain a very small proportion of the variance, they are unlikely to have any substantial effect on second-degree statistics such as correlations and covariances, upon which our subsequent modeling is based.

Developmental Model

The developmental model that was fitted to these data was first proposed by Eaves et al. (1986), and represents a combination of a single general factor present at all ages and a quasi-simplex model of specific effects arising at each age point and their subsequent transmission to later ages. The general factor implies a static developmental process where an influence present at an early age persists across the entire period. The quasi-simplex implies a more dynamic process in which new variation arises at each age, persists to the next age, and is of equal or progressively decreasing importance at subsequent ages. A path diagram of this model appears in Figure 3.7 and the model is described in detail in Chapter 3. Models were fitted using the maximum-likelihood pedigree procedure, also outlined in Chapter 3.

Phenotypic Continuity and Change

The cross-age correlations across all children appear in Table 4.2. As is characteristic of a quasi-simplex pattern of development, these correlations are highest near the diagonal of the correlation matrix, and decrease the further away the correlation is from the diagonal. However, their general large magnitude might also imply a common factor mechanism. The extent to which each process is operating will be examined next.

Table 4.2   Phenotypic correlation matrix

Age	1	2	3	4	7	9
1	155.58	635	593	582	538	435
2	.37	236.46	597	580	523	416
3	.23	.51	226.41	584	511	410
4	.22	.46	.60	158.13	501	406
7	.22	.36	.36	.45	119.56	425
9	.09	.30	.29	.29	.59	1.00

*Note: r*s below diagonal, *N*s above diagonal, and variances on diagonal.
$p < .01$ for all *r*s, except $r_{1,9}$, which was not significant.

For the model-fitting procedures, the data were first standardized within each age, across all individuals as a single group. This standardization procedure effectively eliminates age differences in variances, which most likely are merely a result of using different tests at different ages, while preserving adoptive, nonadoptive, sib 1, and sib 2 variance differences. Resulting parameter estimates were fully standardized to imply unit variance for all latent and observed measures.

Fitting a phenotypic simplex/factor model to the CAP IQ data resulted in the factor structure presented in Figure 4.1. A test of the common factor loadings indicated that those are essential to explain these data (Table 4.3, Model 2), as are the simplex transmission parameters (Model 3). Both processes seem to be operating to produce the age-to-age continuity present in these general cognitive ability measures. The simplex transmission parameters appear stronger between later ages than earlier ages, while the common factor loadings suggest an opposite pattern.

In a phenotypic analysis of Louisville Twin Study data, Humphreys and Davey (1988) found that a simplex adequately accounted for the longitudinal correlations from ages 9 months through 9 years. However, in a recent analysis of CAP sibling

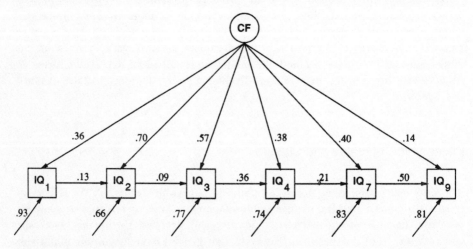

Figure 4.1   Full phenotypic model of IQ from ages 1 through 9 (*CF* represents the phenotypic factor common to the six time points).

data, along with a sample of twins, Cardon, Fulker, DeFries, and Plomin (1992) found that a simplex and common factor were both necessary to explain the phenotypic correlations from ages 1 through 7 years. The nature of the phenotypic continuity will be better understood when separating genetic from environmental influences.

Table 4.3 Phenotypic model comparisons

Model	Form	$-LL^a$	$NPAR^b$	χ^2	df	p
1	Full model	1017.352	17			
2	Model 1, drop common factor	1053.921	11	73.138	6	<.001
3	Model 1, drop simplex	1079.996	12	125.288	5	<.001

[a] Log-likelihood of the data (without the addition of the constant).
[b] Number of free parameters.

Genetic and Environmental Continuity and Change

As discussed in Chapter 3, with data on adoptive and nonadoptive sibling pairs we can examine these developmental structures at both the environmental and the genetic levels, in order to test if similar or different developmental processes are operating at these different levels. Three sources of variation may be observed at each time point, these corresponding to the components of variance in the univariate genetic analysis. The expectation of the correlation among adoptive sibling pairs is c^2, or the proportion of variation ascribable to shared environmental influences. For nonadoptive sibling pairs, the expected correlation is c^2 plus one-half the genetic variance, h^2. A third component, e^2, ascribable to the effects of the specific environment, or influences unique to the individual, is $1 - h^2 - c^2$. These are the same components of variance derivable from the classical twin study. In the sibling adoption design, any effects due to assortative mating will lead to a slight overestimation of h^2 compared to e^2, while c^2 will be unbiased. Combining parent/offspring with sibling data results in unbiased estimates (Phillips & Fulker, 1989). However, the combined analysis is highly complex and is most fruitful when data on the adopted and nonadopted children are obtained when the children are adults. In the present chapter and the next, we will restrict our analyses to siblings. Model fitting to the parent–offspring data for ages 7 and 9 is presented in Chapter 19, in which the effects of the home environment on the child's IQ are explored.

Sibling correlations at each age are presented in Table 4.4. It is clear that in all cases, the nonadoptive correlations are higher than the adoptive correlations, implying genetic influence. At ages 1, 3, and 9, the adoptive correlations are not trivial, which would also imply some shared environmental variance. However, it should be noted that only at age 3 is the adoptive correlation significant. These correlations are consistent with correlations previously reported for ages 1 through 4 years on CAP data (Plomin, DeFries, & Fulker, 1988), although the year 4 adoptive correlation is now somewhat lower. Simply examining sibling correlations at each age does not yield an optimal analysis of longitudinal data, however. Of equal or greater interest are the cross-sibling cross-time correlations, and models of them which will allow further dissection of the developmental process.

Table 4.4   Sibling correlations at each age

Age	Adoptive		Nonadoptive	
	r	N	r	N
1	.12	87	.39**	102
2	.02	80	.35**	91
3	.30*	73	.37**	87
4	.05	74	.24*	88
7	.00	50	.23	65
9	.32	32	.38*	43

$*p < .05, **p < .01.$

Estimates of h^2, c^2, and e^2, obtained from fitting the full simplex/common factor model to the sibling data, are presented in Table 4.5. Heritability appears to decrease from 1 to 4 years of age and then increases at years 7 and 9. The proportions of variance due to shared environmental influences are relatively small, ranging from .07 up to .22. It should be noted that the genetic and shared environmental variance components estimated from fitting the full model to these data are somewhat different from what univariate analyses of the sibling correlations at each age might yield. These differences arise because the multivariate models take into account the cross-sibling cross-time covariance structure which can impact the within-time parameter estimates. For example, even though the sibling correlations at age 9 for nonadoptive and adoptive pairs are highly similar (.38 vs .32, respectively), the estimate of heritability at age 9 is .60 in the multivariate analysis. The (average) cross-sibling correlation between years 7 and 9 is substantially larger for nonadoptive than for adoptive pairs, which accounts for this apparent discrepancy. While such an outcome may on the surface seem inappropriate, it is, in fact, the best estimate of heritability we can obtain from these data. Use of the cross-sibling cross-time information is an inherent advantage of a multivariate longitudinal analysis. When samples sizes are small, such as for our year 9 data, the standard errors of the sibling covariances are relatively large. However, the covariances between one sibling at an earlier age and the other sibling at a later age also contribute information to the within-age sibling covariances. This information would not be used in a univariate analysis at each age; thus, resulting variance component estimates would have larger standard errors and might differ from those obtained from the multivariate analysis.

Table 4.5   Estimates of h^2, c^2, and e^2 at each age

Variance component	Age					
	1	2	3	4	7	9
h^2	.47	.41	.32	.26	.45	.60
c^2	.13	.07	.22	.12	.07	.16
e^2	.40	.52	.47	.62	.48	.24

Tests of the various components of the full model were performed to determine which aspects of the model were essential for explaining these data and which were not. Tests of the unique environmental parameters were conducted first (see Table 4.6). Neither the unique environmental common factor (Model 2) nor the transmission parameters (Model 3) were essential for an adequate model fit. The specific factors were not tested, since measurement error contained in those parameters is essential for the model. The model with only specific factors at the unique environmental level (Model 4) was used as the base model for the next set of tests, those of the shared environmental developmental processes.

Table 4.6 Tests of unique environmental developmental patterns

Model	Form	$-LL^a$	$NPAR^b$	χ^2	df	p
1	Full model	983.183	51			
2	Model 1, drop common factor	984.680	45	2.994	6	>.80
3	Model 1, drop transmission	984.544	46	2.722	5	>.70
4	Model 1, drop transmission & common factor	987.729	40	9.092	11	>.60

[a] Log-likelihood of the data (without the addition of the constant).
[b] Number of free parameters.

At the shared environmental level, none of the parameters was needed to explain these data (see Table 4.7). Tests of the common factor (Model 5) and the transmission parameters (Model 6) indicated neither was necessary. Further tests of the specific factors indicated that those, too, were unnecessary (Model 7). A final test of the common factor against a model with no other shared environmental parameters indicated that it still could be dropped from the model with no significant decrement in fit (Model 8). This final model was then used as the base model in the final set of tests, those of the genetic developmental processes (see Table 4.8).

Table 4.7 Tests of shared environmental developmental patterns

Model	Form	$-LL^a$	$NPAR^b$	χ^2	df	p
	Model 4	987.729	40			
5	Model 4, drop common factor	990.522	34	5.586	6	>.45
6	Model 4, drop transmission	991.401	35	7.344	5	>.15
7	Model 6, drop specifics	991.725	29	0.648	6	>.99
8	Model 7, drop common factor	996.035	23	8.620	6	>.15

[a] Log-likelihood of the data (without the addition of the constant).
[b] Number of free parameters.

At the genetic level, the common factor was found unnecessary in explaining these data (Model 9), but the transmission parameters were necessary (Model 10). Finally, age-specific genetic influences were also necessary (Model 11). When testing each genetic time-specific loading individually, all but the new variation at age 4 were statistically significant.

Table 4.8   Tests of genetic developmental patterns

Model	Form	$-LL^a$	$NPAR^b$	χ^2	df	p
	Model 8	996.035	23			
9	Model 8, drop common factor	1001.827	17	11.584	6	>.05
10	Model 8, drop transmission	1056.713	18	121.356	5	<.001
11	Model 9, drop specifics	1072.051	12	140.448	5	<.001

[a] Log-likelihood of the data (without the addition of the constant).
[b] Number of free parameters.

The final reduced model of general cognitive development is presented in Figure 4.2. There are substantial genetic transmission parameters, along with time-specific genetic influences. The unique environmental influences are only time-specific.

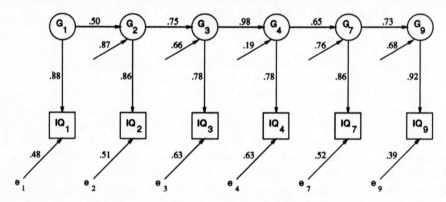

Figure 4.2   Fully standardized final reduced model of the development of IQ (G_i are time-specific genetic factors. Residual effects on the G_i represent new genetic variation arising at each time point. e_i are specific environmental influences at each age. The loadings of the G_i on IQ_i are the h_i, or the square-root of heritability at each age i)

Conclusions

The outcome of these analyses reveals a striking diversity among the genetic and environmental processes that determine continuity and change in individual differences in general cognitive ability during the developmental period from infancy to middle childhood. The nature of these processes could not have been inferred from the phenotypic structure alone.

Although there was insufficient power to detect the small shared environmental influences, they appear to be largely continuous throughout the entire period, as suggested by the full model. That is, shared influences are completely correlated over time. This claim is supported by similar analyses that incorporated twin data along with our sibling sample, thereby increasing the power to detect this small effect (Fulker, Cherny, & Cardon, 1993). This finding suggests that shared environmental influences might be mainly due to stable familial such as socioeconomic status.

While we presently find less shared environmental variance than previous twin studies have reported for these ages (e.g. Wilson, 1983), we must keep in mind that the majority of sibling pairs in the CAP are of opposite sex and that all pairs, necessarily, differ in age. We will, therefore, not detect those shared environmental influences which are only shared by same-sex siblings. Furthermore, siblings differing in age will likely share fewer experiences than dizygotic twin pairs. These factors contribute to reducing the shared environmental variance detected in the sibling adoption design as compared to twin studies.

In the unique environment, the major influences are specific to each time point and there is very little evidence of continuity in the results. That is, influences unique to the individual that either enhance or attenuate cognitive performance are almost entirely transitory and have only a very small tendency to persist from one age to the next. These influences are, in large part, attributable to what psychometricians would refer to as measurement error. However, it is doubtful these influences are entirely due to measurement error, since the unreliability of IQ tests is substantially less than the estimates of the proportions of unique environmental variance, which ranged from .24 to .62. It is comforting to note that environmental influences specific to the individual, such as illness and fluctuation in mood or state, have only transient influences and do not persist over time.

By far the most interesting picture emerges at the level of the genotype, where there is both continuity and change characteristic of a genuine developmental process. There is strong transmission from year 1 through year 9. However, with the exception of year 4, there is also new genetic variation with each age that was not manifested at an earlier point in time and that persists throughout the period of our analysis. A large amount of new genetic variance appears at year 7 and continues to exert its influence at year 9. This influx of new genetic variation from early to middle childhood may represent the genetic underpinnings of the major shift in cognitive processes at this age that has been emphasized in most theories of cognitive development.

While the phenotypic structure suggested both simplex and common factor processes operating, no common factor was found necessary in the full genetic analysis. The common factor process is probably most evident at the shared environmental level, but we had insufficient statistical power to detect it. The genetic and unique environmental levels may also have contributed to the phenotypic common factor, but their effects were not substantial. In the full genetic model, 18 common factor parameters were estimated versus 6 in the phenotypic model. This greater number of parameters would result in less power to detect what is probably a negligible effect.

In conclusion, results of the developmental analysis we have employed suggest that there is not a single developmental process that accounts for individual differences in intellectual ability from ages 1 through 9 years. The three sources of variation we have identified, shared environmental, unique environmental, and genetic influences, each appear to act in a rather different manner, with genetic influences showing the greatest complexity. Those environmental influences shared by siblings have only a small static influence on cognitive ability and do not drive the developmental process. Environmental influences unique to the individual have substantial impact, but their effects are transient and do not persist over time. In

5 Specific Cognitive Abilities

Lon R. Cardon

As explicated in the previous chapter, a developmental genetic analysis of general intelligence is a relatively straightforward process of imposing a developmental model upon the traditional univariate adoption design. The study of the development of specific cognitive abilities adds a further layer of complexity to the problem since we have to define an appropriate psychometric structure for the specific test battery. That is, with multiple specific abilities we must expand the developmental model to account for observed and underlying relationships between different measurements within each occasion.

The study of the psychometric relationships of specific abilities has a long history in psychology, going back to the work of Thurstone (1938), Burt (1939), and Guilford (1967). Following Spearman's (1927) strong statement regarding a general cognitive factor, "g", it became apparent that the tests that assessed specific cognitive domains correlated more highly among themselves than they did across domains, indicating the presence of specific cognitive factors in addition to the pervasive general factor. The psychometric problem was to define these factors and the relationships among them in terms of general intelligence. Thurstone suggested that there were six or seven basic cognitive attributes, whereas later researchers such as Guilford suggested over a hundred.

A basic issue and occasional point of contention among cognitive psychologists, past and present, is whether or not these specific abilities are correlated to any substantial degree or whether they are largely independent of each other. That is, are we dealing with orthogonal or correlated factors? The notion of orthogonal factors is incompatible with the concept of "g" and seems implausible to many psychologists (e.g., Humphreys, 1985; Vernon, 1979). Thus, a model involving correlated factors seems more reasonable to many intelligence researchers.

The notion of correlated factors leads naturally to a hierarchical model of intelligence of the type proposed by Burt (1949), for which there is considerable current support (see Scarr, 1989). The hierarchical intelligence model includes three or more increasingly general levels of abilities, from very specific attributes assessed at the level of the psychometric instrument to one or more superordinate general intelligence factors. The number of specific and general abilities, as well as the number and complexity of the intervening levels, frequently differs among investigators. In the present context there are three levels of the hierarchy we wish to recognize: (1) individual tests, (2) ability groups defined by the tests, and (3) a cluster of ability groups, a single general factor we call general intelligence.

One great advantage of this model is that test-specific variance is separated from that of the factor defining the ability. Thus, idiosyncratic aspects of a particular test, together with unreliability, are in principle separated from our definition of the specific cognitive trait. Consequently, the relationship among these specific cognitive traits is also expected to be free of these artifacts. Of course, the ideal in this respect would involve a very large number of specific tests or indicators of the trait; however, in practice, we are limited to at best three and usually two such indicators.

The test battery that we employ in the CAP is based on the cognitive tests developed for the Hawaii Family Study of Cognition (DeFries, Plomin, Vandenberg, & Kuse, 1981) in which four major specific abilities were defined: verbal ability, spatial ability, perceptual speed, and visual memory. The test battery for adults contains 16 tests tapping these domains. For testing children, test batteries were adapted by CAP investigators in an attempt to assess these same four domains. Unlike general cognitive ability, which can be measured even at the earlier ages, specific cognitive abilities did not seem to emerge as measureable entities until age 3 (Rice, Corley, Fulker, & Plomin, 1986; Singer, Corley, Guiffrida, & Plomin, 1984). Nevertheless, individual differences in several of the abilities measured after age 3 are due substantially to genetic influences (Bergeman, Plomin, DeFries, & Fulker, 1988; Plomin & DeFries, 1985; Rice, Carey, Fulker, & DeFries, 1989). Moreover, these analyses suggest that different abilities are differentially influenced by genetic and environmental factors in early childhood.

The primary issue we address in this chapter is the changing pattern of these differential effects and their emergence and persistence between the ages of 3 and 9 years.

Hierarchical Model of Specific Abilities in the CAP

Cardon, Fulker, DeFries, and Plomin (1992) have developed a genetic hierarchical model for application to CAP sibling data at one time point. This model is designed to estimate genetic and environmental effects which are common to different abilities and to separate these effects from those which impact each ability independently. This separation permits rigorous evaluation of two competing hypotheses: (1) the "genetic 'g' " or "single set" hypothesis (Eaves & Gale, 1974; Vandenberg, 1968), which holds that different abilities are influenced by the same set of genes that create similarities among abilities; and (2) the "specificity" hypothesis, which posits that different sets of genes affect different cognitive abilities. Of course, a third possibility is the presence of both specific and "g" genes, and the hierarchical model can detect this pattern of influence as well. This is an extension of earlier specific-abilities models which discerned the extent to which different abilities are differentially influenced by genetic/environmental components, but did not elucidate the sources of the differential effects.

A path diagram of the hierarchical specification for one time point is presented in Figure 5.1. Although the number of measurements differs at the various child ages in the CAP, the diagram illustrates the typical case of eight measures, of which two each define the Verbal (V), Spatial (S), Perceptual Speed (P), and Memory (M) primary group factors. Correlations among the primary factors are explained by a general common factor (IQ). Factor loadings for the general factor and primary factors are denoted by γ_i and λ_i, respectively, and primary factor residuals and measurement-specific effects are respectively labeled ψ_i and ε_i. A description of the parameter derivations and sibling covariance expectations has been presented by Cardon, Fulker, DeFries, and Plomin (1992).

Schmid-Leiman transformation

Taken at face value, hierarchical models do not yield conclusive evidence for a common factor with independent group factors because all factors are potentially correlated in hierarchical models (Humphreys, 1985, 1989; Humphreys & Davey, 1988). However, hierarchical formulations may be transformed to yield orthogonal common and group factors. One such transformation is the Schmid-Leiman (1957)

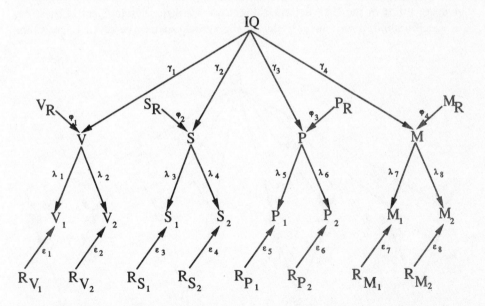

Figure 5.1 Hierarchical path model of specific abilities in the CAP. Symbols V, S, P, and M denote Verbal, Spatial, Perceptual Speed, and Memory abilities, respectively. Residual effects are symbolized as R. General intelligence is represented by the higher-order factor IQ.

procedure, which involves products of hierarchical factor loadings to generate a general factor and uncorrelated primary factors.

Figure 5.2 shows a path diagram of the hierarchical model of Figure 5.1 after Schmid-Leiman transformation. The factor loadings on the general factor are created as the product of the first-order coefficients, λ_i, and the relevant general factor loading, γ_i; the orthogonal primary factor loadings are products of the first-order loadings, λ_i, and the primary factor residual coefficients, ψ_i. Measurement-specific effects are unchanged by the Schmid-Leiman procedure.

Comparison of adoptive and nonadoptive sibling pairs provides a means to assess the extent to which genetic and environmental influences on specific-ability measures correspond to the general factor or specific-ability domains shown in Figures 5.1 and 5.2. This involves estimation of hierarchical parameters for each covariance component, genetic (G), shared environmental (C) and nonshared or unique environmental (SE), and fitting the model according to the general multivariate genetic procedure outlined in Chapter 3. Then, the Schmid-Leiman procedure is employed to separate the genetic and environmental variation of each cognitive ability measure into uncorrelated components relating to general intelligence, primary abilities, and measurement-specific effects. For the ongoing CAP sibling data we employ the pedigree function shown in Eq. 20 of Chapter 3 to make full use of all available data at each age point.

Specific-Ability Measures

A major effort in the CAP has been directed toward the development of measures of specific abilities that broadly define the verbal, spatial, perceptual speed, and

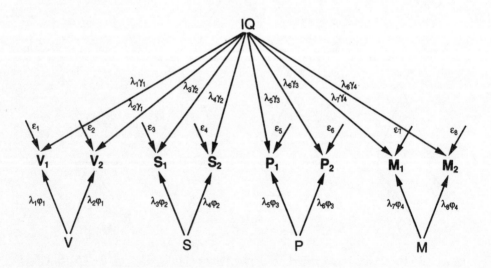

Figure 5.2　Schmid-Leiman (1957) transformation of hierarchical model.

memory domains in early and middle childhood. At age 3 the children are administered seven specific-ability measures constructed by the CAP staff, whereas at 4 years they are administered nine tests. The year 7 test battery consists of two subscales of the WISC-R (Wechsler, 1974) and six tests of specific cognitive abilities constructed by the CAP staff. The tests administered at 9 years of age are modified from the HFSC and CAP adult batteries for telephone administration. Results of

Table 5.1 Factor loadings for child tests of specific cognitive abilities

Age	Measure	Factor			
		Verbal	*Spatial*	*Perceptual Speed*	*Memory*
3	Word Fluency	.97	.11	.03	.01
	Block Design	.11	.93	.02	.04
	Form Discrimination	.10	−.15	.81	.01
	Picture Identification	−.05	.19	.79	−.08
	Figure-Ground	.01	−.04	.48	.40
	Recognition Memory	−.12	.27	.03	.72
	Picture Vocabulary	.16	−.14	.03	.83
4	Word Fluency	.59	.09	.11	−.09
	Picture Vocabulary	.74	.04	−.08	.22
	Vocabulary	.78	−.01	.01	−.09
	Block Design	−.06	.81	.12	.01
	Puzzle Solving	.06	.87	−.11	.02
	Form Discrimination	−.02	.04	.88	.05
	Picture Identification	.06	−.03	.89	.01
	Immediate Memory	.09	.07	−.02	.84
	Delayed Memory	−.08	.02	.08	.88
7	WISC-R Vocabulary	.85	.14	−.14	.04
	Things/Categories	.79	−.13	.18	−.03
	WISC-R Blocks	.05	.81	.11	−.06
	Spatial Relations	−.03	.87	−.04	.06
	Identical Pictures	−.00	.06	.81	.01
	Colo. Perceptual Speed	.01	.01	.86	.03
	Immediate Memory	−.03	.03	.06	.85
	Delayed Memory	.04	−.04	.03	.88
9	Vocabulary	.84	.04	.01	−.02
	Similarities	.89	−.05	−.02	.01
	Hidden Patterns	.07	.55	.35	.04
	Card Rotation	−.00	.96	−.07	.01
	Finding As	−.02	−.07	.93	−.02
	Colo. Perceptual Speed	.04	.09	.78	.04
	Immediate Memory	.01	.06	−.03	.89
	Delayed Memory	−.01	−.06	.02	.92

Note: Measures used to define each factor in the multivariate genetic analyses are in italics. The Figure-Ground test at age 3 is used as an indicator of both Perceptual Speed and Memory.

exploratory and confirmatory factor analyses have indicated that the children's test batteries adequately assess the four cognitive domains and that the Verbal, Spatial, Perceptual Speed, and Memory factors defined by the battery are similar to those assessed by the adult measures, particularly at the later ages (see Table 5.1 and Cyphers, Fulker, Plomin, & DeFries, 1989). Test-retest correlations for the year 9 telephone measures are similar to the in-person adult scales (Kent and Plomin, 1987), and the factor structure is congruent with that observed in the year 7 battery (Cardon, Corley, DeFries, Plomin, & Fulker, 1992).

Table 5.1 shows the factor patterns for the various tests at 3, 4, 7, and 9 years of age. The loadings were obtained by principal-components analysis with oblique rotation to allow for four correlated ability factors at each age. The relevant factor loadings for each measure on the Verbal, Spatial, Perceptual Speed, and Memory factors are italicized, showing a fairly clear separation of abilities at each age. The measures were corrected for sex and age effects by multiple regression prior to the factor analyses. Current sample sizes for which specific ability measurements are available are shown in Table 5.2.

Table 5.2 Number of subjects tested on specific-ability measurements

Type of individuals	Age			
	3	4	7	9
Probands:				
Adopted	226	221	201	191
Nonadopted	232	233	219	178
Siblings:				
Adopted sibs of adoptees	80	77	49	28
Nonadopted sibs of adoptees	23	23	20	17
Sibs of nonadoptees	109	111	74	54

Correlations among the offspring ability factors are presented in Table 5.3. The correlations in the diagonal blocks of this table, typically ranging from .20 to .30, represent ability associations within each measurement occasion, and, thus, are manifestations of the "g" factor at each age. The off-diagonal blocks reflect continuity of the specific abilities over time. The interesting trend in these correlations is the tendency for within-ability correlations to be greater than the cross-trait correlations. For example, the within-ability correlations between ages 4 and 7 are .42, .39, .30, and .16 for V, S, P, and M, respectively, whereas the cross-trait values are generally smaller, and range from .00 to .35. This pattern indicates greater continuity of particular specific abilities than continuity of "g," since the reduced cross-trait, cross-time correlations are representative of overlapping abilities comprising the "g" factor.

Mean differences for gender and adoption status (adopted versus nonadopted probands) were examined with all of the (uncorrected) individual tests at each age serving as dependent variables in a series of MANOVAs. The outcomes revealed adoption status differences at ages 3 ($F_{7,376} = 2.36$, $p = .02$) and 7 ($F_{8,413} = 2.22$,

$p = .03$), and gender differences at ages 4 ($F_{9,369} = 4.02$, $p < .01$), 7 ($F_{8,413} = 8.25$, $p = .00$), and 9 ($F_{8,342} = 4.02$, $p = .02$). Univariate F tests of these significant MANOVAs, although not strictly appropriate due to the lack of independence of the ability tests, reveal that the adoption status differences are uniformly in the direction of higher scores for nonadopted than adopted probands with no consistent relation to ability area. In contrast, the gender differences are domain-specific, with males scoring higher on verbal and spatial tests at ages 7 and 9 and females scoring higher on verbal tests at age 3 and memory tests at ages 3, 7, and 9. In most cases these differences are relatively small and are unlikely to have adverse effects on our model-fitting to covariance structures.

Hierarchical Model-Fitting Results

Hierarchical models including all parameters for genetic, shared environmental, and unique environmental factor structures were fitted to the specific-ability data from the adoptive and nonadoptive siblings at 3, 4, 7, and 9 years. The results from these models are presented separately for each age.

Year 3

The full hierarchical genetic model for the year 3 sibling data has 51 free parameters, comprised of four higher order factor loadings (γ), two primary factor residuals (ψ), four measurement loadings (λ), and seven error terms for each of the three latent covariance components. Application of the model to the seven ability measures at this age yielded a log-likelihood of -1710.23 for the 51 estimated parameters. Standardized parameter estimates from the fit of the full model are presented in Table 5.4. In this table, the "genetic factor" column indicates genetic overlap between "g" and latent group factors for the top portion of the table; in the bottom portion of the table, this column indicates genetic overlap between latent group factors and observed measures. The "genetic residual" column refers to genetic influence independent of "g" (in the top portion of the table) or independent of latent group factors (in the bottom portion). The other columns in the table indicate overlapping ("factor") and independent ("residual") unique environmental and shared environmental influences.

The genetic results in the top portion of the table show moderate to large factor loadings for the four abilities on "g" (range = .26–.74), with additional genetic variation independent of "g" on the Perceptual Speed factor (residual = .53). This outcome indicates that although the genetic "g" factor accounts for the majority of the genetic variance on the four primary abilities, there are heritable effects beyond those impacting general intelligence. The results in the bottom of the table show that much of the genetic influence on each specific test is explained by the latent factors and "g." Most measures show little heritable variation independent of the verbal, spatial, perceptual speed, and memory ability groups.

Table 5.3 Correlations among Verbal (V), Spatial (S), Perceptual Speed (P), and Memory (M) factors in CAP children

Age	Factor	Age 3				Age 4				Age 7				Age 9			
		V	S	P	M	V	S	P	M	V	S	P	M	V	S	P	M
3	V	1.00															
	S	.32	1.00														
	P	.29	.34	1.00													
	M	.31	.32	.28	1.00												
4	V	.38	.31	.26	.33	1.00											
	S	.20	.36	.22	.25	.22	1.00										
	P	.07	.26	.22	.17	.19	.25	1.00									
	M	.16	.14	.12	.29	.16	.19	.20	1.00								
7	V	.22	.27	.11	.27	.42	.10	.02	.10	1.00							
	S	.07	.19	.08	.18	.08	.39	.15	.16	.16	1.00						
	P	.13	.30	.28	.07	.23	.35	.30	-.00	.25	.28	1.00					
	M	.15	.15	.12	.14	.09	.20	.15	.16	.13	.21	.21	1.00				
9	V	.26	.22	.13	.25	.36	.10	.03	.15	.59	.20	.19	.20	1.00			
	S	.12	.27	.16	.23	.18	.34	.22	.10	.19	.45	.31	.11	.31	1.00		
	P	.02	.22	.14	.13	.02	.20	.27	.14	.18	.23	.42	.20	.26	.42	1.00	
	M	.11	.22	.23	.20	.08	.03	.15	.04	.13	.08	.11	.25	.15	.26	.13	1.00

The shared environment findings in Table 5.4 are somewhat similar to the genetic outcomes, but much smaller in magnitude. Most of the shared environmental effects appear general rather than specific at this age, although there may be additional environmental effects specific to memory (residual = .25). Shared environmental variance components of the individual measures also mirror the genetic pattern of little effect independent of that attributable to the ability groups. In contrast, environmental effects unshared by siblings appear to influence each ability differently, as indicated by several large residual estimates (range = .08 to .71) and small or moderate "g" loadings for all ability groups. In addition, the unique-environment measurement residuals are much larger than those estimated for genetic or shared environmental effects, which are likely to reflect to some extent test measurement error which confounds these estimates. Overall, the environmental results suggest that non-genetic factors, particularly unique environmental factors, are strong determinants of specific cognitive abilities at age 3.

Table 5.4 Phenotypically standardized factor loadings and variance components for specific cognitive ability measures at 3 years of age

Measure	Unique environmental		Shared environmental		Genetic		e^2	c^2	h^2
	Fac.	Res.	Fac.	Res.	Fac.	Res.			
Primary factors									
Verbal	.21	.66	.19	.04	.68	.16	.47	.04	.49
Spatial	.45	.66	.53	.04	.26	.16	.63	.28	.09
Perceptual Speed	.14	.08	.38	.00	.74	.53	.03	.14	.83
Memory	.45	.71	.29	.25	.38	.00	.71	.15	.14
Measures									
Word Fluency	.42	.79	.12	.00	.43	.00	.80	.02	.18
Block Design	.49	.79	.32	.00	.19	.00	.86	.11	.04
Form Discrimination	.12	.62	.26	.26	.63	.28	.39	.14	.47
Picture Identification	.24	.81	.09	.00	.52	.10	.71	.01	.28
Figure-Ground[a]	.73	.49	−.40	.01	.37	.01	.69	.15	.16
	−.17			.59		.04			
Recognition Memory	.76	.43	.34	.00	.34	.01	.76	.12	.12
Picture Vocabulary	.13	.76	.21	.00	.60	.01	.59	.04	.37

Note: Residual values ("Res.") represent factor residuals in the context of primary factors and measurement residuals in relation to specific measures. Factor loadings ("Fac.") refer to loadings of primary factors on the common factor for the top section of the table and to observed measures on the primary factors in the bottom section of the table.
[a] Figure-Ground measure loads on Spatial and Perceptual Speed factors. The first row of estimates refers to the Spatial factor; the second row corresponds to Perceptual Speed.

It is important to note that the variance components shown in the top right portion of Table 5.4 warrant somewhat different interpretations than those typically ascribed to h^2, c^2, and e^2. The variance components shown in the table, which are calculated as the sum of the relevant factor loadings squared, reflect proportions of *latent factor* variance. Because unreliability contributes only to the variance of individual tests, estimates of factor heritability and environmentality may be expected to differ from those typically reported, which include measurement error in the e^2 estimates. For example, the Perceptual Speed factor has an estimated h^2 of .83 and e^2 of .03. These indicate that after accounting for the measurement error of the tests defining this factor, the remaining true-score variance is largely familial. In contrast, average h^2 and e^2 estimates for the tests that define the Perceptual Speed factor are .30 and .60, which are fully consistent with those typically reported.

The estimated phenotypic, genetic, and environmental correlations among the ability factors are presented in Table 5.5. These correlations encompass all ability associations explained by "g." The estimated phenotypic correlations between abilities are moderate to large (range = .37–.60) and similar to, albeit somewhat larger than, the observed correlations shown in Table 5.3. The genetic correlations are uniformly large (.69–.97), again emphasizing that the genetic influences on cognitive abilities relate more strongly to "g" than to specific abilities. Shared environmental correlations also are quite large, suggesting that home-environment influences do not differ greatly among abilities. However, these correlations should be interpreted with caution because they are based on very small c^2 effects (shown on the diagonal of the shared environmental correlation matrix). The unique-environment correlations are low to moderate (.16–.48), which is to be expected given that most of these effects appear ability-specific in the parameter estimates in Table 5.4.

Table 5.5   Estimated phenotypic, genetic, and environmental correlations among ability factors at 3 years of age

Factor	Verbal	Spatial	Perceptual Speed	Memory
Phenotypic				
Verbal	1.00			
Spatial	.37	1.00		
Perceptual Speed	.60	.45	1.00	
Memory	.41	.45	.45	1.00
Genetic				
Verbal	.49			
Spatial	.82	.09		
Perceptual Speed	.79	.69	.83	
Memory	.97	.85	.81	.14
Shared environmental				
Verbal	.04			
Spatial	.98	.28		
Perceptual Speed	.98	.99	.14	
Memory	.74	.76	.76	.15
Unique environmental				
Verbal	.47			
Spatial	.17	.63		
Perceptual Speed	.26	.48	.03	
Memory	.16	.30	.45	.71

Note: Variance components are shown on the diagonals of the genetic and environmental matrices.

Year 4

The full hierarchical genetic model for the year 4 sibling data has 66 free parameters, consisting of four higher order factor loadings, four primary factor residuals, five measurement loadings, and nine error terms for each of the three latent covariance components. Application of the model to the nine ability measures at this age yielded a log-likelihood of −2219.18 for the 66 estimated parameters. Table 5.6 shows the parameter estimates from the fit of the full model.

Table 5.6 Phenotypically standardized factor loadings and variance components for specific cognitive ability measures at 4 years of age

Measure	Unique environmental		Shared environmental		Genetic				
	Fac.	Res.	Fac.	Res.	Fac.	Res.	e^2	c^2	h^2
Primary factors									
Verbal	.50	.66	.01	.17	.50	.18	.69	.03	.29
Spatial	.41	.84	.17	.03	.32	.00	.86	.03	.11
Perceptual Speed	.71	.09	.12	.01	.27	.63	.52	.01	.47
Memory	.17	.77	.04	.00	.60	.09	.63	.00	.37
Measures									
Word Fluency	.43	.85	.09	.00	.28	.01	.92	.01	.08
Picture Vocabulary	.09	.60	.49	.01	.46	.43	.37	.24	.39
Vocabulary	.38	.84	.32	.00	.24	.00	.84	.10	.06
Block Design	.79	.52	.15	.00	.28	.00	.90	.02	.08
Puzzle Solving	.35	.76	.41	.00	.36	.03	.70	.17	.13
Form Discrimination	.59	.57	.10	.00	.57	.00	.67	.01	.32
Picture Identification	.67	.63	.12	.13	.36	.00	.84	.03	.13
Immediate Memory	.58	.64	.03	.01	.44	.26	.74	.00	.26
Delayed Memory	.81	.56	.14	.00	.16	.00	.96	.02	.03

Note: Residual values ("Res.") represent factor residuals in the context of primary factors and measurement residuals in relation to specific measures. Factor loadings ("Fac.") refer to loadings of primary factors on the common factor for the top section of the table and to observed measures on the primary factors in the bottom section of the table.

In general, the results for the year 4 data are similar to the year 3 findings. The genetic "g" accounts for most of the genetic effects on the four ability groups (range = .27–.60), with Perceptual Speed ability showing additional specific influence. Heritable variation on each of the nine measures is largely attributable to the four primary abilities, as in the case of the year 3 outcomes. Unique environmental effects also show the year 3 pattern of considerable ability-specific effects on several of the group factors, but reveal an additional general factor emerging at this age. Again, measurement residuals appear quite large for this source of variance, and are likely to reflect test measurement error to some degree. Neither specific nor general abilities seem influenced by shared environmental effects at year 4, as c^2 effects are distinctly absent for the four ability groups (.01–.03) and for most of the individual tests (.00–.24).

The phenotypic, genetic, and unique environmental correlations for the four primary abilities at age 4 are presented in Table 5.7. Shared-environment correlations have been omitted from the table because of the small effects at this age. The patterns of phenotypic and unique-environment correlations closely resemble those observed at 3 years, revealing fairly uniform, moderate correlations at the phenotypic level (range = .27–.49), and low to moderate correlations at the level of the unique environment (.09–.60). The genetic correlations also are similar to the year 3 correlations, showing strong associations due to the impact of "g." Still, several of the genetic correlations are smaller than those noted at age 3, reflecting increased levels of genetic specificity for some abilities at this age. This trend toward increasing specificity is manifest in the average genetic correlations, which change from .87 at age 3 to .67 at age 4.

Table 5.7 Estimated phenotypic, genetic, and environmental correlations among ability factors at 4 years of age

Factor	Verbal	Spatial	Perceptual Speed	Memory
Phenotypic				
Verbal	1.00			
Spatial	.37	1.00		
Perceptual Speed	.49	.40	1.00	
Memory	.39	.27	.29	1.00
Genetic				
Verbal	.29			
Spatial	.94	.11		
Perceptual Speed	.37	.39	.47	
Memory	.93	.99	.39	.37
Unique environmental				
Verbal	.69			
Spatial	.27	.86		
Perceptual Speed	.60	.44	.52	
Memory	.13	.09	.21	.63

Note: Variance components are shown on the diagonals of the genetic and environmental matrices.

Year 7

The year 7 data consist of eight specific-ability measures administered to the adopted and nonadopted siblings in the CAP. The full hierarchical model for the year 7 data has 60 free parameters, including four higher order factor loadings, four primary factor residuals, four measurement loadings, and eight error terms for the genetic, shared environmental, and unique environmental effects. Application of the model to the eight tests administered at age 7 yielded a log-likelihood of −1600.99 for the 60 free parameters. Table 5.8 shows the parameter estimates from the fit of the full year 7 model.

The "genetic factor" column shows moderate to strong loadings for all factors and all measures. This finding indicates that genetic influence on all primary abilities is to some extent shared with genetic influence on "g" and that genetic influence on all observed measures overlaps with genetic influence on the relevant ability group. These outcomes are similar to those observed at earlier ages, but are accompanied by greater proportions of ability-specific genetic variance than at ages 3 or 4 (Verbal = .56(.75²); Spatial = .18; Memory = .71). It should be noted that these results are consistent with those of the hierarchical analyses previously reported by Cardon, Fulker, DeFries, and Plomin (1992), but the parameter estimates are not identical because the present sample is larger than the one examined in the earlier study.

Table 5.8 Phenotypically standardized factor loadings and variance components for specific cognitive ability measures at 7 years of age

Measure	Unique environmental		Shared environmental		Genetic				
	Fac.	Res.	Fac.	Res.	Fac.	Res.	e^2	c^2	h^2
Primary factors									
Verbal	.04	.01	.59	.00	.29	.75	.00	.35	.65
Spatial	.06	.00	.07	.15	.88	.43	.00	.03	.97
Perceptual Speed	.46	.67	.02	.00	.58	.02	.66	.00	.34
Memory	.03	.41	.07	.00	.33	.84	.17	.00	.82
Measures									
Vocabulary	.04	.64	.45	.00	.62	.07	.41	.21	.38
Things	.44	.51	.04	.23	.61	.33	.45	.06	.49
Block Design	.04	.73	.11	.12	.66	.00	.53	.03	.44
Spatial Relations	.29	.50	.10	.32	.74	.00	.34	.11	.55
Identical pictures	.70	.43	.02	.25	.51	.10	.67	.06	.27
Perceptual Speed	.42	.55	.21	.00	.40	.56	.48	.04	.48
Immediate Memory	.40	.18	.07	.12	.88	.13	.20	.02	.78
Delayed Memory	.24	.66	.14	.01	.70	.06	.49	.02	.49

Note: Residual values ("Res.") represent factor residuals in the context of primary factors and measurement residuals in relation to specific measures. Factor loadings ("Fac.") refer to loadings of primary factors on the common factor for the top section of the table and to observed measures on the primary factors in the bottom section of the table.

Unique environmental effects at age 7 closely resemble those apparent at age 3, with two ability factors displaying specific variation and a lack of general factor influence. Shared environmental effects appear substantial for the Verbal ability group, but shared environmental effects do not contribute greatly to either variances or covariances of the other primary abilities or the individual measures which were apparent at 3 and 4 years of age.

The phenotypic and genetic correlations derived from the parameter estimates in Table 5.9 are shown in Table 5.9. The phenotypic correlations are moderate, as are the estimated correlations at the previous occasions. Genetic correlations extend the pattern arising at age 4 toward more moderate genetic associations among the primary abilities. These reduced correlations reflect the greater impact of specific genetic variance at this age. Correlations among shared and unique environmental factors are not presented in Table 5.9 because most shared-environment effects are too small to warrant interpretation and nearly all unique-environment correlations are zero because the effects are specific, rather than general in origin.

Year 9

The year 9 hierarchical model has 60 free parameters, with an identical factor structure to the year 7 model. Parameter estimates from application of the model to the eight measures are given in Table 5.10. The model yielded a log-likelihood of −1476.16.

Table 5.9 Estimated phenotypic and genetic correlations among ability factors at 7 years of age

Factor	Verbal	Spatial	Perceptual Speed	Memory
Phenotypic				
Verbal	1.00			
Spatial	.30	1.00		
Perceptual Speed	.20	.55	1.00	
Memory	.14	.30	.21	1.00
Genetic				
Verbal	.65			
Spatial	.32	.97		
Perceptual Speed	.36	.90	.34	
Memory	.13	.33	.37	.82

Note: Heritabilities are shown on the diagonal of the genetic correlation matrix.

Table 5.10 Phenotypically standardized factor loadings and variance components for specific cognitive ability measures at 9 years of age

Measure	Unique environmental		Shared environmental		Genetic		e^2	c^2	h^2
	Fac.	Res.	Fac.	Res.	Fac.	Res.			
Primary factors									
Verbal	.24	.38	.00	.24	.73	.47	.20	.06	.74
Spatial	.78	.01	.25	.45	.35	.05	.61	.26	.13
Perceptual Speed	.53	.20	.80	.04	.19	.00	.32	.64	.04
Memory	.28	.80	.06	.01	.22	.01	.95	.00	.05
Measures									
Vocabulary	.32	.70	.17	.01	.62	.01	.59	.03	.38
Similarities	.74	.43	.39	.02	.28	.19	.43	.16	.11
Hidden Patterns	.60	.03	.39	.03	.27	.64	.36	.16	.48
Card Rotation	.45	.47	.19	.00	.35	.64	.43	.04	.54
Finding As	.42	.66	.60	.00	.15	.00	.62	.36	.02
Perceptual Speed	.49	.68	.46	.00	.31	.00	.69	.21	.10

Measure	Unique environmental		Shared environmental		Genetic		e^2	c^2	h^2
	Fac.	Res.	Fac.	Res.	Fac.	Res.			
Immediate Memory	.92	.01	.06	.13	.20	.32	.84	.02	.14
Delayed Memory	.68	.68	.02	.00	.26	.00	.93	.00	.07

Note: Residual values ("Res.") represent factor residuals in the context of primary factors and measurement residuals in relation to specific measures. Factor loadings ("Fac.") refer to loadings of primary factors on the common factor for the top section of the table and to observed measures on the primary factors in the bottom section of the table.

The year 9 results differ somewhat from those at earlier ages, possibly due to the small sample sizes at 9 years. At this age nearly all the genetic effects appear in the general factor loadings (range = .19–.73), and only verbal abilities show genetic influence independent of "g" (residual = .47). Shared environmental influences on the four primary abilities are somewhat larger than those observed earlier, particularly with respect to Spatial and Perceptual Speed abilities. Only the unique-environment effects resemble those at 3, 4, and 7 years, revealing substantial environmental influences

Table 5.11 Estimated phenotypic, genetic, and environmental correlations among ability factors at 9 years of age

Factor	Verbal	Spatial	Perceptual Speed	Memory
Phenotypic				
Verbal	1.00			
Spatial	.44	1.00		
Perceptual Speed	.27	.68	1.00	
Memory	.22	.31	.24	1.00
Genetic				
Verbal	.74			
Spatial	.83	.13		
Perceptual Speed	.84	.99	.04	
Memory	.84	.99	.99	.05
Shared environmental				
Verbal	.06			
Spatial	.01	.26		
Perceptual Speed	.01	.49	.64	
Memory	.00	.00	.00	.00
Unique environmental				
Verbal	.20			
Spatial	.54	.61		
Perceptual Speed	.50	.94	.32	
Memory	.15	.28	.26	.95

Note: Variance components are shown on the diagonals of the genetic and environmental matrices.

which are ability-specific for most factors. However, there are unique environmental effects that also are correlated at this age, as "g" loadings range from .24 to .78.

Estimated correlations among the primary ability factors at age 9 are presented in Table 5.11. Again the phenotypic and unique environmental correlations are uniform and moderate, although the correlations between the Spatial and Perceptual Speed factors are increased (phenotypic $r = .68$; unique environmental $r = .94$). The genetic correlations are uniformly quite large, similar to those at ages 3 and 4, but not age 7. It is important to emphasize, however, that several of the genetic correlations are based on very small heritability estimates.

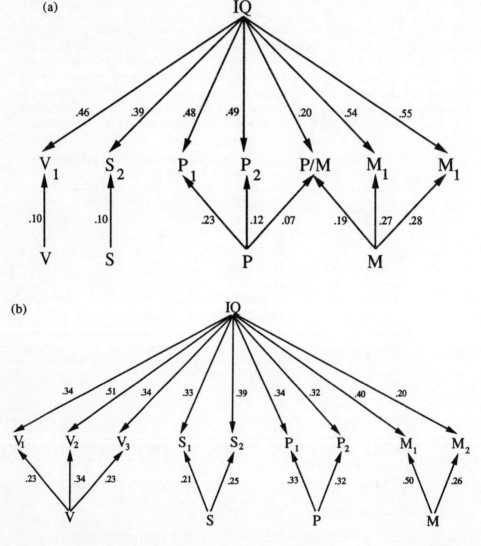

Figure 5.3 Genetic parameter estimates for a) year 3 and b) year 4 after Schmid-Leiman transformation. Residual measurement effects have been omitted from the diagram for clarity.

For comparison of the genetic findings at each of the measurement ages in the CAP children, the genetic factor structures are shown after Schmid–Leiman transformation in Figures 5.3 and 5.4. The results in the figures illustrate some variability over time with respect to genetic influences on primary abilities independent of "g," but some regularity as regards genetic effects relating to "g" and the presence of heritable variation on Verbal abilities beyond that accountable by "g."

The genetic "g" is apparent at all ages, as perhaps are independent Verbal effects, but the other factors do not show these structural similarities across these ages. The Spatial and Memory variables reveal substantial genetic impact during the intermediate ages (years 4 and 7), but these effects are not apparent at ages 3 or 9. The Perceptual Speed tests are influenced by genetic effects independent of "g" at the early ages but disappear almost entirely at the later occasions.

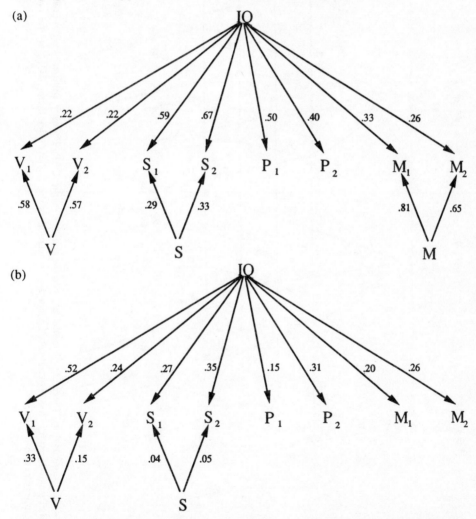

Figure 5.4 Genetic parameter estimates for a) year 7 and b) year 9 after Schmid–Leiman transformation. Residual measurement effects have been omitted from the diagram for clarity.

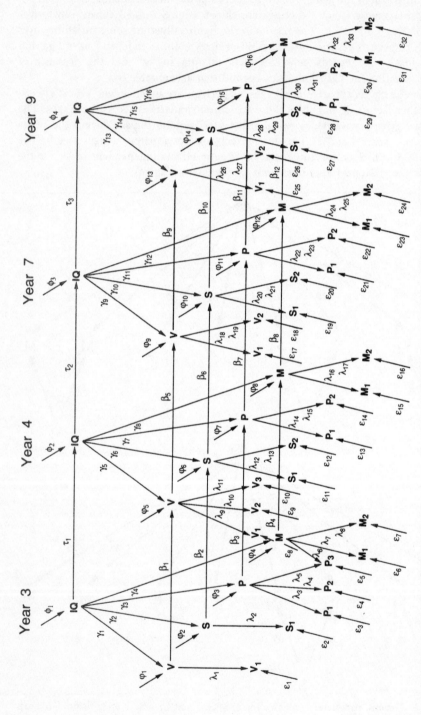

Figure 5.5 Longitudinal hierarchical model of specific abilities in the CAP. Primary abilities are denoted by Verbal: V; Spatial: S; Perceptual Speed: P; and Memory: M. General intelligence is represented as the higher order factor, IQ. Specific measures are subscripted in the order of their appearance in Table 5.1.

Longitudinal Hierachical Model

These hierarchical findings lead to obvious questions concerning the developmental pattern of genetic influences on specific abilities. For example, are the genetic effects that contribute to the apparent "genetic g" the same or different at the four ages? How do the specific genetic factors relate to one another during early and middle childhood? Are any ages particularly important for change in specific abilities, such as that observed in general IQ?

To examine these issues we have recently developed a longitudinal extension of the hierarchical model for application to the CAP sibling data (Cardon & Fulker, in press). A phenotypic path diagram of this model is shown in Figure 5.5. The model consists of hierarchical abilities at each of the four ages as depicted in Figure 5.1, but permits trait variation at each age directly to impact subsequent ages. Developmental stability is modeled by transmission parameters τ_i and β_i, which represent the stability of "g" and the primary abilities, respectively. These stability parameters are allowed to vary between all ages in order to account for differential continuity of specific abilities during early and middle childhood.

Application of this model to the CAP sibling data yielded several interesting results. First, the ability-specific genetic effects show considerable age-to-age stability, indicating that genetic influences on certain abilities at early ages tend to have substantial persistence and impact at later ages. Second, the patterns of longitudinal and hierarchical loadings for "g" and the specific abilities suggest that the genetic effects which influence observed continuity in general intelligence overlap with those influencing specific abilities; that is, the persistent "g" genes are not independent of the persistent specific-ability genes. And third, the residual loadings on "g" and specific abilities strongly suggest a developmental shift in the genetic etiology of individual differences in specific abilities between age 4 and 7 years. In particular, the genetic influences manifested at the first occasion continue to influence abilities at all later occasions, and these effects are augmented by new genetic variation at around age 7, which in turn persists through age 9. This outcome, coupled with a similar finding for general intelligence noted in the previous chapter, points to a genetic basis for the dramatic changes during the transition from early to middle childhood which have long been observed by developmental psychologists (e.g., Piaget, 1962).

Conclusions

The study of individual differences in specific cognitive abilities has inherent complexities beyond those involved in assessments of general mental ability, due to the required treatment of multiple measures and their psychometric foundations. Our recent efforts in the CAP have been devoted to exploring the hierarchical conception of mental abilities in which the psychometric framework is embedded naturally within the theoretical intelligence model. This approach, and its extension to genetic and environmental components, moves beyond evaluation of genetic and/or environmental etiologies for particular cognitive traits to help elucidate the underlying sources contributing to associations among the traits. In this chapter we have discussed applications of the genetic hierarchical model to adoptive/ nonadoptive sibling data in the CAP which were aimed at determining the extent to which Verbal, Spatial, Perceptual Speed, and Memory abilities are genetically and environmentally influenced in early and middle childhood, the extent to which those influences are unique to each ability or common amongst them, and the continuity and change in the genetic and environmental effects from 3 to 9 years of age.

The results present a strong case for the importance of both genetic and environmental effects throughout this period of development. At the early age points, years 3 and 4, the genetic effects appear common to the four abilities, with only Perceptual Speed attributes showing independent genetic influences. Between ages 4 and 7 a trend emerges toward increasing genetic specificity, as different abilities reveal the impact of unique genetic influences. Genetic effects at age 9 seem to revert to the early childhood pattern of overlapping genes for different abilities, but the sample sizes at this age are very small. Longitudinal results further emphasize the childhood shift between 4 and 7 years of age, showing continuity of genetic influences throughout early and middle childhood, with an influx of new genetic variation at age 7. The environmental effects on specific abilities in the CAP are in large part individually specific at all ages examined; environmental influences shared by siblings reared together do not exert substantial effects on these abilities at these ages. There are notable exceptions, however, as spatial abilities exhibit shared environmental influence at two of the ages (3 and 9), as do Verbal abilities and Perceptual Speed at ages 7 and 9, respectively.

In summary, results of the present study suggest that the etiologies of individual differences in specific cognitive abilities differentiate from those of general cognitive ability during middle childhood. However, additional data at 9 years of age will be required to test this hypothesis more rigorously. Future CAP analyses will also assess the extent to which these developmental genetic and environmental influences on specific cognitive abilities change during the dramatic transition from middle childhood to early adolescence.

6 Longitudinal Predictions of School-Age Cognitive Abilities from Infant Novelty Preference

Lee Anne Thompson
Stephen A. Petrill

Earlier theories of individual differences in intelligence that trace intellectual development from the first year of life through childhood were based on change, primarily due to the lack of stability between the first year of life and later years. However, more recent theories have begun to explore the contribution of continuity as well as change to the developmental process. The increased popularity of the information-processing approach which studies intelligence in terms of separable cognitive processes, and the relatively recent discovery of infant measures that are predictive of later intellectual status, have greatly contributed to this theoretical transition (Colombo & Mitchell, 1990).

Infant measures that have been developed over the last 20 years have shown marked improvement in their ability to predict later intelligence when compared to their predecessors, infant sensori-motor tests (Bornstein & Sigman, 1986; Rose & Feldman, 1990). Although some work has been done using auditory stimuli (O'Connor, Cohen, & Parmalee, 1984), most measures have been in the visual domain. One widely used measure involves the rate of habituation to a visual stimulus. Habituation may reflect the speed and efficiency of the infant's ability to encode visual information (Rose, Slater, & Perry, 1986). A second popular measure involves preferential looking at a novel target as assessed through the paired comparisons paradigm (Fagan & Shepherd, 1986) or through dishabituation to a novel stimulus (Colombo & Mitchell, 1990). Both approaches are thought to reflect infants' ability to discriminate between complex stimuli and their ability to retain pertinent information about the stimuli. Newer measures that have been less widely tested show considerable promise and include cross-modal transfer (Rose & Wallace, 1985), learning (Fagen & Ohr, 1990), and visual expectations (DiLalla et al., 1990; Haith, Hazan, & Goodman, 1988).

In the early 1980s, when the first studies reporting significant predictive validity of infant information-processing tests were published (Fagan & McGrath, 1981; Lewis & Brooks-Gunn, 1981; Rose & Wallace, 1988), researchers suggested that continuity was mediated via processes shared by early infant measures and later intelligence tests, for example, encoding, discrimination and retention (Fagan & Singer, 1983); however, these claims were not specifically tested with empirical research. A few years later, researchers were in pursuit of the actual links between infant cognition and later intelligence (*European Bulletin of Cognitive Psychology*, 1988).

Two approaches have been used to validate the utility of an information-processing theory, a theory that postulates separable and independent dimensions of infant intelligence. Concurrent validation can be attained through a series of infant measures that have face validity for tapping different processes. As an example, factor analysis and partial correlations facilitate interpretation of overlapping versus separable processes. Predictive validation can be achieved through longitudinal correlations from infant measures to later indices of specific processing abilities. Examples of studies utilizing each of these approaches are discussed below.

A study involving a sample of high-risk infants (Jacobson, Jacobson, O'Neill, Padgett, Frankowski, & Bihun, 1992) collected several different information-processing measures as well as other developmental measures during the first year of life. Visual recognition memory was assessed at 6.5 and 12 months of age, cross-modal transfer was assessed at 12 months of age and visual expectations were assessed at 6.5 months. When a principal-components analysis with varimax rotation was performed on all of the variables, the information-processing measures loaded most strongly on a factor that reflected processing speed and on a factor that appeared to represent quality of attention and/or memory. This study provides important concurrent validation for an information-processing model of infant intelligence. Based on results from this study, a global unitary process model does not sufficiently account for the data. However, the approach taken by Jacobson et al., while important, does not address the issue of developmental continuity at the process level. Furthermore, while each of the measures used had previously demonstrated predictive validity for later intelligence, it is unclear whether each of the factors from the current analysis was predictive. In other words, while more than one process is operating in infancy, they may or may not contribute to later intellectual ability.

Studies that have used a longitudinal follow-up design can address questions of continuity and predictive validity at the process level. Typically a study will administer one or more information-processing measures during the first year of life and then administer cognitive tests that tap different sets of processes during the preschool years. Fagan and Knevel (1989) found that infant visual recognition memory predicted scores on a verbal factor from the Stanford-Binet administered at 3 years of age but did not relate to the nonverbal factor. Another study (Colombo, Mitchell, Dodd, Coldren, & Horowitz, 1989) found that visual recognition memory at 7 months of age was related to short-term spatial memory at 16 months of age.

Longitudinal data from the CAP provide a unique opportunity to trace the developmental relations between infant novelty preference and later measures of general intelligence, specific cognitive abilities, language development, and academic achievement. The CAP sample is especially important due to its large size, and its use of a diverse battery of cognitive tests across a wide age range spanning infancy and early and middle childhood. A previous paper has explored the relationship between early visual recognition memory and specific cognitive abilities in the CAP at ages 1, 2, and 3 years (Thompson, Fagan, & Fulker, 1991). Visual recognition memory was assessed at 5 and 7 months of age using novelty preferences in a paired comparisons paradigm. Novelty preference scores were significantly related to 3-year-old IQ but not to Bayley Mental Development Index scores at 1 and 2 years of age. Novelty preference was related to language skills at 2 and 3 years, and verbal and memory scales at 3 years. Partial correlations suggested that novelty preference is independently related to language and memory at 3 years of age, again lending support to a multi-process model of infant intelligence.

All of the studies referred to so far have involved longitudinal prediction through the preschool years. Few studies looking at predictive validity have gone beyond the preschool years. When intellectual functioning is studied after a child starts formal schooling, the impact of intelligence on academic achievement becomes an important question. The current chapter extends the Thompson et al. (1991) study to encompass the 4- and 7-year-old cognitive testings in the CAP and also includes measures of academic achievement at 7 years of age. Relating novelty preference scores taken early in life to later scholastic achievement provides additional information for the exploration of developmental precursors of later school achievement reported in the following chapter.

Method

Sample

The current report involves a subsample of infants in the Colorado Adoption Project recruited after 1982, and includes 30 adopted and 81 nonadopted infants. Infants were full term and normal birthweight with no serious perinatal complications.

Measures

Infant novelty preference The infants were tested at 5 and 7 months of age with an early version of the Fagan Test of Infant Intelligence (FTII: Fagan &

Shepherd, 1986). The FTII is a paired comparisons test of visual novelty prefer-ence. The infants were administered four novelty problems at 5 months of age, two problems comparing abstract patterns and two comparing photographs of faces. Each problem consisted of familiarization to a single stimulus where 20 seconds of looking time was required for both the right and left positions, for a total of 40 seconds. Twenty seconds were allowed for the novelty test phase, where novel and familiar stimulus positions were counterbalanced across two 10-second trials. The infants were administered six additional problems consisting of facial photographs at 7 months of age. Familiarization trials involved two identical stimuli presented simultaneously for 20 seconds. Two counterbalanced 5-second trials were allowed for the novelty test phase.

To maximize the sample size and to increase reliability, novelty scores across the 10 problems were averaged. The use of aggregated novelty scores is common practice in the literature (Colombo & Mitchell, 1988; Fagan, Singer, Montie, & Shepherd, 1986; Rose, Feldman, McCarton, & Wofson, 1988). Data from infants who did not complete at least 5 of the 10 problems were excluded from the analysis. A split-half correlation with Spearman-Brown correction of .40 was used as an estimate of reliability for the 10-problem test in the current sample and is comparable to that found in previous studies using the same test (Fagan, 1984; Fagan & McGrath, 1981; Fagan & Singer, 1983). A reliability of .40 is considered low; however, when paired with a later measure with a reliability of .80 or greater, as is typical with standardized intelligence tests, adequate levels of predictive validity are attained (Fagan & Detterman, 1992).

Longitudinal follow-up measures The infants were involved in the regular schedule of CAP testing at 1, 2, 3, and 4 years of age and again at 7 years of age. All follow-up tests were administered by testers who were not aware of the child's performance at 5 and 7 months of age. As previously described in Thompson et al. (1991), at 1 and 2 years of age the infants were administered the Bayley Scales of Infant Development (Bayley, 1969). In addition to the traditional mental and motor scores, MDI and PDI, the Bayley items were also used to form composites scores reflecting specific skills: Means–End, Imitation, and Verbal Skill at 1 year and Spatial, Lexical, Verbal–Symbolic, and Imitation at 2 years. These composite scores have been extensively described and explored in previous studies (Plomin & DeFries, 1985; Thompson, Plomin & DeFries, 1985).

A separate language measure, the Sequenced Inventory of Communication Development (SICD: Hedrick, Prather, & Tobin, 1975) was administered at both the 2- and 3-year testings. Measures of Receptive and Expressive language are provided.

At 3 and 4 years of age the Stanford-Binet Intelligence Scale (Terman & Merrill, 1973) was administered along with a battery of eight specific cognitive abilities tests (SCA). The SCA provides four separate factor scores: Verbal, Spatial, Perceptual Speed, and Memory (Rice, Corley, Fulker, & Plomin, 1986). The first unrotated principal component was used as an additional index of general cognitive ability and accounts for about 30% of the variance (Plomin, DeFries, & Fulker, 1988).

At 7 years of age, the children were administered the Wechsler Intelligence Scale for Children–Revised (WISC-R), the specific cognitive abilities battery, and measures of scholastic achievement. Again the SCA yields four specific factors representing Verbal, Spatial, Perceptual Speed, and Memory abilities, in addition to the first unrotated principal component, which reflects general cognitive ability. In the current sample, the first unrotated principal component of the SCA correlates with WISC-R Full Scale IQ at .70.

Teacher ratings of each child's academic performance with respect to class and grade level were collected during spring semester of the first grade and during the fall semester of the second grade. Several tests of academic achievement were administered during the summer between the first and second grades and include: Reading Recognition from the Peabody Individual Achievement Test (PIAT), KeyMath Numeration, Addition, and Subtraction subtests, and two tests from the Clinical Evaluation of Language Functions (CELF): Producing Model Sentences, and Processing Word and Sentence Structure.

Results and Discussion

Means, standard deviations, and sample sizes for all of the measures at each of the ages are presented in Table 6.1. Novelty preference means are similar to other normal infant samples in the literature and the variances are slightly lower (Fagan & Shepherd, 1986). The current analyses combine adopted and nonadopted infants into one sample to maximize the sample size for longitudinal analyses. The two groups did not differ for mean novelty preference ($F_{1,109} = .003$, $p > .95$). Novelty preference means and variances did not differ for males and females – males: $N = 62$, $M = 61.3$, $SD = 7.4$; females: $N = 49$, $M = 61.6$, $SD = 6.8$; $F_{1,109} = 0.027$, $p > .85$.

Table 6.1 Means and standard deviations for
intelligence and language measures

Variable	M	SD	N
Novelty preference	61.4	7.1	113
Follow-up measures			
Year 1			
Bayley MDI	111.3	11.3	111
Year 2			
Bayley MDI	108.4	11.6	107
Language			
Total	33.9	9.6	107
SICD Receptive	15.4	4.7	107
SICD Expressive	18.5	6.2	107
Year 3			
Stanford–Binet IQ	108.7	14.4	99
Language			
Total	24.7	6.5	99
SICD Receptive	11.8	3.5	99
SICD Expressive	12.9	3.7	99

Table 6.1 (Cont.)

Variable	M	SD	N
Year 4			
Stanford-Binet IQ	109.6	10.8	95
Year 7			
WISC-R IQ	113.3	10.1	97

Note: All results from ages 1, 2, and 3 years are taken from "Longitudinal Prediction of Specific Cognitive Abilities from Infant Novelty Preference" by L. A. Thompson, J. F. Fagan, and D. W. Fulker, 1991, *Child Development 62*, 530–538.

Longitudinal correlations between infant novelty preference and later intelligence scores at 1, 2, 3, 4, and 7 years of age are presented in Table 6.2.

Table 6.2 Longitudinal correlations between infant novelty preference and general cognitive ability at 1, 2, 3, 4, and 7 years of age

Ability	Novelty preference	
	r	N
Year 1		
Bayley MDI	.07	111
Year 2		
Bayley MDI	.09	107
Year 3		
Stanford-Binet IQ	.25*	98
SCA First PC	.21*	97
Year 4		
Binet IQ	.09	94
SCA First PC	.16	89
Year 7		
WISC-R IQ	.13	96
SCA First PC	.12	96

Note: All results from ages 1, 2, and 3 years are taken from "Longitudinal Prediction of Specific Cognitive Abilities from Infant Novelty Preference" by L.A. Thompson, J.F. Fagan, and D.W. Fulker, 1991, *Child Development, 62*, 530–538.
*$p < .05$, one-tailed.

As previously reported, infant novelty preference does not correlate with the Bayley MDI at ages 1 and 2. The Bayley, especially items administered at 1 year of age, is a sensori-motor test and is not predictive of later intelligence. The correlations with both the Stanford-Binet IQ and the SCA first principal component are significant at age 3. The magnitude of these relations is consistent with earlier research. However, while infant novelty preference correlates positively with general cognitive ability at ages 4 and 7, the correlations do not reach significance.

Longitudinal correlations were also calculated between infant novelty preference and measures of later specific cognitive abilities, as seen in Table 6.3. These data indicate that infant novelty preference significantly correlates with Verbal Skill at age 1 and Verbal and Memory ability at age 3. However, infant performance does not predict specific cognitive skills at 4 and 7 years of age. It should be noted, however, that while the correlations are low, they are uniformly positive; relations between novelty preference scores and later cognitive abilities in samples of normal full-term infants are not generally very high (Fagan & Detterman, 1992) and a sample of 100 infants has only about 37% power to detect an effect size of .20 (Cohen, 1977).

Table 6.3 Longitudinal correlations between infant novelty preference and specific cognitive abilities at 1, 2, 3, 4, and 7 years of age

Abilities	Novelty preference	
	r	N
Year 1		
Means–End	.08	111
Imitation	.03	111
Verbal Skill	.22**	111
Year 2		
Spatial	.10	107
Lexical	.16*	107
Verbal–Symbolic	.10	107
Imitation	−.05	107
Year 3		
Verbal	.18**	99
Spatial	.14*	99
Perceptual Speed	.07	97
Memory	.30**	94
Year 4		
Verbal	.18	94
Spatial	.06	94
Perceptual Speed	.15	94
Memory	.00	90
Year 7		
Verbal	.15	96
Spatial	.10	96
Perceptual Speed	.14	96
Memory	.05	96

Note: All results from ages 1, 2, and 3 years are taken from "Longitudinal Prediction of Specific Cognitive Abilities from Infant Novelty Preference" by L. A. Thompson, J. F. Fagan, and D. W. Fulker, 1991, *Child Development*, *62*, 530–538.
*$p < .10$, **$p < .05$, one-tailed.

In addition to later cognitive abilities, infant novelty preference was compared to later language development. As reported in Table 6.4, longitudinal correlations calculated between infant novelty preference and SICD indicate that infant performance positively and significantly relates to both Receptive and Expressive language ability at age 3.

Table 6.4 Longitudinal correlations between infant novelty preference and language at 2 and 3 years of age

Abilities	Novelty preference	
	r	N
Year 2		
Total	.14*	107
Receptive	.15*	107
Expressive	.10	107
Year 3		
Total	.30***	98
Receptive	.33***	98
Expressive	.22**	98

Note: All results from ages 1, 2, and 3 years are taken from "Longitudinal Prediction of Specific Cognitive Abilities from Infant Novelty Preference" by L. A. Thompson, J. F. Fagan, and D. W. Fulker, 1991, *Child Development*, *62*, 530–538.
*$p < .10$, **$p < .05$, ***$p < .01$, one-tailed.

Finally, longitudinal correlations, presented in Table 6.5, were calculated between infant novelty preference and academic achievement at age 7. These data indicate that infant novelty preference significantly correlates with teacher ratings of class performance in reading skills and with achievement-test performance in math.

As previously mentioned, there are few studies in the literature that have reported longitudinal correlations from infant novelty preference into the early school years. To summarize, although novelty preference scores do not relate to Bayley scores at 1 and 2 years of age, they significantly relate to Stanford-Binet IQ, Memory, Verbal Ability and language development at 3 years of age. Cognitive measures at 4 and 7 years of age are not significantly related to novelty preference; however, teacher ratings of class performance in reading and the KeyMath Addition test administered at age 7 are significantly related to novelty preference during the first year of life.

While the pattern of results at age 3 supports information-processing theories of cognitive development, the results for ages 4 and 7 years are not as strong. This may be due to several factors. First, we have limited power to detect small effect sizes in the current sample. The correlations at ages 4 and 7 are not significantly different from those at age 3. Previous studies exploring the predictive validity of novelty preference in normal samples have not found relationships much larger than those reported here. A larger sample may be required to explore the information-processing model into the early school years. Second, the FTII novelty preference test used in the study was an early version of the test and consisted of fewer items than the more widely used current version. Fewer items produce a less reliable test.

Low reliability may have contributed to the relatively low correlations found between novelty preference and cognitive ability at ages 4 and 7.

Table 6.5 Longitudinal correlations between infant novelty preference and achievement at 7 years of age

Abilities	Novelty preference	
	r	N
Teacher ratings		
Grade performance		
Reading	.20	79
Math	.17	79
Class performance		
Reading	.22*	79
Math	.14	78
Achievement tests		
Peabody Individual Achievement		
Test – Reading Recognition	.12	96
KeyMath		
Numeration	.05	96
Addition	.21*	96
Subtraction	.11	96
CELF		
Sentence Structure	.02	94
Model Sentences	.12	95

*$p < .05$, one-tailed.

Conclusions

Infant data from the CAP provide support for an information-processing approach to the study of cognitive development from the first year of life through the early childhood years. The results indicate that infant novelty preference represents a set of independent specific cognitive processes, an important contribution to current theories of cognitive development.

7 School Achievement

Sally J. Wadsworth

Whereas the two previous chapters focus on the etiology of continuity and change in general and specific cognitive abilities during early and middle childhood, the subject of this chapter is educational achievement. Three main issues will be addressed. First, the hypothesis that adopted children are at an increased risk for learning disabilities will be discussed. Secondly, developmental precursors of academic achievement, including performance on measures of general and specific cognitive abilities, will be explored. Finally, by analyzing data from related and unrelated sibling pairs, the etiology of covariation between measures of general cognitive ability and academic achievement will be assessed.

Risk of Learning Disabilities

Results of several previous studies suggest that adopted children may be at a relatively high risk for learning disabilities. In 1970, Silver reported that the frequency of adoption among learning-disabled children was almost four times that expected in a normally-achieving population. In a subsequent study of 225 students from three private schools for learning-disabled children, Silver (1989) found that 17.3% were adopted, as compared to 3.9% of the total population of live births in the United States in 1982.

More recently, Brodzinsky and Steiger (1991) ascertained the prevalence of adopted children in selected special-education populations. In a sample of more than 7,000 students classified as neurologically impaired, perceptually impaired, or emotionally disturbed, the prevalence of adoptees was three to four times that expected in the normal population. However, the authors cautioned that methodological problems may occur when subjects are recruited through clinical settings; thus, more representative target populations are needed in order to determine if adoptees are especially vulnerable to school-related difficulties.

In an attempt to estimate the prevalence of Attention Deficit Disorder (ADD) in adopted children, Deutsch, Swanson, Bruell, Cantwell, Weinberg, and Baren (1982) used Bayesian probability methods. Given an eightfold increase in the rate of non-relative adoption in two clinical populations of children diagnosed with ADD, the authors estimated that 23% of adopted children have ADD. They cautioned, however, that this estimate is sensitive to methodological variation, such as diagnostic criteria, procedures for collection of prevalence data, assumed rate of non-relative adoption in the general population, and the gender ratio in the affected group. In particular, they noted that such estimates may be inflated due to ascertainment bias. Adoptive parents may be more likely than nonadoptive parents to seek medical or psychological help for their children, due to their higher socioeconomic status and increased contact with social agencies.

Although results of these studies suggest that adopted children are over-represented in special-education populations, they do not provide direct evidence of the prevalence of school-related difficulties in such children. In a direct assessment of the vulnerability of adopted children to emotional and academic problems, Brodzinsky, Schecter, Braff, and Singer (1984) analyzed questionnaire data provided by parents and teachers of 130 adopted and 130 nonadopted children. They found that adopted children were rated higher on average for psychological and school-related behavior problems, and lower in social competence and school achievement, than nonadopted children. However, mean ratings of the adopted children were well within the normal range. Although adopted children may be more likely to experience increased stress and poorer prenatal care than nonadopted children, Brodzinsky et al. (1984) concluded that most adopted children adapt successfully both psychologically and academically.

Wadsworth, DeFries, and Fulker (in press) recently analyzed CAP data in order to test the hypothesis that adopted children are at increased risk for learning problems. Results of those analyses will be summarized in this chapter. The average achievement and cognitive test scores of adopted and nonadopted control children tested at age 7 will be compared, and the prevalence of low achievement and placement in special-education classes among these children will be reported.

Precursors of Academic Achievement

In addition to reporting average scores during middle childhood, infant and early childhood precursors of academic achievement are examined in this chapter. Although a number of investigators have reported high correlations between contemporaneous measures of cognitive ability and academic achievement (Brooks, Fulker, & DeFries, 1990; Cardon, DiLalla, Plomin, DeFries, & Fulker, 1990; Dunn & Markwardt, 1970; Jensen, 1969; Thompson, Detterman, & Plomin, 1991), surprisingly few studies have examined the relationship between cognitive ability in infancy and early childhood, and later school achievement. Kagan, Lapidus, and Moore (1978) analyzed data from 35 boys and 33 girls on various infant measures, and on several measures of IQ and achievement at age 10, including eight subtests of the Wechsler Intelligence Scale for Children – Revised (WISC-R) (Wechsler, 1974) and a modification of Spache's Diagnostic Reading Scale. While measures of attentiveness at 4, 8, and 13 months of age predicted IQ and reading ability at age 10, the authors noted that this relationship may have been due at least in part to the correlation of parental social class with the infant and childhood measures.

Analyzing data from 26 subjects, Roe, McClure, and Roe (1983) correlated infant scores on the Gesell Developmental Schedules (Gesell, 1926) obtained at 3, 5, 7, 9, and 15 months of age with WISC-R scores at age 12, as well as with scores on the Peabody Picture Vocabulary Test (PPVT) (Dunn, 1981), and the Wide Range Achievement Test (WRAT) (Jastak and Jastak, 1978). Whereas the Gesell was significantly correlated with later tests of nonverbal ability (e.g., WISC-R Performance IQ), it was not significantly correlated with either reading or arithmetic scores on the WRAT. In addition, the Gesell did not correlate with later tests of verbal intelligence, such as the WISC-R Verbal IQ. These investigators suggested that the failure of the Gesell (and many other infant tests of intelligence) to predict later intelligence and achievement may be due to the fact that the infant tests often measure nonverbal, visual–perceptual, and motor coordination skills, while tests of later intelligence and achievement rely heavily on verbal items.

In this chapter, scores of adopted and nonadopted children on infant measures of general cognitive ability at 1 and 2 years of age, and on measures of general and specific cognitive abilities at ages 3, 4, and 7, will be correlated with measures of school achievement at age 7.

Etiology of Covariation

Although measures of intelligence are correlated with academic achievement, little is known regarding the etiology of their covariation. Numerous studies have provided evidence for significant heritability of general cognitive ability, averaging about .50 by age 7 (Plomin & DeFries, 1985; Plomin, DeFries, & Fulker, 1988). In addition, several studies have investigated genetic and environmental influences on measures of reading and mathematics achievement in school-aged children, resulting in heritability estimates as high as .78 for reading achievement, and .37 for mathematics achievement (Gillis & DeFries, 1991; Stevenson, Graham, Fredman, & McLoughlin, 1987; Thompson, Detterman, & Plomin, 1991). Given the substantial phenotypic correlations between measures of intelligence and academic achievement, and the apparent influence of genetic factors on both, covariation among these measures may be due, at least in part, to heritable influences.

A few recent studies have investigated the etiology of covariation among scores on measures of cognitive abilities and academic achievement. While some have focused on the relationship between achievement and general cognitive ability, others have employed measures of specific cognitive abilities. Brooks et al. (1990) examined the relationship between WISC-R Full Scale IQ scores and performance on three subtests of the Peabody Individual Achievement Test (PIAT) (Dunn & Markwardt, 1970), Reading Recognition, Reading Comprehension, and Spelling. Analyzing data from a sample of 86 monozygotic (MZ) and 60 same-sex dizygotic (DZ) twin pairs participating in the Colorado Reading Project, they assessed the contributions of genetic and environmental influences to the variance in each of the measures, as well as to their covariation. Multivariate genetic analyses yielded heritability estimates of .57 and .45 for IQ and Reading Recognition, respectively, suggesting substantial genetic influences on both of these measures. Moreover, an estimated genetic correlation of .58 accounted for about 77% of their phenotypic correlation, suggesting that the relationship between intelligence and reading recognition is largely due to heritable influences.

Cardon et al. (1990) investigated the etiology of the relationship between WISC-R Full Scale, Verbal, and Performance IQ and PIAT Reading Recognition, using CAP parent–offspring data from 119 adoptive families and 120 nonadoptive families. The phenotypic correlations among these measures were moderate, ranging from .27 between Reading Recognition and Performance IQ to .46 between Reading Recognition and Verbal IQ. Multivariate behavioral genetic analyses yielded moderate heritability estimates of .36 for Full Scale and Verbal IQ, .38 for Reading Recognition, and .41 for Performance IQ, suggesting a moderate influence of genotype on individual differences for each of these measures. Moreover, genetic influences accounted for about 78% of the observed correlation between reading achievement and both Full Scale and Verbal IQ, and about 67% of that between reading achievement and Performance IQ. Thus, these results also suggest that the relationships between reading achievement and both verbal and nonverbal intelligence are largely due to genetic influences.

Subsequently, Thompson et al. (1991) examined scores of 146 MZ and 132 same-sex DZ twin pairs on measures of specific cognitive abilities (verbal, spatial, perceptual speed, and memory) as well as on the Metropolitan Achievement Test (MAT) (Prescott, Barlow, Hogan, & Farr, 1986), including measures of reading, mathematics, and language skills. Both the phenotypic and genotypic relationships among these measures, as well as the genetic and environmental contributions to the variance of each measure, were investigated. Phenotypic correlations among the individual measures of specific cognitive abilities and those of scholastic achievement ranged between .22 and .40, with reading correlating .40 to both verbal and spatial ability, and math correlating .32 with both of these measures. Genetic influences accounted for more than 80% of the phenotypic correlations between verbal ability and each of three measures of scholastic achievement. In addition, more than 80% of the observed correlation between spatial ability and language, and greater than 90% of that between spatial ability and both reading and mathematics achievement, was found to be due to genetic influences. These findings suggest that the observed correlations among measures of specific cognitive abilities and scholastic achievement are also, to a large extent, genetically mediated.

These previous studies analyzed either twin data or parent–offspring data. Moreover, whereas the studies of both Brooks et al. (1990) and Cardon et al. (1990) focused on the relationship between general cognitive ability and reading achievement, the study by Thompson et al. (1991) concerned specific cognitive abilities and both reading and mathematics achievement. The results presented in this chapter complement and extend these previous studies by analyzing data from adoptive and nonadoptive sibling pairs participating in the CAP at age 7. The contributions of genetic and environmental influences to the variance in Verbal and Performance IQ and to measures of both reading and mathematics achievement will be assessed, and the etiology of the relationship among these measures will also be examined.

Method

In order to assess academic achievement and general cognitive ability of CAP participants at age 7, the Reading Recognition (REC) subtest of the Peabody Individual Achievement Test (PIAT), the Numeration, Addition, and Subtraction subtests of the KeyMath Diagnostic Arithmetic Test (Connolly, Nachtman, & Pritchett, 1976), and the Wechsler Intelligence Scale for Children – Revised (WISC-R), excluding the Digit Span subtest, were administered to adopted and nonadopted children during the summer following first grade (average age of 7.4 years). Based upon the results of a principal-components analysis, a composite mathematics measure (MATH) was computed. Information concerning placement in special classes for academic or emotional/behavioral problems (including counseling, referral to resource rooms, special reading classes, tutoring, and classes for children with emotional/behavioral disorders) was obtained by parental report using the Child Behavior Checklist (CBCL: Achenbach & Edelbrock, 1983). For the present analysis, data were available from 199 adopted and 216 nonadopted children.

To assess general cognitive ability in infancy and early childhood, the Bayley Scales of Infant Development (Bayley, 1969) were administered at 1 and 2 years of age, and Form L-M of the Stanford-Binet Intelligence Scale (Terman & Merrill, 1973) was administered during the home visit at ages 3 and 4. Tests of specific cognitive abilities at ages 3, 4, and 7 comprise four factors, based on the results of factor analyses: Verbal, Spatial, Perceptual Speed, and Memory. Detailed descriptions of the methods and measures employed at ages 1–4 have been provided by Plomin et al. (1988), and those for age 7 have been described in chapter 2. Currently, data are available for 207 adopted and 230 nonadopted children at age 1, 192 adopted and 218 nonadopted children at age 2, 190 adopted and 206 nonadopted children at age 3, 182 adopted and 206 nonadopted children at age 4, and 199 adopted and 216 nonadopted children at age 7.

In order to assess the etiology of covariation between Verbal and Performance IQ and measures of academic achievement at 7 years of age, data were analyzed from 60 pairs of adopted children and their unrelated siblings, as well as from 71 pairs of nonadopted control children and their related siblings. Scores on the Reading Recognition (REC) subtest of the PIAT, a MATH composite, and the WISC-R Verbal IQ and Performance IQ were analyzed using multivariate behavioral genetic methods.

Results

Risk of learning disabilities

Table 7.1 presents means, standard deviations, and effect sizes for the achievement (REC and MATH), Verbal IQ (VIQ) and Performance IQ (PIQ) test scores of adopted and nonadopted children at age 7. Effect sizes were computed as the difference between the two means divided by the pooled standard deviation (Cohen, 1977). The data were subjected to a multivariate analysis of variance (MANOVA) (SPSS-X, 1988), with the model including adoptive status and gender, as well as their interaction. Means for adopted children were slightly lower than those for nonadopted control children on each of the four measures, resulting in a significant multivariate F test ($F_{4,398} = 3.12, p \leq .02$). With regard to the univariate comparisons, a Bonferroni adjustment (Judd & McClelland, 1989) indicated that an alpha level of .01 should be used to determine the significance of the difference between the two groups ($\alpha / number\ of\ comparisons = .05/4 = .0125$). Based on this criterion, adopted and nonadopted children differed significantly only for VIQ ($F_{1,401} = 10.67$, $p \leq .001$). Furthermore, this difference accounts for less than 3% of the observed variance ($SS_{between}/SS_{total} = R^2$ for each comparison).

Table 7.1 Means, standard deviations, and effect sizes on achievement and IQ measures for adopted and control children at age 7

Measure	Adopted	Control	Effect
REC	31.31 ± 7.52	31.71 ± 8.38	.05
MATH	10.91 ± 1.26	11.16 ± 1.42	.19
VIQ	108.54 ± 11.57	112.30 ± 12.22	.32*
PIQ	113.15 ± 11.55	114.18 ± 11.68	.09
(N)	195–199	212–216	
Multivariate test $F_{4,398} = 3.12,\ p \leq .02$			

Note: After "Cognitive Abilities of Children at 7 and 12 Years of Age in the Colorado Adoption Project," by S. J. Wadsworth, J. C. DeFries, and D. W. Fulker (in press), *Journal of Learning Disabilities*. By PRO-ED Inc. Reprinted by permission.
*$p \leq .01$

Table 7.2 presents means, standard deviations, and effect sizes for the individual subtests of the WISC-R for the adopted and nonadopted groups. Adopted children obtained slightly lower mean scores than controls on 9 of the 10 subtests. However, the multivariate F test was nonsignificant ($F_{10,398} = 1.58, p \geq .1$). The univariate tests are significant only for Similarities ($F_{1,407} = 11.07, p \leq .001$), and the adopted versus control comparison accounts for less than 3% of the observed variance.

Gender differences were significant for the MATH composite, PIQ, and three of the WISC-R subtests (Block Design, Object Assembly, and Mazes), with males scoring higher than females for each measure. However, these differences account for only 2–3% of the observed variance. The interaction between gender and adoptive status was significant only for Information, resulting in a nonsignificant multivariate F test ($F_{10,398} = 1.66, p \geqslant .09$).

Table 7.2 Means, standard deviations, and effect sizes on WISC-R subtests for adopted and control children at age 7

Measure	Adopted	Control	Effect
Information	11.54 ± 2.36	11.92 ± 2.35	.16
Similarities	12.14 ± 3.04	13.08 ± 2.96	.31*
Arithmetic	10.44 ± 2.67	10.89 ± 2.82	.16
Vocabulary	11.53 ± 2.89	12.24 ± 2.94	.24
Comprehension	11.62 ± 2.86	12.15 ± 2.74	.19
Picture Completion	12.00 ± 2.41	12.14 ± 2.34	.17
Picture Arrangement	12.79 ± 2.28	13.11 ± 2.66	.13
Block Design	11.68 ± 2.81	11.90 ± 2.76	.08
Object Assembly	11.04 ± 2.71	11.26 ± 2.27	.09
Mazes[a]	12.04 ± 2.91	11.79 ± 2.80	−.09
(N)	195–199	215–216	
Multivariate test $F_{10,398} = 1.58, p \geqslant .1$			

Note: After "Cognitive Abilities of Children at 7 and 12 Years of Age in the Colorado Adoption Project," by S. J. Wadsworth, J. C. DeFries, and D. W. Fulker (in press), *Journal of Learning Disabilities*. By PRO-ED Inc. Reprinted by permission.
[a] The Mazes subtest was administered at age 7 in place of the Coding subtest.
*$p \leqslant .005$.

Although adopted children tested in the Colorado Adoption Project at age 7 have slightly lower mean scores than controls on measures of general cognitive ability and academic achievement, all group differences are relatively small, accounting for less than 3% of the variance of any measure. Moreover, both groups have average scores that are well within the normal range.

Although differences between the average test scores of adopted and nonadopted children in the CAP are not large, the proportion of adopted children exhibiting learning problems could exceed that of controls. In order to assess the prevalence of possible learning disabilities in the adopted and nonadopted CAP children, IQ-achievement discrepancy scores were computed and the proportion of children in each group scoring at least 1.5 standard deviations below expected based on the regression of achievement on IQ was determined. The proportions of adopted and nonadopted children placed in special-education classes due to academic or emotional/behavioral problems were also compared. Chi-square tests, using continuity correction when cells contained less than 10 individuals (CROSSTABS/CHISQ) (SPSS-X, 1988) were performed to determine the significance of the difference between the proportions of adopted and nonadopted children whose academic achievement was below that predicted, based on IQ, and who were placed in special-education classes. Results indicated that there were no significant differences for either the proportions of adopted and nonadopted children with low achievement scores relative to their IQ, or the proportion placed in special education classes. In this sample, only 1.0% of the adopted group scored 1.5 standard deviations or more below expected on REC, versus 1.9% for the control group ($\chi^2 = .09$, $p \geq .7$). For MATH, 7.7% of the adopted children and 5.7% of the control children scored at least 1.5 standard deviations below expected ($\chi^2 = .70$, $p \geq .4$). With regard to the proportions of adopted and nonadopted control children placed in special classes for academic or emotional/behavioral problems, 2.8% of adopted and 4.8% of non-adopted children required special placement ($\chi^2 = 0.54$, $p \geq .4$).

Precursors of academic achievement

Correlations of early and middle childhood measures of general and specific cognitive abilities with measures of academic achievement are presented in Table 7.3. Although there is some suggestion of higher correlations of general and specific cognitive abilities with the achievement measures in nonadopted than in adopted children, the general pattern is similar for the two groups. As expected, correlations of measures of general cognitive ability with measures of reading and mathematics achievement tend to increase with age. This may be due, in part, to the use of tests which rely more heavily on verbal ability at ages 3, 4, and 7, which would be consistent with the suggestion of Roe et al. (1983) that infant tests which rely on nonverbal measures correlate less with later measures of intelligence and achievement.

Table 7.3 Correlations of measures of general and specific cognitive abilities with measures of academic achievement

Measure	Adopted		Control	
	REC	MATH	REC	MATH
General cognitive ability				
Age 1	.08	.13	−.08	.14
Age 2	.20*	.21*	.20*	.30*
Age 3	.21*	.20*	.13	.15
Age 4	.27*	.19*	.28*	.17
Age 7	.37*	.33*	.40*	.50*
Verbal ability				
Age 3	.18	.07	.03	.04
Age 4	.14	.15	.25*	.17
Age 7	.26*	.16	.32*	.31*
Spatial ability				
Age 3	.13	.15	.03	.19*
Age 4	.13	.30*	.24*	.36*
Age 7	.19*	.18*	.18*	.28*
Perceptual speed				
Age 3	−.07	−.06	.04	.18
Age 4	.05	.21*	.20*	.14
Age 7	.13	.24*	.20*	.15

Table 7.3 (Cont.)

Measure	Adopted		Control	
	REC	MATH	REC	MATH
Memory				
Age 3	.02	.09	.13	.20*
Age 4	.03	.03	.06	.07
Age 7	.20*	.11	.20*	.15

*$p \leq .01$.

While measures of verbal ability tend to be more highly correlated with reading achievement, and measures of spatial ability are more highly correlated with mathematics achievement, the similarities among these correlations are striking, suggesting the presence of a general factor influencing both achievement measures. However, results of hierarchical multiple regression analyses suggested that Perceptual Speed at age 7 contributed significantly ($p \leq .0001$) to the prediction of both achievement measures in nonadopted children when IQ was held constant. Moreover, Spatial ability at age 4 contributed significantly ($p \leq .001$) to the prediction of mathematics achievement in both groups. In agreement with the results of previous studies (e.g. Pennington, Van Orden, Kirson, & Haith, 1991), Memory at age 7 was significantly correlated with reading achievement in both groups.

Etiology of covariation

Phenotypic analysis   Phenotypic covariance matrices were computed separately for adopted probands, unrelated siblings, nonadopted control probands, and related siblings, and subjected to a Cholesky decomposition analysis in order to examine their factor structures. A multivariate path model including VIQ, PIQ, REC, and MATH (Figure 7.1) was fitted to the data by the method of maximum likelihood estimation using LISREL 7 (Jöreskog & Sörbom, 1989). Data from the four groups were analyzed simultaneously, by fitting first a general model allowing separate solutions, followed by a nested submodel equating covariance structures among the groups. The difference between the goodness of fit χ^2 obtained for the general model, and that obtained for the submodel, provided a test of the equality of covariance structures among the groups.

Results of the phenotypic model, equating covariance structures for the four groups, are presented in Figure 7.2. Phenotypic correlations among the measures are moderate (.32 between VIQ and PIQ, .40 between VIQ and REC, .18 between PIQ and REC, .43 between VIQ and MATH, .31 between PIQ and MATH, and .41 between REC and MATH), with a large proportion of variance specific to each measure. However, 44% of the observed correlation between REC and MATH is mediated by IQ, with most of this being accounted for by VIQ. This provides further support for the suggestion of a general factor influencing both reading and mathematics achievement.

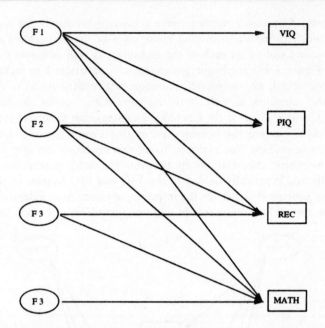

Figure 7.1 Path diagram illustrating Cholesky decomposition of factor structure among measures of Verbal IQ (VIQ), Performance IQ (PIQ), Reading Recognition (REC), and Mathematics achievement (MATH).

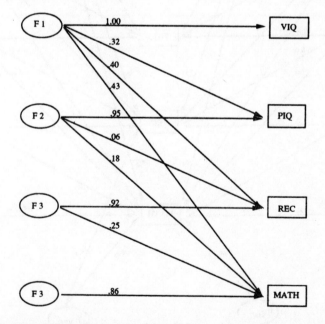

Figure 7.2 Results of Cholesky decomposition of factor structure among measures of Verbal IQ (VIQ), Performance IQ (PIQ), Reading Recognition (REC), and Mathematics achievement (MATH) ($\chi^2_{30} = 21.64$ ($p > .8$)).

Genetic analysis Covariance matrices were computed separately for unrelated and related sib pairs yielding variances and covariances of the first sibling's score with that of the second sibling on each of the measures, as well as across measures. As depicted in Figure 7.3, the phenotypic model was partitioned to include genetic, shared environmental, and nonshared environmental contributions to the variance in each of the measures, as well as to the correlations among the measures. As discussed in Chapter 3, use of the Cholesky decomposition facilitates exploration of the factor structure among the variables, permitting a more thorough interpretation of their interrelationships. For example, using this model we are able to determine the extent to which the relationship between the achievement variables, both phenotypically and genetically, is due to the VIQ and PIQ factors. In this manner, the model can provide evidence for the influence of common or independent genetic influences on the various measures. In addition, estimates of heritability, environ-

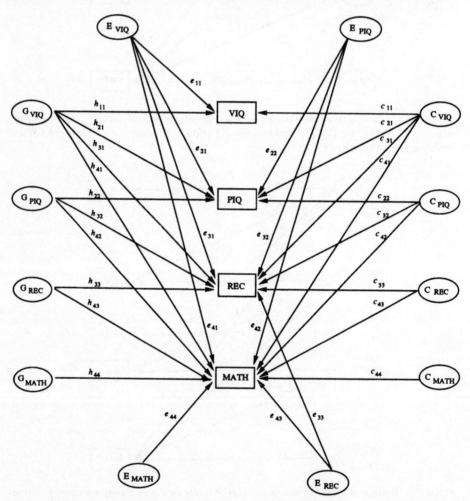

Figure 7.3 Path diagram illustrating partitioning of phenotypic model into genetic (h), shared environmental (c), and nonshared environmental (e) components of variance.

mentality, and genetic and environmental correlations among the measures are computed with relative ease.

Proportions of variance due to genetic and environmental effects for each of the measures were calculated from the sum of the squared standardized path coefficients from common and specific factors to each measure. For example, the heritability of VIQ is simply the square of the path from G_{VIQ} to VIQ, i.e. h_{11}^2, whereas that for PIQ equals $h_{21}^2 + h_{22}^2$, etc. (Figure 7.3). A similar procedure is used to obtain estimates of shared (c^2) and nonshared (e^2) environmental influences.

Estimates of the genetic, shared environmental, and nonshared environmental correlations among the measures are also obtained from the standardized path coefficients. For example, $h_{11} \times h_{21} = h_{VIQ} \times r_G \times h_{PIQ}$, where h equals the square root of the heritability of the measure, and r_G equals the genetic correlation between the two measures. That portion of the phenotypic correlation which is due to shared

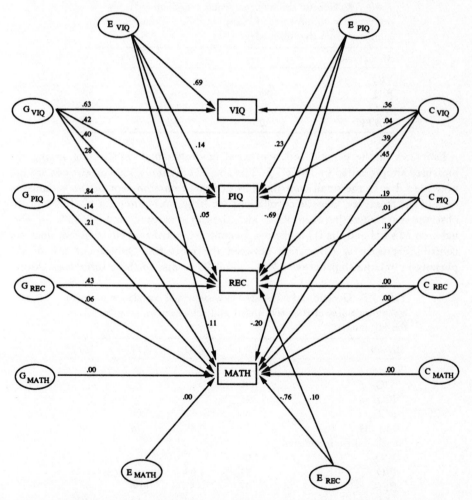

Figure 7.4 Results of genetic/environmental Cholesky decomposition ($\chi_{42}^2 = 38.66$ ($p^2 > .5$)).

genetic influences is obtained from $h_{11} \times h_{21}/r_P$, where $h_{11} \times h_{21}$ estimates the phenotypically standardized genetic correlation.

Results of the genetic analysis are presented in Figure 7.4. The results suggest one moderate genetic common factor. In addition, there remains a strong genetic influence specific to PIQ, and a moderately strong genetic influence specific to REC, but not to MATH. There is one shared environmental common factor, but little or no additional influence of shared environment specific to any of the measures.

Table 7.4 summarizes the contributions of genetic and environmental influences to the variances of each of the measures, estimated from the full model, with substantial heritability indicated for VIQ (.40), PIQ (.89), and REC (.36), but not for MATH (.12). Similarly, nonshared environmental influences are substantial for all measures except PIQ. In contrast, there is very little influence of shared environment on any of these measures.

Table 7.4 Genetic and environmental contributions to performance on measures of cognitive ability and achievement, estimated from the full model

Measure	h^2	c^2	e^2
VIQ	.40	.13	.47
PIQ	.89	.04	.08
REC	.36	.15	.49
MATH	.12	.24	.63

Estimates of the genetic and nonshared environmental correlations among the measures are presented in Table 7.5. The shared environmental correlations are not presented due to the small contributions of shared environment to the variance in each of the measures. Genetic correlations are large, accounting for 62–100% of the observed covariance between REC and the IQ measures, and 40–93% of that between MATH and the IQ measures. In contrast, nonshared environment does not contribute greatly to covariation between the measures. Only about 8% of the phenotypic relationship between REC and VIQ is due to these influences. More-

Table 7.5 Genetic and nonshared environmental correlation matrices among measures of cognitive ability and achievement, estimated from the full model

Measure	VIQ	PIQ	REC	MATH
Genetic				
VIQ	1.00			
PIQ	.45	1.00		
REC	.66	.52	1.00	
MATH	.79	.88	.78	1.00
Nonshared environmental				
VIQ	1.00			
PIQ	.51	1.00		
REC	.07	−.81	1.00	
MATH	.13	−.14	.12	1.00

over, the nonshared environmental correlation between REC and PIQ is negative, suggesting that genes and nonshared environment may influence these measures in different ways. Similarly, nonshared environment accounts for only about 17% of the observed correlation between MATH and VIQ, but does not appear to contribute to covariation between MATH and PIQ. Finally, only about 16% of the phenotypic relationship between REC and MATH can be attributed to nonshared environmental influences.

Discussion

In this chapter, three issues relating to school achievement in the CAP were discussed. In the first section, the prevalences of learning problems in adopted and nonadopted children were compared. Although adopted children obtained slightly lower mean scores on tests of cognitive ability and academic achievement than nonadopted children, differences were small, accounting for less than 3% of the variance in any measure, and average scores of both groups were well within the normal range. Moreover, proportions of individuals scoring below expected, based on IQ, did not differ significantly for the two groups, and there was no significant difference in frequency of special-education placement.

These results are consistent with those of Brodzinsky et al. (1984) who found that adopted children had somewhat more behavioral and academic problems than nonadopted children, but were within the normal range of variation. However, the results appear to differ from those of Silver (1970, 1989), Deutsch et al. (1982), and Brodzinsky and Steiger (1991), who reported that adopted children are over-represented in special-education populations. These apparently contradictory results may be due to differences in ascertainment procedure or measures used. For example, in the previous studies, children were ascertained from clinic and special-education populations. As noted by Deutsch et al. (1982), adoptees may be over-represented in these populations because adoptive parents may be more likely to refer their children to medical and special-education programs. Moreover, children placed at later ages may be more likely to have academic and behavioral problems. Silver (1989) and Deutsch et al. (1982) did not report age at placement for adoptees, but Brodzinsky et al. (1984) reported an age range at placement from 3 days to over 3 years. In the more recent study by Brodzinsky and Steiger (1991), most of the children were placed within the first year after birth. In contrast, adopted children in the CAP were placed within the first month after birth.

Although results of several previous studies suggest that the prevalence of learning problems may be relatively high in adopted children, the evidence is indirect and possibly subject to ascertainment biases and other methodological problems. Thus, additional research comparing adopted and nonadopted children with regard to behavioral problems, school success, and performance on school-administered achievement tests is clearly warranted. Nevertheless, the present results suggest that "easily placed" adopted children are not at a substantially increased risk for learning disabilities or school achievement problems (Wadsworth et al., in press).

In the second section of this chapter, developmental precursors of academic achievement were explored. As expected, correlations between measures of general cognitive ability and measures of reading and mathematics achievement tended to increase with age, as well as with reliance on verbal content of the measures. Furthermore, although there was some tendency toward higher correlations between measures of early cognitive ability and academic achievement in nonadopted children, the correlational pattern was similar for the two groups.

While at some ages verbal measures were more highly correlated with reading achievement than with mathematics achievement, and spatial measures were more highly correlated with mathematics achievement, there were striking similarities among the correlations at the various ages, suggesting the existence of a general intelligence factor influencing performance on the achievement measures. However, independent of the effects of IQ, perceptual speed at age 7 contributed significantly to the prediction of both achievement measures in the control group, and spatial ability at age 4 contributed significantly to the prediction of mathematics achievement in both adopted and control groups.

These results suggest that some early childhood measures of general and specific cognitive abilities may be useful for predicting later school achievement. Furthermore, it may be possible to facilitate academic achievement by encouraging the development of those skills which are most strongly related to performance on measures of achievement. However, it is important to note that this study has focused on only two measures of academic achievement collected at one time point. Further research into the relationships among cognitive abilities and measures of school achievement, such as school-administered achievement tests and subject-area grades, is warranted.

Finally, the etiology of covariation among scores on measures of cognitive ability and academic achievement was examined. Consistent with earlier studies, the phenotypic analysis revealed moderate correlations among the IQ and achievement measures. Furthermore, although results suggest that a large proportion of the observed covariance between the measures of academic achievement is due to general cognitive ability, over half of their covariation is independent of IQ.

Results of the genetic analysis support previous findings that genetic influences contribute substantially to individual differences in VIQ, PIQ, and reading achievement, as well as to their covariation. However, the measure of mathematics achievement used in this study does not appear to be substantially influenced by genetic factors, and it is only slightly influenced by shared environmental factors. In contrast, twin studies by Thompson et al. (1991) and by Gillis and DeFries (1991) found substantial influences of shared environment on individual differences in mathematics performance. The absence of shared environmental influences on mathematics achievement in the present study could be due to the age at which participants were tested, or to the fact that the siblings are tested on different occasions. Whereas the CAP siblings were tested at age 7, participants in the Thompson et al. (1991) study were 6–12 years of age, and those in the Gillis and DeFries (1991) study were 8–20 years of age. Alternatively, individual differences in mathematics achievement at age 7 may be due in part to differences in exposure to math at school; this would be expressed as nonshared environmental influence in the CAP sib pairs, who differ in age by 3.2 years, on average.

The present results conform closely to those obtained by Cardon et al. (1990) and by Brooks et al. (1990). The phenotypic correlation between Verbal IQ and reading achievement is largely mediated by genetic factors. In addition, while performance on the mathematics achievement measure was not found to be highly heritable, a strong genetic correlation between reading and mathematics achievement accounted for more than 45% of their phenotypic correlation. This is also consistent with the results of Gillis and DeFries (1991), who found that genetic influences accounted for 55% of the observed correlation between reading and mathematics achievement in a twin sample.

Results of these analyses suggest that measures of IQ and reading recognition are moderately heritable, with genetic influences accounting for more than one-third of the variance in Reading Recognition and Verbal IQ, and two-thirds of the variance in Performance IQ. Moreover, the moderate phenotypic correlations among the measures appear to be mediated largely by genetic influences. Commonalities between "verbal" and "nonverbal" measures both phenotypically and genotypically suggest the presence of a general intelligence factor and/or a general achievement factor. As the CAP sample size increases, more rigorous analyses of the etiology of the relationships among cognitive abilities and academic achievement will be undertaken.

Results of the present study have only begun to elucidate the genetic and environmental correlates of school achievement during middle childhood. Moreover, a different pattern of results may emerge as the CAP children approach early adolescence. Thus, additional research with other measures of cognitive abilities and school achievement in this unique sample is clearly warranted.

8 Developmental Speech and Language Disorders

Susan Felsenfeld

"I can't talk wight. I don't know why but I tink I was born-did dat way."
An 8-year-old boy during a speech evaluation

This chapter provides an overview of behavioral genetic studies of developmental speech and language disorders. The chapter begins with a classification taxonomy and description of speech disorder phenotypes. Following this, pertinent twin and family studies of these conditions are reviewed. The chapter concludes with a description of the *Colorado Adoption Project Speech Disorders Study*, the first investigation to date to use an adoption database to examine the etiology of developmental speech and language disorders. As part of this discussion, preliminary data from the *Speech Disorders* project are presented, and ongoing speech analyses using the CAP database are described.

Developmental speech and language disorders (hereafter referred to as DSLDs) are a relatively common cluster of related disabilities that affect between 4 and 8% of all young children. These disorders occur in the absence of organic factors that are known to interfere with speech development (e.g., mental retardation, hearing loss, neurological impairment, cleft palate, etc.). Most often, children affected with a DSLD differ from other children only in their communication skills. As a group, these children can be expected to perform in the average to low-average range on nonverbal tests of intelligence (Bloodstein, 1981; Leonard, 1987; Winitz, 1969), and come from home environments that do not differ in significant ways from the rearing environments of other children (Bloodstein, 1981; Leonard, 1987; Parlour & Broen, 1991; Schery, 1985). As with many developmental disabilities (e.g., dyslexia, hyperactivity), more boys are affected than girls, with reported ratios ranging from 2:1 to 3:1 (Kidd, Kidd, & Records, 1978; Ludlow & Cooper, 1983; Van Riper, 1971). The causes of DSLD are unknown. Systematic attempts to identify prenatal, birth, medical, or social variables that are related to DSLD have been generally unsuccessful (Bloodstein, 1981; Schery, 1985; Shriberg, Kwiatkowski, Best, Hengst, & Terselic-Weber, 1986), with the exception of the robust and well-replicated clinical observation of familial aggregation (to be discussed later).

With some exceptions (notably, chronic stuttering), the manifestations of DSLD are most pronounced in childhood, with symptoms appearing to remit by adolescence. Because of this, it has been suggested that these disorders are "outgrown" either with or without direct speech intervention and are therefore of little consequence. Recently, however, follow-up studies have demonstrated that adolescents and adults with a history of moderate speech and language disability are distinguishable from control subjects across a variety of receptive and expressive speech and language tasks (Aram, Ekelman, & Nation, 1984; Felsenfeld, Broen, & McGue, 1993; Lewis & Freebairn-Tarr, 1992) and may experience educational and occupational outcomes that are less satisfactory (Felsenfeld, Broen, & McGue, 1993). This information has created a renewed interest in identifying etiological factors that are related to DSLD, including the possibility of a genetic contribution.

Classification Taxonomy

"Developmental speech and language disorders" actually refers to a cluster of related but phenotypically distinguishable conditions that affect children as they acquire their native language. At a superordinate level, it is important to differentiate between children whose disorders are of unknown origin and those whose disorders can be attributed to known organic pathologies, either congenital or acquired.

Organic communication disorders in children result from structural, cognitive, or sensory constraints that adversely affect either the speech or hearing mechanisms (as in cleft lip/palate, syndromic deafness, etc.) and/or the information-processing and abstraction systems (as in syndromic mental retardation). There are well over 100 major gene and chromosomal syndromes that have speech or language deficits as a major clinical feature (cf. Siegel-Sadewitz & Shprintzen, 1982). These include single-gene conditions with both dominant (e.g., Apert's Syndrome) and recessive X-linked (e.g., Hunter's Syndrome) inheritance, as well as disorders involving sex chromosome aneuploidey (e.g., Kleinfelter's Syndrome) and trisomy and monosomy conditions (e.g., Down's Syndrome).

In addition, children with developmentally intact speech systems can (and unfortunately do) acquire communication disorders, usually as the result of

traumatic or degenerative events such as closed head injuries, near-drowning, progressive neuromuscular diseases, etc. While the speech and language characteristics
of these populations are essential to study for both rehabilitative and theoretical
reasons, results are not generalizable to the (larger) functionally impaired population
since these conditions tend to involve multiple intrinsic systems and/or diffuse
organismic damage.

Figure 8.1 presents a traditional taxonomy of developmental speech and language
disorders of unknown origin. While each of these disorders has the effect of
reducing the amount, complexity, intelligibility, and/or transmission efficiency of
the speaker's message, there are significant differences in phenotype.

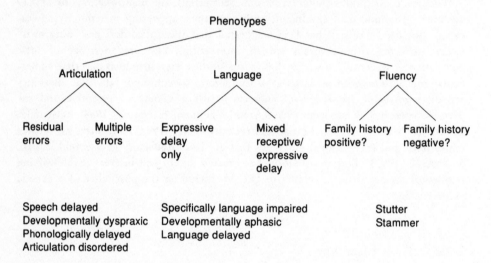

Figure 8.1   Taxonomy of Developmental Speech and Language Disorders.

At a broad level, distinctions are made between those disorders that primarily
affect the articulation system, those that affect the language system, and those that
adversely affect speech fluency (and create stuttering). Although this taxonomy
presents articulation, language, and fluency disorders as independent, it must be
pointed out that there is a much greater than chance likelihood that these disorders
will co-exist within an affected individual. This is particularly true for language and
(multiple error) articulation disorders. Shriberg et al. (1986), for example, found
that in a large sample of children whose primary diagnosis was articulation disorder,
75% also displayed clinically significant expressive language problems. In addition,
there is evidence that developmental articulation disorders and stuttering co-occur
at a high rate among children, with estimates ranging from 16% (Blood & Seider,
1981), to 22% (Homzie, Lindsay, Simpson, & Hasenstab, 1988). Regardless of
diagnosis, severity of involvement appears to be a good predictor of the probability
of co-morbidity, with children who manifest a moderate to severe communication
disorder (of whatever form) more likely to receive secondary and tertiary speech
diagnoses.

Characteristics of Disorder Phenotypes

Articulation disorders

Children who receive a primary diagnosis of developmental articulation disorder continue to misarticulate phonemes (usually consonants) beyond the age when mastery of the sounds is expected. As illustrated in Figure 8.1, there are at least two major subgroups of articulation-disordered children: those who make "residual errors" and those who produce "multiple errors" (Shriberg & Kwiatkowski, 1982). These subgroups differ from one another in terms of both number and types of misarticulations. The residual errors subgroup is the larger of the two, and contains children who substitute or distort a small number of consonant sounds. Readers who themselves attended speech therapy to correct an /r/ sound or a "lisp" were undoubtedly members of this group. In contrast, children in the "multiple errors subgroup" are those who persistently substitute and/or omit a large number of sounds. These errors are not random, but follow systematic simplification patterns that are fairly predictable across children. The following sample transcript from a 6-year-old child with a moderate multiple articulation disorder may be a helpful illustration. This child was shown a picture of two children having a picnic, and was asked to tell a story about it.

CHILD'S PRODUCTION: De tids are eating pood. Dey're on a bantit. One tid is eating tate.
TRANSLATION: The kids are eating food. They're on a blanket. One kid is eating steak.

Not surprisingly, there are subgroup differences in prognosis. While children in the residual errors subgroup do not appear to be at significant risk for related expressive language or reading problems, children in the multiple errors subgroup display these concomitant deficits at a high rate (Bishop & Adams, 1990). At present, it is not known if heritabilities differ between these two articulation subgroups, although there is speculation that they will be higher for children with multiple errors (Felsenfeld, McGue, & Broen, 1993).

Language disorders

Children who receive a primary diagnosis of developmental language disorder are those who fail to acquire the structure and lexicon (vocabulary) of their native language at age-appropriate times for reasons that cannot be readily identified. Typically, these children demonstrate some (or all) of the following clinical signs in language expression: delayed onset of speech, delays in the production of word combinations, difficulty in the correct production of morphemes (grammatical markers), reduced vocabulary size, difficulty in generating idea units, and difficulty in producing utterances that are organized and conversationally appropriate. For example, a 6-year-old child with a developmental language disorder, upon viewing the same picture as before, might produce the following narrative:

CHILD'S PRODUCTION: Them eating. Sitting on a thing. Having meat.

Clinically, distinctions are often made between children whose deficits are primarily in the production (expression) of language and those whose deficits involve not only language production but difficulties in auditory comprehension of verbal material as well. These comprehension deficits may be manifested as poor performance on tasks requiring short-term auditory memory (as in sentence or digit repetition), demonstration of vocabulary or concept knowledge, and comprehension of grammatically complex sentence structures. Again, little is known about subgroup differences in genetic etiology, although there is some preliminary evidence to suggest that the heritability of mixed expressive-receptive disorders may be higher (Whitehurst, Arnold, Smith, Fischel, Lonigan, & Valdez-Menchaca, 1991).

Fluency disorders

Children who receive a primary diagnosis of developmental fluency disorder (or stuttering) demonstrate chronic disturbances in their ability to produce (but not conceive) smooth, effortless, forward-moving speech. According to Van Riper (1971), the essence of stuttering is the "fracturing and disruption of the motor sequence of the word," almost as though a temporary loss of control over the movement of the speech musculature had occurred (Perkins, 1990). Although there is still some controversy about the single most appropriate "definition" of stuttering, the clinical diagnosis is usually made by counting the frequency of occurrence of a small number of discriminating speech behaviors: the repetition of sounds or syllables, the prolongation of sounds, and the presence of articulatory "blocks" in which the speech musculature appears to be frozen in position. Although these behaviors do occasionally occur in the speech of nonstuttering individuals, their frequency of occurrence is far greater among those classified as stutterers.

For purposes of comparison, a narrative of the "picnic scene" is included for a hypothetical 6-year-old stutterer. All three core stuttering topographies (syllable repetitions, prolongations, and blocks) have been included in the example.

CHILD'S PRODUCTION: The two ki-ki-ki-kids are having a fffffun time at the p. . . .ark. They're ea-eating sssteak and c. . . .ookies.

Among adults, an accurate diagnosis of chronic stuttering is usually not difficult, except in rare cases in which the stuttering is extremely mild or the speaker demonstrates excellent stuttering avoidance skills. For young children, however, the diagnosis is more problematic, since many normally developing youngsters experience periods of nonfluency that may be misidentified as incipient stuttering. For this reason, some behavioral genetic researchers will not consider a child to be affected with stuttering unless core behaviors have been present for at least six months (Kidd, 1983), or will not identify affected cases that are under the age of 5 (Barnes MacFarlane, Hanson, Walton, & Mellon, 1991).

As with other speech and language disorders, there is considerable speculation that etiologically distinct subgroups of developmental stutterers exist, although there is no consensus about what these subgroups should be. One of the most promising proposals for subgrouping stutterers is the presence or absence of family history (where positive family history is loosely defined as the presence of stuttering in a first-degree relative of a proband). Recently, a small number of investigators have examined data obtained from heterogeneous samples of stutterers grouped according to family history status (Cooper, 1972; Janssen, Kraaimaat & Brutten, 1990; Kidd, Heimbuch, Records, Oehlert, & Webster, 1980; Poulos & Webster, 1991; Seider, Gladstein, & Kidd, 1982). Although many of the studies in this area are preliminary, these analyses have revealed interesting similarities and differences between stutterers who report a positive family history (FHP) and those who report that no other immediate relatives ever stuttered (FHN). A partial list of these comparisons is presented in Table 8.1.

Table 8.1 Comparison of family history negative (FHN) and family history positive (FHP) subgroups of stutterers

Investigators	Comparison variable
	Non-significant subgroup differences
Kidd, Heimbuch, Records, Oehlert, and Webster (1980)	Severity of stuttering
Janssen, Kraaimaat, and Brutten (1990)	Speech-associated anxiety
Janssen, Kraaimaat, and Brutten (1990)	Reading ability
Janssen, Kraaimaat, and Brutten (1990)	Responsiveness to treatment
Seider, Gladstein, and Kidd (1982)	Reports of other speech problems
	Significant subgroup differences
Poulos and Webster (1991)	Prenatal, birth, or medical event (FHN more)
Cooper (1972)	Frequency of spontaneous recovery (FHN more)
Janssen, Kraaimaat, and Brutten (1990)	Frequency of prolongations (FHP more)
Janssen, Kraaimaat, and Brutten (1990)	Abnormal speech-motor behavior when fluent (FHP more)

If this subgroup classification does reflect true etiological differences, several predictions can be made. For example, both incipient and advanced stutterers with a negative family history should be more likely than FHP stutterers to report the presence of a prenatal, birth, or medical event that could reasonably have precipitated the stuttering behavior, and may be more likely to experience spontaneous recovery, perhaps as the effects of the initial precipitating event diminish. Conversely, stutterers with a positive family history should have prenatal, birth, and early developmental histories that are unremarkable; in fact, this appears to be the case (Cox, Seider, & Kidd, 1984).

In contrast to stutterers with a negative family history, FHP stutterers should display speech production breakdowns that reflect the effects of a fairly isolated and

Table 8.2 Summary of methodology for family studies of DAD/SLI

| Investigators | Primary selection phenotype | Index cases | | Who assessed | Relatives |
		Number of index cases	Ages at time of testing		Primary selection criteria
Neils and Aram (1986)	Specific language impairment	74 probands 36 controls	4–6 years	Parents & siblings	Reports of problems with pronounciation, stuttering, reading, or language
Lewis, Ekelman, and Aram (1989)	Moderate/severe articulation disorder	20 probands 20 controls	4–6 years	Parents & siblings	Reports of speech, reading, or learning disorder
Tallal, Ross, and Curtiss (1989)	Specific language impairment	62 probands 50 controls	4 years	Parents & siblings	Reports of language problems, below average school achievement in reading or writing, grade repetition, or learning disability
Tomblin (1989)	Specific language impairment	51 probands 136 controls	7–9 years	Parents & siblings	Receipt of speech-language therapy
Whitehurst, Arnold, Smith, Fischel, Lonigan, and Valdez-Menchaca (1991)	Expressive language delay	62 probands 55 controls	2–3 years	Parents, siblings, grandparents, aunts, uncles, & cousins	Report of late talking, speech problems, or school problems
Felsenfeld, McGue, and Broen (1993)	Moderate/severe articulation disorder	24 probands 28 controls	30–32 years	Offspring	Receipt of speech-language therapy

persistent inherited "defect" in the speech-motor control system. More specifically, members of the FHP subgroup might be expected to produce a higher frequency of certain types of dysfluencies that reflect problems in precise timing of speech-motor (particularly laryngeal) action (i.e., prolongations and blocks), and should display atypical speech-motor responses that are independent of the stuttering itself (as appears to be the case). Finally, the family members of FHP but not FHN stutterers should evidence atypical speech-motor behaviors that may be considered subclinical forms of the inherited defect. Clearly, until more work is done in this area, the subgroup classification proposed here for stuttering must be considered preliminary (albeit provocative).

Genetic Studies of DSLD

As noted previously, one long-standing observation about developmental speech and language disorders is that they "run in families." Within the last 15 years, there has been increasing interest in examining this familiality from a biological (genetic) perspective, using family and twin study designs. Although many of these recent studies are of modest size, their results have provided the foundation needed to justify more sophisticated quantitative and molecular genetic studies of communication disorders.

Issues of phenotype definition and the presence of co-morbidity have been problematic for researchers studying the transmission of communication disorders, particularly those involving deficits in articulation and language development. Because so little is known about the etiology of these phenotypes, most studies have elected to adopt broad criteria for the identification of affected probands and relatives, so that individuals with multiple phenotypes (e.g., articulation and language disorders) and those with only one diagnosis (e.g., expressive language disorder) have been collapsed for analysis. Because studies have not controlled for the frequent co-morbidity of these disorders in their samples, the acronym DAD/SLI (developmental articulation disorder/specific language impairment) has been adopted here to represent the broad phenotype that includes language disorder with or without co-occurring deficits in articulation. Clearly, future studies should select affected cases more narrowly so that differences in etiology among diagnostic categories and for subgroups within a category can be identified.

Family studies of DAD/SLI

Table 8.2 provides a summary of the methodologies of the principal family studies of developmental articulation and language disability. As can be seen, data were usually obtained from the first-degree relatives of individuals identified as articulation and/or language disordered, and were compared to data from control families with negative speech disorder histories (Byrne, Willerman, & Ashmore, 1974; Felsenfeld, McGue, & Broen, 1993; Lewis, 1992; Lewis, Ekelman, & Aram, 1989;

Neils & Aram, 1986; Tallal, Ross, & Curtiss, 1989; Tomblin, 1989; Whitehurst et al., 1991). In most of these studies, young affected children were identified as index cases, and data were collected about the speech, language, and/or learning status of their first-degree relatives (usually parents and siblings).

The results of these family studies are summarized in Table 8.3. In describing the degree of familiality, two types of analyses are reported: (1) the percentage of proband subjects with one or more affected first-degree family members; and (2) the percentage of affected first-degree relatives of proband and control subjects.

Table 8.3 Rates of disability among relatives as reported in family studies of DAD/SLI

Investigators	% of probands with affected relative(s)	% of relatives who are affected	
		Proband	Control
Byrne, Willerman, and Ashmore (1974)	55%	—	—
Neils and Aram (1986)	—	20%[a]	3%
Tallal, Ross, and Curtiss (1989)	77%	41%[a]	18%
Tomblin (1989)	53%	23%[a]	3%
Lewis (1990)	—	33%[a]	5%
Whitehurst, Arnold, Smith, Fischel, Lonigan, and Valdez-Menchaca (1991)	—	12%[a]	7%
Felsenfeld, McGue, and Broen (1993)	53%	33%[b]	0%

[a] = parents and siblings
[b] = offspring

With the exception of one study (Whitehurst et al., 1991) these investigations have demonstrated that relatives of individuals affected with a DSLD are more likely to display speech, language, or learning disabilities than are the relatives of control subjects. Specifically, results of these studies suggest that: (1) approximately 50% of the families of probands will contain at least one additional first-degree relative who is affected; and (2) between 20 and 33% of the first-degree relatives of a proband subject will report the presence or history of similar disorders, in comparison to an average of 6% of the relatives of controls. Not surprisingly, when very broad criteria for affection are employed (as in the study by Tallal et al., 1989), the absolute rates of reporting are elevated for both proband and control subjects, although proband families still report a significantly greater number of problems. Only one family study has failed to find increased rates of DSLD among the relatives of language disordered children (Whitehurst et al., 1991). Unlike previous family studies, the probands in this investigation displayed a narrow disorder phenotype: all were 2-year-old children with delayed onset of speech and normal comprehension skills. According to the investigators, this negative result suggests that disorders limited to language expression may be less heritable than are disorders that are more moderate or are mixed (i.e., involve both expressive and receptive deficits).

Family studies of stuttering

References to the familiality of stuttering and to "stuttering families" have appeared in the literature for more than 50 years (Andrews & Harris, 1964; Gray, 1940; Johnson, 1959; Wepman, 1939; West, Nelson, & Berry, 1939). Sheehan and Costley (1977) reviewed many of these early studies, and concluded that, on average, about one-third of all stutterers sampled reported the presence of "other stutterers" in their families. Although these findings are informative, there are problems with many of these early studies (for example, there were no controls for family size, particularly high-density families may have been selectively recruited) that limit their usefulness for future quantitative genetic analyses.

Much of our recent knowledge about the transmission of stuttering has come from the Yale Family Study of Stuttering, the most comprehensive behavioral genetic study of this disorder performed to date (see Kidd, 1983 for a description). In this large project, data were collected on nearly 600 stutterers and more than 2,000 of their first-degree relatives. To be included in the study as a proband, potential subjects were required to have received a diagnosis of stuttering by a speech pathologist at some time in their lives ("recovered" stutterers were therefore included). Relatives were considered to be affected if: (1) they had received a stuttering diagnosis, or (2) they were reported to have ever had excessive repetitions or prolongations in their speech that lasted at least six months. In most cases, information about the speech status of relatives was obtained by interviewing the proband (or the proband's parents); that is, direct testing of relatives to determine or verify speech status was not performed. To prevent a bias toward the ascertainment of particularly large or high-density families, probands were admitted into the study without regard to family size or knowledge about family speech history.

The results of the Yale Family Study verified that stuttering aggregates within families and is transmitted vertically from parent to child (Kidd, 1983; Kidd, Heimbuch, & Records, 1981; Seider, Gladstein, & Kidd, 1983). Among the first-degree relatives of chronic stutterers, approximately 15% reported that they had stuttered at some point in their life: of these, about one-half were considered to be persistent (chronic) stutterers and one-half were reported to be recovered, with female relatives somewhat more likely than male relatives to report spontaneous improvement (Seider et al., 1983). When compared to rates of "ever stuttered" in the population (estimated to be about 5% in males and 1–2% in females), these investigators concluded that a rate of 15% for family members of probands was significant. Interestingly, severity of stuttering in the proband was not a good predictor of family history status; that is, the more severe stutterers were not necessarily those who had a positive family history (Kidd et al., 1980).

Gender analyses: Sex ratios, and risk to relatives

Many of the current family studies of DSLDs have reported gender differences among affected index cases and their relatives. Two useful observations can be derived from these data: (1) the proportion of males who are affected; and (2) the risk to family members of male versus female probands.

Table 8.4 presents the gender distribution of affected probands and relatives across family studies of DSLD. As can be seen, the data tend to confirm the historical reports of elevated rates of disability among males. Regardless of phenotype, it appears that the fathers, brothers, and sons of affected index cases are more likely to be affected with DSLD than are mothers, sisters, and daughters. Moreover, there is some suggestion that the sex ratios may not be the same for all primary selection phenotypes, with moderate articulation disorder showing a less pronounced gender ratio than stuttering or developmental language disorder.

Table 8.4 Gender distribution of probands and relatives affected with a DSLD

Investigators	Primary selection phenotype	% of affected cases who are male	
		Probands	Relatives
Kidd, Kidd, and Records (1978)	Stuttering	75%	75% (284/370)
Tallal, Ross, and Curtiss (1989)	Specific language impairment	71%	73% (27/37)
Tomblin (1989)	Specific language impairment	74%	—
Whitehurst, Arnold, Smith, Fischel, Lonigan, and Valdez-Menchaca (1991)	Expressive language delay	85%	51% (23/43)
Lewis, Ekelman, and Aram (1989)	Moderate/severe articulation disorder	65%	63% (35/56)
Felsenfeld, McGue, and Broen (1993)	Moderate/severe articulation disorder	63%	36% (5/14)

Although more males than females in the population are affected with DSLDs, a small number of family and pedigree studies have provided evidence that the risk for disability is greater for relatives of *female* probands, both for stuttering (Andrews & Harris, 1964; Barnes MacFarlane, Hanson, Walton, & Mellon, 1991; Kidd et al., 1978) and for DAD/SLI (Lewis, 1990). In examining the risk to relatives of stutterers, for example, Kidd et al. (1978) noted that approximately 13% of the first-degree relatives of male probands were affected with stuttering in comparison to 18% of the relatives of female probands, a difference that is statistically significant. As pointed out by these investigators, an observation of unequal risk to relatives based on proband gender is difficult to explain on purely cultural or environmental grounds. Rather, they suggest that a more plausible explanation is that stuttering liability may be transmitted as a sex-modified trait in which genotypes are expressed as different susceptibilities depending on gender, with males having a lower liability threshold.

Twin studies of DSLD

Only two moderately-sized twin studies (in which one (or both) members of a twin pair had a reported history) of DSLD have been completed within the last 30 years

(Howie, 1981; Lewis & Thompson, 1992). Table 8.5 presents the probandwise concordance rates for speech disorder from both of these investigations.

Table 8.5 Probandwise concordance rates as reported in twin studies of DSLDs

Phenotype	# Pairs		Concordance rates	
	MZ	DZ	MZ	DZ
Stuttering[a]	17	13	.77	.32
DAD/SLI[b]	32	25	.86	.48

[a] Howie (1981).
[b] Lewis and Thompson (1992).

As can be seen, the obtained twin concordances not only complement the findings of family studies, but provide evidence that genetic factors are important contributors to familial resemblance. Monozygotic twins were found to be highly concordant for the presence of a DSLD (.77–.86), although the phenotype and severity of the disorder were not necessarily identical. In comparison, the concordances obtained for the same-sex dizygotic twins were more moderate (.32–.48) but were somewhat greater than sibling risk estimates obtained from family studies (.20–.33). In addition, Lewis & Thompson (1992) noted that more first-degree relatives of concordant than discordant twin pairs reported speech or language problems (25% versus 13% respectively), a finding that is more readily explained by a genetic than an environmental transmission hypothesis.

Only one study (Andrews, Morris-Yates, Howie, & Martin, 1991) has estimated the heritability of a DSLD (stuttering) by applying quantitative model-fitting procedures to twin data. In this study, an item about stuttering was inserted into a lengthy questionnaire mailed to several thousand Australian twins. Results revealed that the best fit to the data was provided by a model in which 71% of the variance in liability to stuttering was attributed to additive genetic variance, with the remaining 29% attributed to the individual's unique environment. Interestingly, the shared environmental effects parameter was estimated to be zero, a finding of particular interest in light of the importance that has been paid in the stuttering literature to between-family factors such as parental behavior and modeling.

Nongenetic Transmission of DSLD

The twin and family studies of DSLD performed to date have provided consistent evidence that the etiology of these disorders may be largely genetic. However, because these studies have almost exclusively studied intact nuclear families, cultural transmission factors cannot be dismissed. The role of cultural transmission is particularly important in the etiology of speech and language disorders, since the development of communication is so unambiguously dependent on environmental input. It has been pointed out that parents with a history of DSLD might be expected to provide a language environment that is less than optimal for speech development (for example, they may provide poor speech models, may be overly concerned about speech performance, etc.); as such, the vertical transmission of DSLD in these homes may be explained in purely environmental terms. Although this explanation does not address the differences in concordance between MZ and DZ twins, there have been other environmental hypotheses (for example, encouragement of twin secret languages, more limited mother–child dyadic interactions) that have been invoked to interpret this phenomenon.

Two general hypotheses have been proposed to explain how cultural transmission of DSLDs might occur: (1) children who manifest DSLDs come from homes that are intellectually or linguistically impoverished; and/or (2) parents of children with DSLDs create an emotional climate that interferes with normal communication development (e.g., they set standards for speech that are unrealistic, they are not verbally responsive, etc.).

To test these hypotheses, the "linguistic environments" of children with and without DSLD have been compared, both from a macro level (e.g., by examining factors related to socioeconomic status) and from a micro level (e.g., assessing the quality, complexity, or emotional tone of maternal speech input). Regardless of the level of analysis, the results of these studies have been negative or inconclusive, and, in some cases, are actually contradictory. For example, mothers of children who stutter have been characterized as perfectionistic, achievement-oriented, and verbally dictatorial in one set of studies (Johnson, 1959; Kasprisin-Burrelli, Egolf, & Shames, 1972) and as passive low-achievers in another (Andrews & Harris, 1964). Thus, despite many attempts to identify differences, most researchers have concluded that aberrant verbal behaviors or patterns of interaction are not characteristic of parents of DSLD children (see Bloodstein, 1981; Conti-Ramsden, 1985; Leonard, 1987 for reviews).

Similarly, although there is a weak but positive association between low socioeconomic status and the presence of a DSLD (Templin, 1957; Vetter, 1980), studies that have compared the environments of affected and control families along a variety of speech-specific dimensions (e.g., availability of stimulating toys, variety in experience, family attitudes towards speech) have failed to find robust group differences (Cox et al., 1984; Parlour & Broen, 1991). Despite these negative findings, there continues to be a great deal of emphasis upon environmental etiologies for DSLD; in some cases, treatment programs have been developed to modify parental behaviors and attitudes that are presumed to cause and/or exacerbate these conditions.

The most effective way to assess the relative importance of genetic and between-family environmental factors to the etiology of a familial condition is through an adoption design (Plomin, DeFries, & Fulker, 1988). While desirable, it is unlikely that an adoption study of DSLD will ever be completed given both the difficulty and expense associated with adoption studies, and the relatively low frequency of DSLD within the population. Fortunately, although the project is not specifically designed to assess speech disorders, data obtained as part of the ongoing Colorado Adoption Project can be used to address questions pertaining to the etiology of developmental speech and language disorders.

The Speech Disorders Study

Using data from the Colorado Adoption Project, a preliminary investigation addressing the etiology of developmental communication disorders has been initiated at the University of Pittsburgh. At the time of this writing, most of the analyses of interest are still in progress, and will not be reported here. The purpose and design of this study will, however, be described, and some early comparative data will be presented.

As discussed in Chapter 2, all biological, adoptive, and control parents who participated in the CAP completed an extensive personal history questionnaire which included items relating to history of speech disorder. For the *CAP Speech Disorders Study*, the responses to these particular questions have been used to identify affected and unaffected parental index cases (and their associated offspring). Thus, although the index cases for this study are parents, the principal analyses will be performed on the at-risk offspring and a matched group of offspring with negative parental history. When this study is completed, it will provide the first systematic evidence detailing the independent contributions of genes and environment to the etiology of developmental speech disorders. Specifically, the following questions will be addressed in the *CAP Speech Disorders Study*:

1 What are the differences in articulation, language, and fluency performance between:
   (a) Adopted children with a biological parent who reports a history of speech disorder and children raised by an adoptive parent who reports such a history;
   (b) Adopted children with a positive biological *or* rearing history and adopted children with no parental history;
   (c) High-risk adopted children and high-risk control (nonadopted) children;
   (d) High-risk nonadopted children and matched low-risk nonadopted children?
2 What are the differences in recovery rates for affected children in these groups?
3 Are adoptive and control *mothers* who report a history of speech disorder rated differently than mothers with a negative speech disorder history on a scale of maternal verbal behavior?
4 What is the gender distribution of children who are judged to be affected with a DSLD?

Selection of parental index cases

For the *CAP Speech Disorders Study*, any parent in the CAP sample (biological, adoptive, or control) who responded affirmatively to a select set of speech history questions on the CAP questionnaire was considered an index parent. Once a parent was identified as affected, that individual's child was considered to be at risk for a

DSLD, and was included in the offspring analysis. It should be noted that the questionnaire items were written to identify individuals with a stuttering problem in particular, although some items may also have been endorsed by parents with a history of articulation disorder. For this reason, the offspring analyses will focus on the articulation and fluency domains.

It would, of course, have been desirable to obtain validation of speech disorder status for all of the parents identified as "affected" in the present study. When using unvalidated questionnaire responses, there is the danger that clinically significant speech problems will be either under-reported (if they remitted early) or over-reported (if normal developmental articulation errors or periods of nonfluency were misidentified as disorders). However, Parlour (1991) found that adults with a documented history of moderate articulation disorder endorsed questionnaire items about speech problems far more often than did unaffected control adults. Thus, although the sensitivity and specificity of the speech items used in the CAP study are unknown, they were judged to be acceptable for identifying adults who experienced clinically significant problems with stuttering (and possibly articulation performance) at some point in their life.

Specifically, in order to be identified as an index parent, an individual was required to have:

1 Responded in the affirmative to at least two of the five following speech-related questions:
   (a) "Do you have a history of a stuttering or stammering problem?"
   (b) "Do/did you have a definite problem saying certain words or sounds?"
   (c) "Do/did you have a definite problem getting words out or getting started speaking?"
   (d) "Has there ever been a problem with gaps or hesitancy in your speech?"
   (e) "Was a speech therapist ever seen for an evaluation or therapy?"
2 Reported the age of onset of the speech problem to be less than age 13 (or unspecified). (This criterion was included to maximize the likelihood that the reported problem was indeed developmental rather than acquired).

When these criteria were applied to the CAP database, the following three groups of affected parents were identified: 23 of the 244 biological parents (9%); 32 of the 451 adoptive parents (7%); and 40 of the 480 control (nonadoptive) parents (8%). In addition, two comparison groups of parents with negative speech disorder history were matched to the proband groups on the basis of socioeconomic status and offspring age and gender: 57 unaffected adoptive parents, and 41 unaffected control parents. Data from the offspring of all of these subjects were reviewed to ensure that information was available at ages 3 and 7. In addition, adopted children who had both a positive biological and adoptive history of speech disorder were excluded. There were 93 high-risk and 98 comparison (low-risk) children who met these criteria. These 191 children formed the following five groups:

1 Adopted children with a positive biological history of DSLD ($N = 21$);
2 Adopted children with a positive adoptive history of DSLD ($N = 32$);

3 Nonadopted children with both a positive biological and environmental history of DSLD (at-risk nuclear families) ($N = 40$);

4 Adopted comparison (low-risk) children with both negative biological and adoptive histories ($N = 57$);

5 Nonadopted comparison (low-risk) children ($N = 41$).

Depending on the nature of the assessment measure, offspring performance will be evaluated both continuously (e.g., by comparing mean differences for standard scores) and categorically (e.g., as a pass versus a fail on scales assessing fluency and intelligibility of speech). Since longitudinal data are available for children at ages 3 and 7, it will be possible to examine patterns of recovery among children who are considered to be "affected" with a DSLD at age 3, and determine if there are group differences in recovery profiles.

Measures

Although several assessment instruments were included in the CAP battery, only a subset were selected for analysis in the *CAP Speech Disorders Study*. Most of the selected measures are norm-referenced tests of speech and language ability or general cognitive functioning. In addition, audio (age 7) and video (age 3) samples of the children's conversational speech are available for analysis. These tapes are invaluable since they permit an examination of important aspects of speech production not assessed in standardized tests (i.e., fluency, intelligibility, prosody, and articulation) and allow an examination of the mother's verbal interactions with her child at age 3. In addition to evaluating the performance differences across the five groups, the offspring speech and language measures will be correlated with general cognitive and language measures obtained from the biological and/or adoptive parents, as well as with measures of the home environment and ratings of maternal interactions. Taken together, the outcome of this study will provide a comprehensive assessment of both genetic and nongenetic factors associated with atypical communication development.

Preliminary analyses

The conversational speech and maternal interaction analyses are in progress, and will be presented in forthcoming reports. At present, a few preliminary analyses have been completed to examine the general comparability of the five offspring groups on selected language and environmental variables. These comparisons are summarized in Table 8.6.

As can be seen, the five offspring groups do not differ significantly in Receptive or Expressive language development as measured by the *Sequenced Inventory of Communication Development* (SICD) at age 3, total score on an interview and observational scale of the home environment (the HOME Scale), or WISC Full Scale IQ at age 7. Given these findings, any significant differences in speech outcome that may be observed across groups cannot be readily attributed to factors such as IQ or general adequacy of the home environment.

Table 8.6   Means and standard deviations for the preliminary analyses from the Colorado Adoption Project Speech Disorders Study

Test	Group					P-value
	1	*2*	*3*	*4*	*5*	
Age 3						
SICD Receptive	10.4	9.2	9.7	10.3	10.4	
	(4.2)	(4.7)	(4.0)	(4.0)	(4.3)	.741
SICD Expressive	10.7	10.8	11.7	12.0	12.9	
	(3.9)	(3.9)	(3.7)	(3.4)	(2.8)	.095
HOME Scale total	31.5	30.9	31.1	31.4	31.3	
	(2.0)	(2.5)	(2.3)	(2.2)	(2.2)	.867
Age 7						
WISC-R Full Scale	109	111	113	110	113	
	(10.3)	(11.2)	(9.6)	(10.4)	(12.8)	.659

Note: Group 1 = at-risk; positive biological history (N = 21);
Group 2 = at-risk; positive adoptive history (N = 32);
Group 3 = at-risk control; positive nuclear family (N = 40);
Group 4 = comparison low-risk adoptive (N = 57);
Group 5 = comparison low-risk control (N = 41).

Although not statistically significant, the results from the expressive portion of the SICD are worthy of comment. When compared to the adopted children at low risk for disorder (group 4), the mean scores for the two at-risk adoptive groups (groups 1 and 2) were noted to be approximately one-third of a standard deviation lower. This trend toward poorer expressive language performance among the at-risk adoptees provides the first evidence of depressed verbal production abilities among these children, and motivates our ongoing analyses of parameters such as articulation and fluency that are more likely to distinguish the groups.

Conclusions

Both family and twin studies have demonstrated that developmental articulation, language, and fluency disorders are familial and are probably strongly influenced by genetic factors. Specifically, between 20 and 33% of the first-degree relatives of probands with developmental articulation/language disorders and 18% of the relatives of stuttering probands report a positive history of similar disorders, in comparison to rates within control groups of about 3–4%. Although only two twin studies of speech disorders have been reported, they show much higher concordance of disability among monozygotic (.77–.86) than dizygotic (.32–.48) twins. However, because communication development is clearly dependent on environmental input, the possibility of cultural transmission cannot be rejected.

Using CAP data, a preliminary risk analysis for speech disorder is being performed. This project (the *Colorado Adoption Project Speech Disorders Study*) will assess the speech of five groups of children: three groups that are considered to be "at-risk" for speech disorder because of a positive parental history (biological, adoptive, and nonadoptive), and two groups that are considered to be "low-risk" for these conditions (adoptive and nonadoptive comparison groups). Preliminary analyses of test data have demonstrated a general comparability across groups for variables that have been historically associated with speech outcome, such as IQ, language comprehension, and general adequacy of the home environment. In addition, these results have shown a trend toward poorer performance in language expression for the two adoptive groups at risk for DSLD, a finding that is consistent with predictions and will be explicated further through the analyses that are in progress. Unlike any previous study, results of this project will help to establish the independent contributions of genes and family environment to the etiology of developmental speech disorders of unknown origin, information that is of considerable interest to both theoreticians and practitioners who study and treat abnormal communication development.

9 Personality and Temperament

Stephanie Schmitz

Although definitions of temperament vary widely (Goldsmith et al., 1987), temperament usually refers to a subset of early-appearing personality traits that display biological origins and are consistent across situations and time. Behavioral genetic research has led the way toward understanding the origins of individual differences in personality and temperament. It brings to bear the armamentarium of quantitative genetics, such as twin and adoption studies, to investigate empirically the relative influence of an important class of biological factors: genes.

Research on adolescent and adult personality has been a major focus of behavioral genetic research (Plomin, Chipuer, & Loehlin, 1990). A landmark in this field was the 1976 book by Loehlin and Nichols, *Heredity, Environment, and Personality*, a book that set out the agenda for research. A second landmark is a 1989 book by Eaves, Eysenck, and Martin, *Genes, Culture, and Personality*, which summarizes two decades of research by the authors and their advances in model-fitting. Another recent book by Loehlin (1992), *Genes and Environment in Personality Development*, is likely to become a classic in this area.

Most of this research is based on self-report questionnaire data. Surprisingly similar results emerge from twin analyses of most of the myriad self-report measures of personality. For example, in a model-fitting meta-analysis of personality data from over 30,000 pairs of twins in four studies conducted in four countries, Loehlin (1989) reported heritability estimates in excess of .50 and negligible shared-environment estimates for extraversion and neuroticism. A few new issues have since arisen, such as gender differences, nonadditive genetic variance, and assimilation effects for identical twins reared together (Plomin et al., 1990).

When attention turned to the study of the temperamental foundations of personality in early childhood, parental rating questionnaires were employed widely, modeled after the self-report questionnaires used in studies of adolescents and adults. This was a reasonable strategy since parents see their children in a variety of situations over time. Nearly all of the dozens of behavioral genetic studies of temperament in childhood have relied on parental rating questionnaires and the twin method. Like the adult twin studies using self-report questionnaires, twin studies of parental ratings of children's temperament yielded evidence for ubiquitous genetic and negligible shared environmental influence.

However, unlike self-report questionnaire studies of adults, twin studies using parental rating questionnaires produced an odd pattern of results. Identical twin correlations were moderate, typically about .50, but fraternal twin correlations were zero or even negative, especially for traits like activity, and for questionnaires that ask about broad aspects of behavior. For example, twin studies of infants and young children that used the Emotionality–Activity–Sociability–Impulsivity (EASI) rating scales of Buss and Plomin (1975) yielded average identical and fraternal twin correlations of .59 and −.01 respectively (Buss & Plomin, 1984). Similarly, a twin study in England yielded average EASI correlations of .25 for identical and −.06 for fraternal twins in infancy and early childhood (Stevenson & Fielding, 1985). Studies using other parental rating instruments yielded similar results (Plomin et al., 1990). The low fraternal twin correlation is a problem because, if a trait is heritable, fraternal twins who share half of their segregating genes should resemble each other, even in the presence of epistasis.

Uneasiness about twin analyses based on parental rating questionnaires has in part been responsible for motivating researchers to undertake more difficult assessments, such as observational measures, and more difficult designs, such as adoption designs. The relatively few attempts to conduct studies of this kind are promising. Bayley's Infant Behavior Record (IBR: Bayley, 1969) was designed to assess infants' temperamental responses to a standardized and somewhat stressful mental test session. For the first two years of life, these observational data displayed reasonable patterns of twin correlations indicative of moderate genetic influence in the Louisville Twin Study (Matheny, 1980). In the first report using the sibling adoption design, CAP data confirmed these findings in infancy (Braungart, Plomin, DeFries, & Fulker, 1992). The MacArthur Longitudinal Twin Study has incorporated a wide variety of observational measures in the second year of life, ranging from observational measures to parent and tester ratings (Emde et al., 1992; Plomin, Chipuer, & Neiderhiser, 1993).

However, CAP data in infancy and early childhood yield strikingly different results for parental rating data in comparison to twin studies. Although twin studies consistently find high correlations for identical twins and negligible or even negative correlations for fraternal twins, results of both parent–offspring and sibling adoption analyses provide little or no evidence for genetic influence on parental ratings of temperament (Plomin, Coon, Carey, DeFries, & Fulker, 1991).

Another important source of information are teachers who have as their own standard of comparison the behavior of many children seen in the school situation. However, to our knowledge, teacher ratings have not previously been reported in behavioral genetic research on temperament.

Finally, self-report questionnaires can be administered to children toward the end of middle childhood. The Minnesota and Texas adoption studies included children whose average age was about 9 years, although there was a wide age range (Loehlin, Horn, & Willerman, 1981, Loehlin, Willerman, & Horn, 1982; Scarr, Webber, Weinberg, & Wittig, 1981). These studies found little evidence for genetic influence in parent–offspring analyses. Adoption studies using self-report questionnaires for older subjects suggest heritabilities of about .20 to .30 rather than heritabilities of .40 or .50 usually found in twin studies.

The purpose of this chapter is to present results of CAP temperament analyses from early through middle childhood employing tester and teacher ratings in addition to parental ratings and self-reports. Because the children are tested as children and their parents as adults, this age disparity and the consequent low magnitude of parent–offspring resemblance led us to focus on the contemporaneous sibling adoption design that compares nonadoptive siblings (biological siblings reared together with their biological parents) and adoptive siblings (either genetically unrelated children adopted into the same adoptive family or adopted children and the adoptive parents' biological children).

We expected to confirm for middle childhood our earlier findings that parental rating questionnaires yield no systematic evidence for genetic influence in parent–offspring and sibling adoption designs. We were curious to see whether tester ratings on a version of the IBR modified for use with older children (Modified Behavior Record, MBR) would continue to show genetic influence past infancy. We were especially interested in teacher ratings, which have not been previously examined in a behavioral genetic study. Finally, although the sibling sample is small at 9 and 10, at these ages the CAP employed self-report questionnaires for the children. These data make it possible to assess parent–offspring and sibling resemblance for self-report questionnaires toward the end of middle childhood, although results from the Texas and Minnesota adoption studies did not lead us to expect to find genetic influence at this age.

Concerning environmental influence, we expected to find that shared family environment would not play much of a role in personality development. This is the result typically found in twin studies of self-report personality questionnaires (Plomin, Chipuer, & Neiderhiser, 1993; Plomin & Daniels, 1987). The adoption design provides a much more powerful test, however, because resemblance for genetically unrelated individuals brought together by adoption directly tests the importance of shared environment.

Method

CAP personality data for parental ratings up to the age of 7 years have previously been reported (see Plomin et al., 1991). The current analyses utilize data starting at age 7, continuing through annual testings to age 10. The most important addition to previous publications is that starting at age 7, the children's temperament is assessed not only by their parents, but also by their teachers. Additionally, trained examiners rate the children at age 7, and the children's self-reports are available at ages 9 and 10.

Sample

The numbers of adoptive and nonadoptive families with completed data for parent–offspring comparisons are 187 and 198 at age 7, 160 and 153 at age 8, 170 and 161 at age 9, and 164 and 145 at age 10, but the number of biological fathers with complete data is considerably lower (between 34 and 40, depending on the measure).

The number of adoptive and nonadoptive sibling pairs ranges from 60 to 61 and 66 to 68, respectively, at 7 years of age. Because the CAP is an ongoing longitudinal study, the sample of sibling pairs is progressively smaller at 8 to 10 years of age. The gender distribution was similar in the sample for adopted and control children. Originally the control children were matched to adopted children of the same gender; however, because inclusion in our analysis required existing data for their siblings, the gender distribution differs slightly, but not significantly, in the present sample. The adoptive sample consists of 33 boys and 28 girls, while there are 34 boys and 34 girls in the control sample. The gender composition of a sibling pair, however, differs in control families from that in adoptive families. There is a preponderance of mixed gender pairs in the adoptive families (76% boy–girl sibling pairs), whereas the control sample consists of more same-gender pairings (61%). Obviously, the adoption agencies considered the gender of second placements in order to favor brother–sister pairs.

Procedure and measures

The parents' temperament was assessed using the EASI (Buss & Plomin, 1975). Although several instruments were used for the children, the measure employed at each age was the Colorado Childhood Temperament Inventory (CCTI: Rowe & Plomin, 1977). The CCTI was used for parent ratings at all four testing dates (7 through 10 years, averaged over mother's and father's ratings), teacher ratings for the same age points, tester ratings at age 7, and children's self-ratings at ages 9 and 10. Four scales (Sociability, Emotionality, Activity, Attention Span) are common to parent, teacher, tester, and self-ratings. Additionally, parents rate their children on two CCTI scales that represent dimensions of Chess and Thomas' (1984) New York Longitudinal Study, namely Reactions to Foods, and Soothability. Since parental ratings on these last two scales cannot be compared to ratings by others, we will report results on the first four scales.

According to Buss and Plomin (1984), the CCTI Emotionality scale is a measure of distress, while the Activity scale is homologous to the adult EASI Activity scale, and the CCTI Sociability scale measures a mixture of sociability and shyness. Exploratory factor analysis (Plomin & DeFries, 1985) showed the adult EASI measures to yield factors comparable to the child CCTI measures.

Additionally, the examiner used a modified version of the Infant Behavior Record (IBR: Bayley, 1969), the Modified Behavior Record (MBR), with more age-appropriate items (Plomin, DeFries, & Fulker, 1988). Exploratory principal-components analysis of the MBR items, with varimax rotation, yielded four factors which were used to construct scales, namely Attention Span, Fear (of strangers), Sociability, and Impulsivity.

At age 7, the children were tested in the CAP laboratories at IBG, University of Colorado, Boulder. Testing for ages 8, 9, and 10 was completed during a telephone interview, and teacher ratings were obtained by mail.

Design and analysis

To assess genetic and environmental etiologies of temperamental traits, we will present both parent–offspring and sibling correlations. With regard to the parent–offspring adoption design, we compare correlations between adopted children and their biological and adoptive parents to those between nonadoptive parents and their offspring. For reasons mentioned earlier, we will focus on the sibling adoption design, which compares correlations for adoptive siblings (whose resemblance is due only to their shared environment) and nonadoptive siblings (who share heredity as well as family environment). For a more detailed discussion of the designs employed and the model-fitting expectations, see Chapter 3.

Power considerations

Since the samples available for sibling comparisons are small, particularly at later ages, we are only able to detect "large effects." At age 10, we only have 80% power to detect even large effects (Judd & McClelland, 1989). Also, assuming 50% heritability in childhood and in adulthood, as well as a genetic correlation of .50 from childhood to adulthood, the CAP sample would only have 50% power to detect the resulting expected correlation of .125 (Plomin, 1986). Clearly, the results reported in this chapter must be regarded as preliminary.

Results

As indicated earlier, the present chapter reports CAP temperament results for teacher and tester ratings in addition to parent ratings and self-reports. Although parent–offspring correlations for the EASI traits are included, our focus is on the sibling adoption design, which will be applied to CCTI traits as well as other variables. We begin, however, with preliminary issues such as mean comparisons between adoptive and nonadoptive families and between boys and girls. Individual differences analyses concerning rater agreement, longitudinal stabilities, and selective placement, as well as assortative mating, are also described.

Means analyses

Table 9.1 shows the means, standard deviations, and effect sizes on all measures by gender and Table 9.2 by family status. Effect size is a standardized measure of the difference between the means of two groups. Tester ratings are given as mean ratings per scale. Entries for year 7 under the column heading 'Tester/Self' refer to tester ratings, while entries for year 9 and year 10 under the same heading refer to self-ratings. (Neither tester nor self-ratings are available for year 8.)

Effect sizes over .26 are considered "large" by Cohen (1977), whereas those over .13 are termed "medium." It is remarkable that the children's self-reports show effect sizes of .17 and above in regard to gender. During the ages of 9 and 10 years, children perceive considerably more differences related to gender than their parents or teachers do.

Multivariate repeated-measures analyses of variance across all scales and time points were performed to test whether there were systematic differences in the observed temperament ratings due to gender or to family status, i.e. adopted versus nonadopted children. Table 9.3 shows significant differences for these analyses, as well as for those at age 7, the only time point for which tester data were available.

Adult raters (parents, teachers, and testers) perceive differences between nonadopted and adopted children. Contrary to prevailing views about adoption, adopted children are not uniformly rated more negatively. Parents rate adopted children as more sociable than control children. However, parents rate adopted children as more active, and teachers rate them as lower in attention span. Moreover, adopted children do not rate themselves differently on average from nonadopted children.

Parents and teachers report typical gender differences. Parents rate boys as more active and teachers rate girls as more attentive. Boys themselves report that they are more active *and* more sociable than girls; girls rate themselves as more emotional.

Although these reported differences are statistically significant, they only account for between 1% and maximally 5% of the observed variance. No significant interactions emerged between adoptive status and gender. The age variable in Table 9.3 refers to mean changes across the four age points. The results indicate that the children are seen, at least by their parents and by themselves, as increasingly sociable and less active as middle childhood progresses.

For the multivariate analyses of tester ratings at age 7, shown in the bottom part of Table 9.3, results are generally similar to those for parents and teachers. Testers rate adopted children as more emotional, more active, and less attentive than children from control families.

Interrater agreement

Since temperament ratings of others are the result of the "real" temperament and the rater's perception or bias, the agreement among different raters can be viewed as an indication of the extent to which the same trait is being measured in the

Table 9.1 Means ± standard deviations and effect sizes of CCTI ratings by gender

Measure	Parents			Teachers			Tester/Self		
	Boys	Girls	Effect	Boys	Girls	Effect	Boys	Girls	Effect
Year 7									
Sociability	19.12 ± 3.78	19.22 ± 4.24	.02	18.02 ± 3.89	18.51 ± 4.24	.12	3.69 ± 0.90	3.72 ± 0.91	.03
Emotionality	13.62 ± 3.63	14.23 ± 4.20	.16	11.04 ± 4.38	10.26 ± 4.03	.18	1.43 ± 0.68	1.50 ± 0.58	.11
Activity	19.93 ± 3.02	18.75 ± 3.77	.35	19.13 ± 3.82	18.03 ± 4.01	.28	3.72 ± 0.69	3.47 ± 0.70	.36
Attention Span	18.64 ± 3.16	18.20 ± 3.23	.14	17.49 ± 3.98	18.80 ± 3.85	.34	3.39 ± 0.93	3.35 ± 0.90	.04
Year 8									
Sociability	19.02 ± 3.96	19.58 ± 3.59	.15	18.19 ± 4.19	18.54 ± 4.33	.08			
Emotionality	13.85 ± 3.31	14.34 ± 3.49	.14	10.58 ± 4.52	11.04 ± 4.43	.10			
Activity	19.46 ± 2.90	18.96 ± 3.24	.16	18.75 ± 3.90	18.00 ± 4.11	.19			
Attention Span	18.29 ± 3.31	18.40 ± 2.95	.04	17.63 ± 4.28	18.19 ± 3.89	.14			
Year 9									
Sociability	19.52 ± 4.11	19.68 ± 4.17	.04	18.68 ± 3.87	18.25 ± 4.08	.11	15.63 ± 3.04	14.56 ± 3.28	.29
Emotionality	13.46 ± 3.98	13.99 ± 4.41	.13	11.11 ± 4.45	10.50 ± 4.37	.14	9.79 ± 3.91	10.49 ± 4.20	.17
Activity	19.32 ± 3.69	18.31 ± 3.81	.27	18.78 ± 4.07	17.93 ± 3.89	.21	15.56 ± 3.17	14.23 ± 2.67	.46
Attention Span	18.09 ± 3.92	18.41 ± 3.65	.08	17.39 ± 4.29	17.87 ± 4.08	.11	15.70 ± 3.09	15.09 ± 3.13	.20
Year 10									
Sociability	19.17 ± 4.26	19.80 ± 4.25	.15	18.62 ± 3.87	17.95 ± 4.60	.16	16.22 ± 3.03	15.20 ± 3.03	.34
Emotionality	13.34 ± 3.92	14.15 ± 4.42	.19	11.04 ± 4.49	10.91 ± 4.05	.03	9.14 ± 3.56	10.26 ± 4.12	.29
Activity	18.21 ± 3.95	18.08 ± 4.04	.03	18.54 ± 4.12	17.99 ± 4.39	.13	15.70 ± 3.18	14.75 ± 2.94	.31
Attention Span	18.17 ± 3.96	18.15 ± 4.14	.00	17.25 ± 4.26	18.57 ± 3.84	.33	15.72 ± 3.03	15.12 ± 3.08	.20

Table 9.2 Means ± standard deviations and effect sizes of CCTI ratings by family status

Measure	Parents			Teachers			Tester/Self		
	Adopted	Control	Effect	Adopted	Control	Effect	Adopted	Control	Effect
Year 7									
Sociability	19.85 ± 4.04	18.51 ± 3.84	.27	18.70 ± 3.86	17.82 ± 4.20	.22	3.72 ± 0.91	3.69 ± 0.90	.03
Emotionality	13.71 ± 4.03	14.08 ± 3.79	.09	10.59 ± 4.04	10.75 ± 4.42	.04	1.57 ± 0.64	1.36 ± 0.62	.33
Activity	19.62 ± 3.68	19.16 ± 3.17	.13	19.13 ± 3.98	18.17 ± 3.87	.24	3.69 ± 0.70	3.53 ± 0.70	.23
Attention Span	18.57 ± 3.26	18.31 ± 3.14	.08	17.49 ± 4.11	18.63 ± 3.77	.29	3.23 ± 0.84	3.49 ± 0.96	.29
Year 8									
Sociability	19.55 ± 4.11	18.97 ± 3.44	.15	18.42 ± 4.05	18.26 ± 4.47	.04			
Emotionality	14.07 ± 3.49	14.07 ± 3.30	.00	10.49 ± 3.99	11.09 ± 4.93	.13			
Activity	19.43 ± 3.26	19.04 ± 2.84	.13	18.35 ± 4.11	18.49 ± 3.91	.03			
Attention Span	18.24 ± 3.26	18.45 ± 3.04	.07	17.21 ± 4.29	18.59 ± 3.82	.34			
Year 9									
Sociability	20.15 ± 4.10	19.00 ± 4.10	.28	18.68 ± 3.87	18.25 ± 4.08	.11	15.08 ± 3.19	15.21 ± 3.20	.04
Emotionality	13.57 ± 4.15	13.82 ± 4.22	.06	10.94 ± 4.32	10.70 ± 4.53	.05	10.02 ± 4.24	10.21 ± 3.86	.05
Activity	19.35 ± 3.89	18.37 ± 3.58	.26	18.63 ± 3.95	18.11 ± 4.05	.13	15.35 ± 3.00	14.53 ± 3.00	.27
Attention Span	18.19 ± 4.06	18.28 ± 3.52	.02	17.06 ± 4.36	18.22 ± 3.92	.28	15.35 ± 3.13	15.51 ± 3.11	.05
Year 10									
Sociability	19.91 ± 4.48	18.92 ± 3.94	.23	18.67 ± 4.00	17.96 ± 4.41	.17	15.81 ± 3.01	15.72 ± 3.14	.03
Emotionality	13.71 ± 4.08	13.67 ± 4.26	.01	11.25 ± 4.03	10.71 ± 4.55	.13	9.77 ± 4.05	9.48 ± 3.62	.08
Activity	19.11 ± 3.76	17.99 ± 3.64	.30	18.64 ± 4.19	17.94 ± 4.29	.17	15.33 ± 3.16	15.10 ± 3.14	.07
Attention Span	18.14 ± 4.20	18.17 ± 3.73	.01	16.92 ± 4.05	18.78 ± 4.00	.46	15.71 ± 3.28	15.16 ± 2.79	.18

different context of home, laboratory, and school. Correlations of ratings by persons that have a different role for the child, such as parents and teachers, have been reported to average .28, whereas correlations between self-ratings and ratings by others averaged .22 (Achenbach, McConaughty, & Howell, 1987). Table 9.4 lists the interrater agreement for our sample for all possible combinations.

Table 9.3　Multivariate analyses of CCTI scores

Source	Overall	Sociability	Emotionality	Activity	Attention Span
Parents (ages 7 to 10)					
Gender				.03	
BAC status[a]		.01		.04	
Age	.04	.05		.01	
Teachers (ages 7 to 10)					
Gender					.01
BAC status[a]	.02				.00
Age					
Self (ages 9 and 10)					
Gender	.00	.00	.02	.00	
BAC status[a]					.00
Age		.00		.04	
Tester (age 7)					
Gender				.00	
BAC status[a]			.00	.02	.00

Note: Significant probability levels are shown.
[a] BAC status: adopted versus nonadopted.

Table 9.4　Interrater correlations of CCTI ratings

Measure	PT	PR	PS	TR	TS
Year 7					
Sociability	.46***	.45***		.33***	
Emotionality	.07	.08		.16	
Activity	.30***	.28***		.16*	
Attention Span	.26***	.15*		.28***	
Min. *N* used	151	144		130	
Year 8					
Sociability	.37***				
Emotionality	.26**				
Activity	.40***				
Attention Span	.38***				
Min. *N* used	135				
Year 9					
Sociability	.40***		.23**		.13
Emotionality	.23**		.14		.07
Activity	.24**		.19*		.19*
Attention Span	.35***		.15		.05
Min. *N* used	156		176		162

Measure	PT	PR	PS	TR	TS
Year 10					
Sociability	.45***		.38***		.34***
Emotionality	.12		.25***		.05
Activity	.28***		.27***		.18*
Attention Span	.35***		.16*		.12
Min. *N* used	135		173		138

Note: PT = parent–teacher correlation; PR = parent–tester correlation; PS = parent–self correlation; TR = teacher–tester correlation; TS = teacher–self correlation.
*$p < .05$, **$p < .01$, *** $p < .001$.

Generally, all raters show significant agreement on their perception of the child's sociability, activity level, and, with the exception of ages 9 and 10, its attention span/persistence. However, emotionality showed less evidence of rater agreement, which might be indicative of the different contexts or of a rating bias. Coon and Carey (1988) found evidence for a weak rating bias for temperament data on the same sample at younger ages. The magnitude of the correlation between parent and teacher ratings is similar to that reported by Goldsmith, Rieser-Danner, and Briggs (1991) for preschoolers.

Stability of measures over time

Longitudinal correlations indicate how stable a particular temperament trait is over time. Table 9.5 lists same-trait longitudinal correlations from 7 to 10 years of age for parent, teacher, and self-ratings.

Table 9.5 Longitudinal correlations of CCTI ratings by rater

Measure	Age					
	7–8	7–9	7–10	8–9	8–10	9–10
Sociability						
Parent rating	.80***	.79***	.74***	.79***	.77***	.82***
Teacher rating	.35***	.43***	.44***	.48***	.38***	.48***
Self-rating						.54***
Emotionality						
Parent rating	.71***	.66***	.61***	.65***	.71***	.68***
Teacher rating	.24**	.17*	.18*	.23**	.27**	.23**
Self-rating						.44***
Activity						
Parent rating	.76***	.72***	.62***	.77***	.74***	.76***
Teacher rating	.41***	.41***	.26**	.41***	.33***	.33***
Self-rating						.47***
Attention Span						
Parent rating	.73***	.60***	.62***	.66***	.66***	.70***
Teacher rating	.34***	.51***	.35***	.47***	.38***	.42***
Self-rating						.22**

Table 9.5 (Cont.)

Measure	Age					
	7–8	*7–9*	*7–10*	*8–9*	*8–10*	*9–10*
Min. *N* used for						
Parent rating	163	170	167	156	154	161
Teacher rating	125	133	116	143	123	132
Self-rating						175

$*p < .05, **p < .01, ***p < .001.$

Parental ratings show high stability as they did in early childhood (Plomin et al., 1988). The teacher ratings show less stability. Parental rating correlations may be higher because parents spend more time with their children and thus see more consistency across time since they average their children's behavior over much time and many situations. Alternatively, it is possible that parents' perception of their children's temperament does not change over time. It should be noted that the source of parental ratings is the same persons each year, whereas different teachers rate the children each year. Also noteworthy is the finding that self-report ratings show significant stability from 9 to 10 years. Finally, sociability ratings show slightly greater stability for all four raters than do the other CCTI scales. Thus, aspects of the children's temperament show some stability over time, as longitudinal correlations are highly significant for the various raters and scales.

Selective placement and assortative mating

Since selective placement can inflate estimates of both genetic and environmental parameters (see Plomin & DeFries, 1985), correlations between biological and adoptive parents were used to detect selective placement. Table 9.6 lists selective placement correlations.

Table 9.6   Selective placement correlations for parental EASI scales

Scale	Biological mother– adoptive mother	Biological mother– adoptive father	Biological father– adoptive mother	Biological father– adoptive father
Emotionality				
Fear	.06	.07	.09	.35*
Anger	−.05	.02	−.07	.08
Activity	.01	−.05	−.10	.04
Sociability	.01	.11	−.30*	−.14
Impulsivity	−.03	.08	−.05	−.04
(*N*)	226–229	220–222	49	48

$*p < .05.$

Although two of the 20 EASI correlations are statistically significant, one of these is negative and the average correlation is close to zero; thus, selective placement is minimal for these personality measures.

Another influence on observed familial correlations is assortative mating. When the child's biological or adoptive parents are similar, this will increase the parent–offspring and sibling resemblance. Table 9.7 lists assortative mating correlations for the three types of parents.

Table 9.7 Assortative mating correlations for parental EASI scales

Scale	Biological parents	Adoptive parents	Control parents
Emotionality			
Fear	.10	−.14*	−.07
Anger	−.09	−.02	−.09
Activity	−.09	.08	.06
Sociability	.08	.06	.21***
Impulsivity	−.26	.05	.15*
(N)	47–49	227–229	234–236

*$p < .05$, ***$p < .001$.

Correlations between biological fathers and mothers and between adoptive parents provide little or no evidence for assortative mating. For control parents, spouse correlations are significant for two of the five traits. However, as indicated in the following section, this suggestion of assortative mating for control parents does not result in greater parent–offspring similarity for control families.

Parent–offspring correlations

Parent–offspring correlations of mean ratings averaged over the middle childhood years are presented in Table 9.8. The correlations involve parental self-report ratings and ratings of the children by four sources: parents, teacher, tester, and self. We present average parent–offspring correlations for three reasons: there are many correlations; few of them are statistically significant; and no age trends were observed. In Table 9.8, the number of statistically significant correlations (9) is only slightly greater than the number expected by chance (6). Moreover, two of the significant correlations are negative in sign. This finding of negligible familial resemblance suggests that neither heredity nor shared environment contributes importantly to individual differences in personality measures.

Table 9.8 Parent–offspring correlations based on children's CCTI ratings and parents' EASI self-ratings

Measure	BF	BM	AF	AM	CF	CM
Sociability						
Parent rating	.12	.09	.14	.20*	.12	.11
Teacher rating	.18	.15	.07	.18	.11	.02
Self-rating	.15	−.06	.11	.02	.20*	.14
Tester rating	.13	.07	−.03	−.04	.00	−.08
Min. N used	18	95	94	95	79	79

Table 9.8 (Cont.)

Measure	BF	BM	AF	AM	CF	CM
Emotionality – Fear						
Parent rating	−.06	.10	−.16	−.01	.02	.01
Teacher rating	.12	.01	.01	−.01	−.02	.11
Self-rating	.04	.05	−.05	−.01	−.02	.04
Tester rating	−.05	.00	.30***	−.14	.16*	−.08
Min. *N* used	22	110	106	108	95	96
Emotionality – Anger						
Parent rating	.17	.05	.07	−.08	.12	.09
Teacher rating	−.19	−.01	.12	−.11	−.13	.21*
Self-rating	.02	.03	.04	.02	.08	.02
Tester rating	.29	−.05	.19*	−.05	−.01	−.04
Min. *N* used	22	110	106	108	95	96
Activity						
Parent rating	.20	−.01	.17	.02	.17	−.04
Teacher rating	.02	.03	.06	.14	.10	−.12
Self-rating	.16	.09	−.05	.01	.03	.03
Tester rating	.02	−.08	.00	−.02	.12	.00
Min. *N* used	24	115	113	115	99	100
Impulsivity						
Parent rating	.07	.05	−.15	−.03	−.02	.05
Teacher rating	.13	−.06	.01	.08	−.02	−.05
Self-rating	.03	−.03	.05	−.23**	−.17*	.00
Tester rating	.10	−.06	−.11	−.05	.02	.14*
Min. *N* used	24	115	113	115	102	103

Note: Correlations averaged over the two to four ages for which data available; tester ratings (one time point only) are given for completeness.
BF = birth father; BM = birth mother; AF = adoptive father; AM = adoptive mother; CF = control father; CM = control mother.
*$p < .05$, **$p < .01$, ***$p < .001$

However, two aspects of the design warrant caution in this interpretation. First, the results are based on the parents' self-report, whereas the data for the children come from various sources. Second, the parents are adults, whereas the offspring data were obtained during middle childhood. Thus, genetic and environmental influences that differ between children and adults will not contribute to parent–offspring resemblance.

Sibling correlations

As discussed in Chapter 3, the sibling adoption design can be used to assess genetic and environmental influences because the correlation for adoptive siblings directly estimates the proportion of variance due to shared environment, whereas the difference between nonadoptive and adoptive sibling correlations estimates half the variance due to heredity. In the present chapter, the results are presented for genders combined because the sample size does not provide adequate power to detect significant differences in sibling correlations for boys and girls separately. Table 9.9 lists sibling intraclass correlations for adoptive and nonadoptive siblings for parents' ratings on the CCTI at 7, 8, 9, and 10 years of age. Sociability at 7 and 8 years yields significantly greater nonadoptive than adoptive sibling correlations; however, because the correlations for adoptive siblings are negative, these differences are exaggerated.

Table 9.9 Sibling intraclass correlations for parent CCTI ratings at years 7–10

Measure	Adoptive	N	Control	N
Year 7				
Sociability	−.20	61	.21*	66
Emotionality	.03	60	−.24*	68
Activity	−.29*	61	−.12	67
Attention Span	−.09	60	−.06	68
Year 8				
Sociability	−.21	32	.37**	45
Emotionality	.16	32	.07	44
Activity	−.06	33	.06	43
Attention Span	−.33*	32	.13	44
Year 9				
Sociability	−.27	32	.01	48
Emotionality	.14	33	.10	48
Activity	−.11	33	−.20	47
Attention Span	.06	32	−.03	48
Year 10				
Sociability	−.22	22	−.06	31
Emotionality	.10	22	−.10	31
Activity	−.11	22	−.46**	29
Attention Span	−.48**	22	.20	32

Note: Intraclass correlations are based on double-entered data, and the N for significance testing has been adjusted accordingly; those families have been excluded from analyses who had been identified as bivariate outliers.
*$p < .05$, **$p < .01$

Averaging over the four years, the adoptive and nonadoptive sibling correlations for Sociability are −.23 and .13, respectively; .11 and −.04 for Emotionality; −.14 and −.18 for Activity; and −.21 and .06 for Attention Span. The negative correlations for Activity are interesting in that Activity also yields negative correlations for fraternal twins, who are genetically as similar as non-twin siblings (Buss & Plomin, 1984).

The results in Table 9.9 clearly confirm CAP results in early childhood, indicating that parental ratings on the CCTI provide no evidence for genetic influence either from parent–offspring or sibling adoption analyses, which is in contrast to results from twin studies. However, DZ correlations are often low and sometimes negative, which is similar to the results for non-twin siblings in Table 9.9.

Table 9.10   Sibling intraclass correlations for tester CCTI
ratings at year 7

Measure	Adoptive	N	Control	N
Sociability	−.02	63	.30*	72
Emotionality	.06	36	−.08	38
Activity	.03	64	.23*	73
Attention Span	−.08	63	.11	74

Note: Intraclass correlations are based on double-entered data, and the N
for significance testing has been adjusted accordingly; those families have
been excluded from analyses who had been identified as bivariate outliers.
*$p < .05$

In contrast, tester ratings on the CCTI at 7 years of age, as shown in Table 9.10
provide evidence of genetic influence for Sociability and Activity. For all four
scales, adoptive sibling correlations are near zero, whereas nonadoptive sibling
correlations are significant for Sociability and Activity. As shown in Table 9.11,
tester ratings on the MBR also suggest some genetic influence for Fear and possibly
for Attention Span and Impulsivity.

Table 9.11   Sibling intraclass correlations for tester
MBR ratings at year 7

Measure	Adoptive	N	Control	N
Attention Span	−.09	63	.17	74
Fear	−.05	63	.30**	71
Sociability	.10	61	.06	73
Impulsivity	.06	63	.14	73

Note: Intraclass correlations are based on double-entered data, and
the N for significance testing has been adjusted accordingly; those
families have been excluded from analyses who had been identified
as bivariate outliers.
**$p < .01$

These results are important because the siblings were rated when each child was
7 years old, i.e., they were rated in the laboratory several years apart. Moreover,
unlike parents, testers have had considerable experience rating children of that age.

Teacher ratings are interesting for similar reasons, although it cannot be assumed
that temperament in the classroom is characteristic of temperament in the home
with parents or in a session with a tester in a laboratory. Table 9.12 lists adoptive
and nonadoptive sibling correlations based on teacher CCTI ratings.

At 7 years of age, when the sample is the largest, teacher ratings suggest genetic
influence for Sociability, Emotionality, and Activity. Some hint of genetic influence
can be seen at age 10, although none of the four CCTI scales shows a consistent
pattern of correlations across the years from 7 to 10. The average adoptive and
nonadoptive sibling correlations are .04 and .23 for Sociability; −.06 and .12 for
Emotionality; .09 and .35 for Activity; and −.04 and .20 for Attention Span.

Table 9.12 Sibling intraclass correlations for teacher CCTI ratings at years 7–10

Measure	Adoptive	N	Control	N
Year 7				
Sociability	−.18	40	.14	41
Emotionality	−.18	52	.17	60
Activity	.08	53	.42***	60
Attention Span	.25*	54	.13	61
Year 8				
Sociability	.46**	30	.09	37
Emotionality	−.07	30	.04	35
Activity	.18	30	.26	36
Attention Span	.07	30	−.07	36
Year 9				
Sociability	−.22	31	.18	42
Emotionality	.05	31	.21	41
Activity	.32*	32	.33*	42
Attention Span	−.08	32	.22	42
Year 10				
Sociability	.09	16	.50**	23
Emotionality	−.02	17	.07	25
Activity	−.23	17	.38*	23
Attention Span	−.41*	17	.52**	25

Note: Intraclass correlations are based on double-entered data, and the N for significance testing has been adjusted accordingly; those families have been excluded from analyses who had been identified as bivariate outliers.
*$p < .05$, **$p < .01$, ***$p < .001$.

Finally, Table 9.13 lists sibling correlations for the self-report version of the CCTI for the sample at 9 and 10. Genetic influence is suggested for Sociability at 9 and Activity at both 9 and 10.

Table 9.13 Sibling intraclass correlations for self CCTI ratings at years 9 and 10

Measure	Adoptive	N	Control	N
Year 9				
Sociability	.00	40	.26*	51
Emotionality	.28*	40	.14	53
Activity	.28*	38	.45***	51
Attention Span	−.07	39	−.05	52
Year 10				
Sociability	.04	27	−.01	35
Emotionality	.53**	26	−.27	36
Activity	−.26	26	.27	35
Attention Span	.16	27	−.12	36

Note: Intraclass correlations are based on double-entered data, and the N for significance testing has been adjusted accordingly; those families have been excluded from analyses who had been identified as bivariate outliers.
*$p < .05$, **$p < .01$, ***$p < .001$

Model fits

Several model-fitting methods have been developed in behavioral genetics to assess the relative contributions of genetic and environmental influences. As discussed in more detail in Chapter 3, one such method is the multiple regression approach of DeFries & Fulker (1985) that can easily be used with any statistical software package.

The model for analyzing sibling data, which has mainly been applied to twin data but can be fitted to any genetic relationship, uses the following formula:

$$C = B_1P + B_2R + B_3PR + A \tag{1}$$

where C is the predicted score for a sibling, P is the child's score, R is the coefficient of their relationship ($R = 0.5$ for biological siblings and 0.0 for adoptive siblings), PR is the product of the child's score and the coefficient of the relationship, and A is the regression constant. B_1 is a direct estimate of the variance due to the shared environment (c^2), while B_3 is a direct estimate of the heritability (h^2). As gender differences exist for some scales in our sample (see Tables 9.1 and 9.3), the above model was extended to include gender as a covariate. The model used is:

$$C = B_4G + B_5P + B_6R + B_7PR + A \tag{2}$$

where B_7 estimates h^2 and B_5 estimates c^2 independent of gender (G). This flexibility of the model to incorporate covariates and interaction terms is one of its advantages over other modeling approaches.

Since regression coefficients are not constrained to be between zero and one, this model can yield "non-sensible" solutions, for example by giving negative estimates of h^2 or c^2. When this occurs, a reduced model is fitted, which constrains the estimate to be zero (Cherny, DeFries, & Fulker, 1992; Cyphers, Phillips, Fulker, & Mrazek, 1990).

The significance of h^2 or c^2 can be tested by an F test of the difference in R^2, the squared multiple correlation, between the full and the reduced model (Cherny et al., 1992). Compared to maximum-likelihood estimates, the multiple regression model has been shown to be somewhat more conservative, although the difference is negligible when the correlations are small (Cherny et al., 1992), as is the case for most of our analyses. In most cases the probability associated with the F test of the difference in R^2 is similar to the probability of the parameter estimate in the parsimonious model. However, if the underlying model is violated, the significance associated with the parameter estimate in the constrained model will be different from that of the F test of the difference in R^2. Such violations of model assumptions occur, e.g., when large and positive correlations are observed for adoptive siblings, while the correlations for control siblings are negative.

Resulting parameter estimates are presented in Table 9.14. Similar results were obtained when a LISREL 7 model (Jöreskog & Sörbom, 1989) was fitted to the same data.

Table 9.14 Parameter estimates using DF model

Measure	Parents		Teacher		Tester		Self	
	h^2	c^2	h^2	c^2	h^2	c^2	h^2	c^2
CCTI year 7								
Sociability	.48**	.00	.25	.00	.60***	.00		
Emotionality	.00	.00	.33	.00	.00	.00		
Activity	.00	.00	.67**	.05	.33	.04		
Attention Span	.00	.00	.00	.21**	.22	.00		
MBR year 7								
Attention					.31*	.00		
Fear					.77***	.00		
Sociability					.04	.07		
Impulsivity					.16	.04		
CCTI year 8								
Sociability	.72***	.00	.00	.26**				
Emotionality	.00	.04	.29	.00				
Activity	.00	.00	.00	.15				
Attention Span	.38	.00	.09	.06				
CCTI year 9								
Sociability	.00	.00	.35	.00			.32	.00
Emotionality	.00	.10	.31	.06			.00	.18*
Activity	.00	.00	.09	.28*			.35	.24*
Attention Span	.00	.02	.44*	.00			.00	.00
CCTI year 10								
Sociability	.20	.00	.25	.11			.00	.00
Emotionality	.00	.00	.10	.02			.00	.04
Activity	.00	.00	1.0 + ***	.00			.39	.00
Attention Span	.47	.00	1.0 + ***	.00			.00	.00

Note: Bivariate outliers have been excluded from the computations. Constrained estimates are given for those scales where only a reduced model was appropriate.
*p < .05, **p < .01, ***p < .001

These model-fitting results correspond well to the comparisons between nonadoptive and adoptive sibling correlations in Tables 9.8 to 9.12. For the CCTI ratings at age 7, the ratings by two of the three groups (parents and testers) show significant heritability for Sociability. Teacher ratings at 7 also show genetic influence for Activity. The tester ratings of the MBR suggest significant genetic influence on the children's Attention Span and Fear.

At age 8, parent CCTI ratings again give evidence for significant genetic influence on Sociability. At ages 9 and 10, genetic influence approaches significance for teacher ratings of Sociability. At age 10, teachers' reports indicate that Activity and Attention Span are highly heritable. These estimates of heritability of Activity and Attention Span derived from teacher ratings at age 10 should be treated with caution, however, since even the constrained models resulted in heritability estimates greater than unity. This result is almost certainly due to the very small sample of adopted children tested at this age (17 pairs).

A few estimates of shared environmental influence were significant as well: teacher ratings of Attention Span at age 7, teacher ratings of Sociability at age 8, self-reports at age 9 on the Emotionality and Activity scales, and teacher ratings for Activity also at age 9. However, none of these shared environmental influences is consistent over time.

Conclusions

In agreement with the conclusions of Plomin et al. (1988), CAP sibling data in middle childhood do not show important influences of either genetic or shared environmental origins for parental ratings. CAP results are consistent with previous adoption studies like the Texas Adoption Project (Loehlin, 1979), which obtained parent–offspring correlations for personality of .08 for adopted and .09 for natural children, and sibling correlations for the Cattell 16PF (Cattell, Eber, & Tatsuoka, 1970) of .05 for biological and .04 for adoptive siblings. In the Minnesota Transracial Study (Scarr et al., 1981), correlations of .01 for natural and .05 for adopted children were reported. Supplementing parental ratings with those from other sources yielded new insight into the etiology of temperament. Teacher and tester ratings provided evidence for genetic influence on Sociability and, at certain ages, Activity.

Discrepancies between parent and teacher/tester ratings

Although twin studies consistently yield identical twin correlations that are much greater than fraternal twin correlations for parental ratings of temperament as assessed by the EASI questionnaires, both the parent–offspring and the sibling adoption design show little evidence of genetic influence. The results reported in this chapter suggest that, despite their reasonableness and their evidence for genetic influence in twin research, parental rating questionnaires do not assess the genetic origins of temperament.

However, evidence for some genetic effect emerges from tester and teacher ratings. Tester ratings revealed that two of the four behaviors measured by scales of the MBR, namely Fear and Attention Span as well as Sociability at age 7, showed significant genetic influence. Teacher CCTI ratings appear to show genetic influence for Activity at age 7 and for both Activity and Attention Span at age 10, although the latter finding needs to be confirmed with a larger sample size. The results for teacher ratings are particularly surprising because teacher ratings have not previously been employed in behavioral genetic studies of temperament, and because each member of the sibling pair is rated by a different teacher.

Discrepancies between parent and teacher or teacher and tester ratings on the same scales might be attributed to differences in the situations that elicit different aspects of the children's behavior, the school environment being more structured than the home. Generally, the parent–teacher agreement is greater than the average correlation of .27 reported by Achenbach et al. (1981) in their meta-analysis of 119 studies. In the CAP, the average correlation between parents and teachers is .42 for the Sociability scale and .31 for Activity and Attention Span. The lower agreement on Emotionality (.17) might result from it being a less easily observed facet of temperament.

Discrepancies between twin and adoption results

Although the discrepancy between results from twin studies and those from family and adoption studies has been noted previously, additional discussion of this apparent paradox is clearly warranted. Heritability estimates from twin studies are of such magnitude that parent–offspring correlations should be reasonably large, but no study has found more than negligible parent–offspring correlations when the offspring were children. One possibility is that genetic effects on a phenotype differ in childhood and adulthood (Plomin et al., 1988). The expectation of $\frac{1}{2} h^2$ for the correlation between biological parents and their adopted-away children applies only to characters measured at the same age (Plomin, 1986). The parent–offspring design also assumes that shared environmental influences stay the same across generations. However, environmental changes across time (cohort effects) and secular trends might change the etiology of observed phenotypic variations (Coon, Carey, & Fulker, 1990).

Another issue concerns the age differences at the time of testing between biological parents, and adoptive and control parents. While biological parents of adopted children were tested when they were in their early twenties, on average, adoptive and control parents were about a decade older when they entered the study and were administered the questionnaires. This does not seem to be an important factor, however, since the correlations in our sample do not differ systematically by parent type.

Correlations between related siblings would also be expected to be substantial, if heritabilities estimated from twin studies were valid. Loehlin (1992) observes that, whether adopted or biologically related, siblings grow to be less similar over time. It seems as if there is a tendency for the nonshared environmental factor to be the most influential component of behavioral variation, as had already been noted by Plomin and Daniels (1987).

The influence of rating bias on the CAP data was previously found to be insufficient to explain low parent–offspring correlations regarding temperament at younger ages (Coon & Carey, 1988). With regard to sibling correlations, parents seem to contrast their children, while 'outside raters' see their similarities. For example (see Tables 9.9 and 9.12), parents in adoptive families rate their children as being rather dissimilar in their attention span at age 8; their teachers, however, rate the children more similarly. Verhulst and Koot (1992) suggest that whoever rates somebody else's personality might be projecting from their own self-perception. While we cannot reject this possible explanation, it does not seem to have influenced parental ratings of their children. The parent–offspring correlations are too low to support the projection explanation.

Non-additive genetic variance (dominance and epistasis), rather than additive gene action, could explain the unexpectedly high correlations between MZ twin pairs. While the correlation of both DZ twins and non-twin siblings include a quarter of the dominance variance, MZ twins, being genetically identical, share all of the dominance variance. Epistatic interaction might play an even more important role in creating similarities between MZ twins. These additional sources of variance might help to explain why observed MZ twin correlations were higher than expected. Dominance effects could also explain why sibling correlations are slightly higher than parent–offspring correlations. Unfortunately, to detect these effects, a rather large sample size would be needed.

Since the CAP study is ongoing, comparing these results to those obtained when the children enter adolescence will provide valuable information on change and continuity in personality development. In a few years, when the CAP children will be adults themselves, their self-ratings will be more comparable to the data that their parents provided as young adults. Additional sibling data during middle childhood will also be obtained during the next few years and should provide more power to detect the influences of genes and the environment during this important developmental epoch.

10 Competence During Middle Childhood

Jenae M. Neiderhiser
Shirley McGuire

Behavioral competence and a sense of self-worth are especially important for children during middle childhood (Damon & Hart, 1982; Erickson, 1950; Harter, 1990). While much research has been devoted to understanding the antecedents and correlates of children's sense of competence, very few behavioral genetic analyses of such outcomes have been previously reported. The little work which has been done using behavioral genetic designs has focused on children's incompetence (see Rende & Plomin, 1990 for a review) or childhood delinquency and criminality (e.g., Rowe, 1986; Rowe & Rogers, 1985). Investigating the etiology of individual differences in positive outcomes is especially important given the recent assertion by Hoffman (1991) that behavioral genetic studies have not examined outcomes such as social competence which are thought to be heavily influenced by the family environment.

The purpose of this chapter is to examine genetic and environmental influences on positive dimensions of children's competence. While research in child development and family studies has concentrated on the family environment as a primary source of children's self-worth and competence (see Harter, 1983; Wylie, 1979 for reviews), these studies examined only one child per family, thereby not exploring possible differences within families. In addition, most theoretical perspectives concerning the development of social and behavioral competence have focused exclusively on environmental influences (Bandura, 1977; Cooley, 1902; Mead, 1934). For example, one line of research has examined the relationships among parenting styles and children's social responsibility and achievement (e.g., Baumrind, 1971; Steinberg, Mounts, Lamborn, & Dornbusch, 1991). Although there may be links between the parenting styles and children's outcomes, these associations may be genetically mediated or may differ between the parent and each child within the family. Thus, it is important to investigate both genetic and environmental influences on children's social competence. Sibling studies, such as the Colorado Adoption Project (CAP), make it possible to disentangle environmental influences shared by siblings from those specific to one child. Furthermore, sibling pairs who vary in genetic relatedness (nonadoptive versus adoptive pairs) can be used to separate sibling similarity due to shared genetic heritage from that due to shared environment.

In the only study that has explored genetic and environmental components of children's perceived self-competence, a twin/stepfamily analysis of Harter's Self-Perception Profile for Adolescents (1988), genetic influences emerged while shared environmental influences had very little impact on ratings of perceived competence (McGuire, Neiderhiser, Reiss, Hetherington, & Plomin, in press). Specifically, individual differences for the dimensions of Scholastic, Athletic, and Social Acceptance and Physical Appearance were significantly heritable, whereas Self-worth, Friendship, and Behavior Conduct were primarily influenced by nonshared environment. Genetic influences on these measures may be the result of genetic contributions to cognitive development and personality. However, McGuire et al. (1993) found that most of the genetic influences on Scholastic Competence and Social Acceptance were unique to those measures and were not due to genetic influences on verbal ability and sociability, respectively.

Change and continuity in competence over time is another focus of research in this area (Bandura, 1977; Damon & Hart, 1982; Eisenberg & Harris, 1984; Kurdek & Krile, 1983; Nottelmann, 1987). Most of this research has emphasized "universal" developmental stages rather than individual differences. For example, Harter (1990) suggests that the dimensions of competence change across the life-span, with five distinct developmental stages: early childhood, middle childhood, adolescence, young adulthood, and later adulthood. Less is known, however, about the stability over time of individual differences in competence, as well as genetic and environmental contributions to stability.

CAP sibling adoption data were used to address two general issues for this chapter: (1) genetic and environmental influences on measures of self-worth assessed during middle childhood; and (2) genetic and environmental contributions to developmental change and continuity for these measures.

Method

Sample

The present analyses included children participating in the CAP tested at ages 7, 9, and 10 years. Because the CAP is an ongoing longitudinal study, the sample sizes decrease across time, ranging from 59 nonadoptive and 50 adoptive sibling pairs at 7 years to 33 nonadoptive and 25 adoptive sibling pairs at 10 years. There are approximately as many girls as there are boys in this sample (52% boys, 48% girls).

Measures

Competence data were obtained from children participating in the CAP at ages 7, 9, and 10 years of age. Two measures of competence were used: the children's self-report of their competence using Harter's (1982) Self-Perception Profile for Children, and mothers' and teachers' reports of the children's social competence using the CAP Social Competence Scale (CSCS).

Harter's (1982) Self-Perception Profile for Children consists of six subscales: Self-Worth, Behavior Conduct, Athletic Competence, Scholastic Competence, Physical Appearance and Social Acceptance. The version of the Harter used in the CAP is administered as part of a battery of questions about "feelings" during a telephone interview when the children are 9 and 10 years of age.

The CAP Social Competence Scale is based on the Walker-McConnell Scale of Social Competence and School Adjustment (Walker & McConnell, 1988). Mother and teacher reports on the dimensions of Leadership, Confidence, and Popularity were measured when the children were 7 years of age. Teacher reports were also obtained when the children were 9 years of age. Confirmatory factor analysis using LISREL 7 (Jöreskog & Sörbom, 1989) was performed in which the number of factors and the items on each scale were equated for mother and teacher reports.

Results

Mean differences

In order to assess mean gender differences and mean differences between adoptive and nonadoptive families, 2 (boys and girls) × 2 (adoptive and nonadoptive) analyses of variance were conducted. Significant gender differences were identified for 7 of the 24 measures (see Table 10.1). Thus, the effects of gender were regressed out of the data for the remainder of the analyses. Significant mean differences for adoptive status were found for 3 of the 24 measures, but accounted for only 12% of the variance on average.

Univariate results

Intraclass sibling correlations were calculated in order to assess sibling similarity for nonadoptive and adoptive sibling pairs on the self-reports on the Harter, and mother and teacher reports on the CAP Social Competence Scale. Table 10.2 contains the intraclass sibling correlations for the competence measures. The greater correlations for nonadoptive than adoptive siblings on scholastic competence suggest genetic influence at both 9 and 10 years of age. In addition, evidence for genetic influence emerged at either age 9 or 10 for the Harter scales of Self-Worth, Physical Appearance, and Athletic Competence.

Table 10.1 Means for competence measures by adoptive status and gender

Measure	Nonadopted		Adopted	
	Boys	Girls	Boys	Girls
Mother report of CAP				
Social Competence Scale				
Year 7				
Popularity	21.2	21.5	21.9	21.3
Confidence	43.9	46.0	45.1	46.1[a]
Leadership	32.9	32.5	34.0	33.6
Teacher report of CAP				
Social Competence Scale				
Year 7				
Popularity	20.3	19.6	19.9	19.6
Confidence	40.9	43.1	40.1	40.7[a,b]
Leadership	31.1	31.1	29.7	29.8
Year 9				
Popularity	20.1	19.9	19.5	19.5
Confidence	39.2	43.1	38.8	40.4[a,b]
Leadership	30.4	30.6	28.8	28.9
Harter self-report				
Year 9				
Physical Appearance	15.0	13.9	15.6	14.4
Athletic Competence	12.5	11.9	13.1	11.7[a]
Behavior Conduct	14.2	15.5	13.9	14.6[a]
Self-Worth	16.5	17.1	16.8	16.6
Scholastic Competence	13.9	14.4	13.9	13.6
Social Acceptance	15.4	14.5	16.0	15.0
Year 10				
Physical Appearance	15.6	14.6	16.0	14.4[a]
Athletic Competence	12.5	11.6	13.4	11.5[a]
Behavior Conduct	15.2	15.5	14.9	16.0
Self-Worth	17.3	16.8	17.3	17.3
Scholastic Competence	14.7	15.7	15.0	15.0
Social Acceptance	15.2	15.3	16.7	15.7[b]

[a] Indicates significant gender differences.
[b] Indicates significant differences for adoptive status.

Even more interesting than the self-report data on the Harter are the results of mother and teacher ratings of social competence. For mother ratings at age 7 and teacher ratings at both 7 and 9 years, genetic influence is suggested for both Popularity and Confidence. In addition, teacher ratings of Leadership also suggested genetic influence at both age 7 and 9. Ratings of Leadership by mothers, who might not be expected to be as well versed in rating their children's leadership competence as are teachers, did not suggest genetic influence.

In order to estimate genetic and environmental parameters for these measures, maximum-likelihood model-fitting analyses were performed using LISREL VII

Table 10.2 Intraclass sibling correlations for competence measures

	Nonadopted	Adopted
Harter self-report:		
Year 9		
Physical Appearance	.39**	−.05
Athletic Competence	.06	.12
Behavior Conduct	−.01	.15
Self-Worth	.29*	.08
Scholastic Competence	.41**	.16
Social Acceptance	−.13	.22
Year 10		
Physical Appearance	−.07	.31*
Athletic Competence	.20	−.07
Behavior Conduct	.08	.03
Self-Worth	.08	.09
Scholastic Competence	.22	.15
Social Acceptance	.10	.20
Mother report of CAP		
Social Competence Scale		
Year 7		
Popularity	.29*	.05
Confidence	.41*	.02
Leadership	.10	.13
Teacher report of CAP		
Social Competence Scale		
Year 7		
Popularity	.36**	.07
Confidence	.42*	.00
Leadership	.37**	−.01
Year 9		
Popularity	.16	.00
Confidence	.26	−.21
Leadership	.30*	−.10

*$p < .05$, **$p < .01$.

(Jöreskog & Sörbom, 1989). The results of these analyses are illustrated in Figures 10.1 and 10.2. Significance of the genetic and shared environmental parameters was tested by comparing nested models. When a model is compared to a reduced one, the difference in χ^2 between the two models can be used to determine the importance of a parameter to the model. In this case either the genetic or shared environmental parameter was dropped from the full model to test the significance of those parameters. For the Harter, Physical Appearance at age 9 was the only area of competence to show a significant genetic influence (Figure 10.1). However, approximately half of the total phenotypic variance of Self-Worth and Scholastic Competence at age 9 and slightly less than half of the total variance of Athletic and

Scholastic Competence at age 10 can be explained by genetic influences, although none of these parameter estimates reached significance. Shared environmental influences showed no evidence for either significant or substantial influence on any of the Harter measures at age 9 or 10.

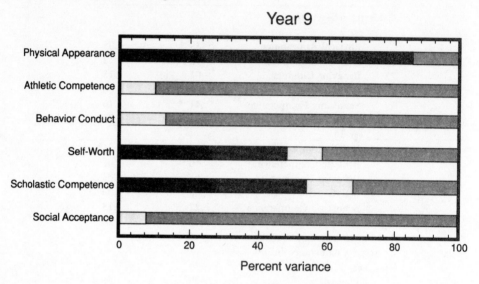

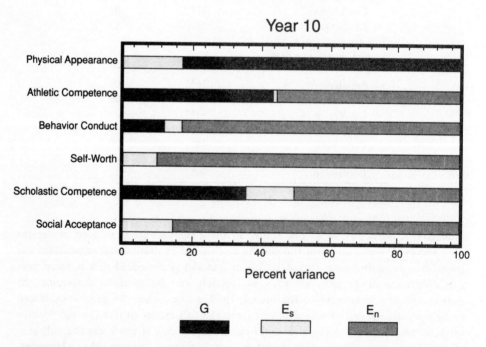

Figure 10.1 Univariate model-fitting results for the Harter (1982) Self-Perception Profile for Children, showing percentage of total phenotypic variance accounted for by genetic and environmental parameters.

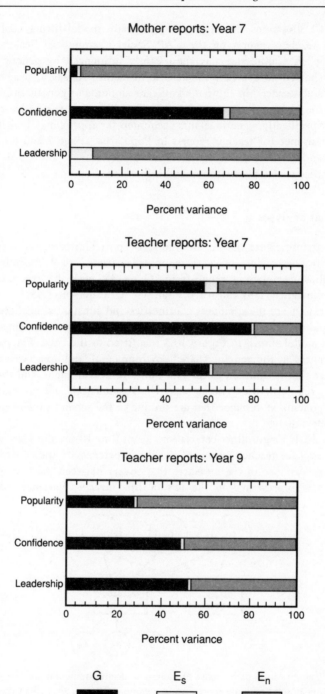

Figure 10.2 Univariate model-fitting results for the CAP Social Competence Scale show-ing percentage of total phenotypic variance accounted for by genetic and environmental parameters.

Figure 10.2 illustrates the results of univariate model-fitting analysis for the mother and teacher reports on the CAP Social Competence Scale. For mother reports at age 7, Confidence was significantly influenced by genetic factors and Leadership showed no evidence for genetic influence. At age 7, teacher reports of Confidence and Leadership showed significant amounts of genetic influence. As we found with the Harter, in several cases genetic influences were substantial but not significant. Specifically, genetic factors accounted for approximately half of the total phenotypic variance for teacher reports of Popularity at age 7 and for Leadership, Popularity, and Confidence at age 9. None of the teacher ratings of competence showed significant shared environmental influences.

Longitudinal analyses

Phenotypic stabilities are modest for the self-report Harter measure from 9 to 10 and for the teacher ratings of social competence from 7 to 9. As indicated below, the longitudinal correlations range from .11 to .37, with the exception of Harter Athletic Competence (.58) and teacher ratings of Leadership (.45).

To what extent are these modest continuities and substantial changes during this time of rapid development mediated by genetic and environmental factors? A longitudinal model shown in Figure 10.3 was fitted to the data. The parameters G, E_s, and E_n represent the genetic, shared environmental, and nonshared environmental covariance that is common to the two times of measurement, and the parameters g, e_s, and e_n represent the genetic, shared environmental, and nonshared environmental components of variance that are unique to the second time of measurement, and thus index change.

In Figure 10.4, longitudinal correlations from 9 to 10 for the Harter self-ratings and from 7 to 9 for teacher ratings of social competence are indicated by the heavy vertical lines. Variance in the measures that covary between the two ages – that is, continuity – is represented to the left of this marker. Variance due to change

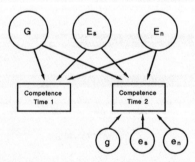

Figure 10.3 Longitudinal path model of genetic and environmental sources of change and continuity. G is a latent variable representing genetic influences that affect the phenotype at both times of assessment. E_s and E_n are latent variables representing shared and nonshared environmental influences common to the two ages. The three latent variables – g, e_s and e_n – represent genetic, shared environmental, and nonshared environmental influences that are specific to the phenotype at Time 2.

(including error of measurement) is to the right of the marker. Longitudinal genetic analysis decomposes these variances into genetic and environmental components, as indicated by the key at the bottom of Figure 10.4. Stability in the Harter Self-ratings

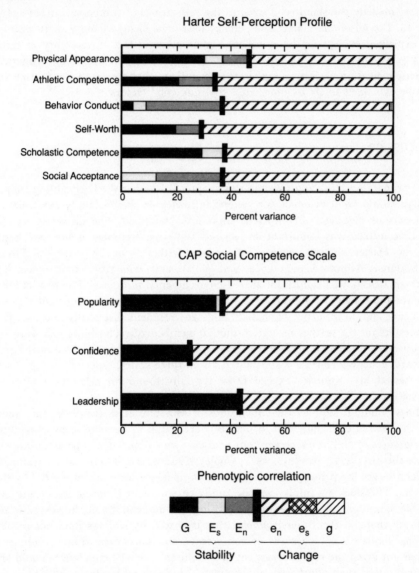

Figure 10.4 Summary of genetic and environmental components of continuity and change in the Harter Self-ratings of competence from 9 to 10 years of age and in the Teacher rating of self-competence from 7 to 9 years. As indicated by the key at the bottom of the figure, the vertical bar indicates the phenotypic correlation at the two ages. To the left of the bar are the genetic, shared environmental, and nonshared environmental components of the phenotypic correlation; to the right of the bar are the components of change.

of competence from 9 to 10 years is mediated by genetic factors for all of the measures except Social Acceptance. Shared environment is also a mediating factor in stability for self-ratings of Physical Appearance, Behavior Conduct, and Scholastic and Social Competence. It is important to note that nonshared environmental factors mediate the stability in most of the self-ratings of competence from age 9 to age 10. For all of the Harter competence measures, change from 9 to 10 years can be explained exclusively by nonshared environmental factors. Stability in teacher ratings of children's competence from 7 to 9 years can be attributed primarily to genetic mediation. Again, change in teacher ratings of competence from 7 to 9 years can be explained solely by nonshared environmental factors.

Discussion

The primary question addressed in this chapter is whether the CAP sibling adoption design would yield evidence for genetic influence or shared family environmental influence on measures of competence in middle childhood. The answer seems clear: Genetic influence is important and shared family environment is not. Self-ratings on the Harter showed genetic influence at either 9 or 10 years for Physical Appearance, Athletic Competence, Self-Worth, and Scholastic Competence. None of the scales showed significant shared environmental influence. The results for the mother and teacher ratings are even more impressive in showing substantial genetic influence and no influence of shared family environment. Especially noteworthy are the results for the teacher ratings. Although members of each sibling pair were rated by different teachers at different measurement occasions separated by two years on average, genetic influence is substantial for all three scales, especially at age 7 when the heritability estimates exceed those typically found for self-report ratings of personality.

These findings support the conclusions of McGuire et al. (in press) that genetic and nonshared environmental influences were found to be significant contributors to measures of children's perceptions of their own competence. The assumptions of some theorists (e.g., Bandura, 1977; Cooley, 1902; Mead, 1934) that environmental influences are the primary determinants of competence were not supported by these results. This does not imply that the family environment is not an important factor in the development of competence during middle childhood; however, it does indicate that the family environment that is shared by siblings does not result in sibling similarity on measures of competence and self-worth. These findings are important given the recent argument that children's social competence should show shared, rather than nonshared, environmental influences (Hoffman, 1991).

Although longitudinal stability of competence from age 9 to 10 for self-ratings and from 7 to 9 for teacher ratings is only modest, genetic factors appear to contribute to this stability for self-reports of Physical Appearance, Self-Worth, and Athletic and Scholastic Competence, and for all three teacher ratings of competence. Genetic factors do not appear to be involved in change across these relatively short intervals.

These findings suggest several avenues for future research. First, all of these measures showed substantial nonshared environmental influences. Studies have examined links between sibling differential experiences and children's adjustment and achievement (e.g., Daniels, Dunn, Furstenberg, & Plomin, 1985; McHale & Pawletko, 1992); however, this research could be extended to include other areas of children's competence including peer and athletic competence. Second, work in this area should be extended from middle childhood to adolescence when there may be important changes in genetic and environmental influences.

11 The Stress of First Grade and its Relation to Behavior Problems in School

Richard Rende

Over the past decade there has been considerable interest in stress as experienced by children and adolescents. This interest has followed in part from research on adults in which stress has been shown to be an important factor relating to both physical health and psychological well-being. An emerging literature has revealed that stressful experiences may also have adverse consequences for children and adolescents (see Compas, 1987; Johnson, 1986). This chapter will focus on the stress of first grade, and its relation to behavior problems, as assessed in the CAP.

Why First-Grade Stress?

Many studies relating stress to behavior problems in childhood have focused on the impact of major life events (see Compas, 1987; Johnson, 1986). A general conclusion from this work is that major life events show a moderate relationship (correlations ranging between .20 and .30) with negative outcomes such as behavior problems. While this research supports the contention that stress in childhood may be linked with negative outcomes, there are other potential sources of stress for children (Compas, 1987). In particular, there have been numerous suggestions that developmentally relevant life transitions may be stressful for some children, and hence may be associated with negative outcomes (e.g., Dunn, 1988; Humphrey, 1984).

One developmentally relevant life transition for children is the beginning of elementary school. Studies of older children enrolled in higher grades in elementary school (third to sixth graders) have documented that school-related events are a source of stress (e.g., Colton, 1985; Humphrey, 1984; Phillips, 1978; Schultz & Heuchurt, 1983). However, the stress of the first year of elementary school has received little direct study, and consequences of stressful experiences in beginning school have yet to be explored.

First-grade stress may be especially salient to children because of the changes experienced in the transition to elementary school. For example, children are required for the first time to be in school for a full day; they begin to experience more academic demands; and they also have increasing demands to get along with peers, as well as to negotiate interactions with older children in the school. It is possible that children will vary in their adaptation to these new demands, and also that individual differences in adaptation will correspond to differences in outcomes, such as behavior problems.

Research issues

This chapter will focus on three research issues concerning first-grade stress: (1) assessing the stressfulness of events related to beginning first grade, (2) finding contemporaneous and longitudinal relations between first-grade stress and behavior problems in school, and (3) estimating the relative contributions of genes and environment to individual differences in stressful experiences in first grade.

Assessing the stress of first grade

As noted earlier, relatively little systematic work has been done on quantifying the stressfulness of events experienced in the first grade. An important methodological issue in this work concerns the informant, or person providing the information on stressful events. Recent research on childhood stress has incorporated child as well as parent reports of stressful events and the distress experienced by children (e.g., Brown & Cowen, 1988; Colton, 1985; Compas, Howell, Phares, Williams, & Ledoux, 1989; Rende & Plomin, 1991; Sandler, Wolchik, Braver, & Fogas, 1986). An important conclusion from this work is that children's own perceptions of the distress due to stressful events often differ from adult ratings of the children's stress. Researchers in the field have strongly recommended that children's own perceptions of events be collected (Compas, 1987).

In the CAP, both child and parent reports of the occurrence and upsettingness of 18 potentially stressful events experienced in the first grade were collected. The perspective of both child and parent is represented, allowing direct comparison of their reports.

Relations between stress and behavior problems

There is a growing literature documenting how stressors such as major life events influence maladaptation in middle childhood (see Compas, 1987). Furthermore, there is evidence that children's self-reports of the upsettingness of major life events correlate with school adjustment (Pryor-Brown & Cowen, 1990). Given this, it is reasonable to speculate that first-grade stress may also be related to behavior problems in school. Such information would not only extend our knowledge about behavioral consequences of stress in childhood, but also help to identify children who have difficulties adjusting to elementary school.

A more complex issue concerns longitudinal relations between stress and behavior problems. Traditionally, most studies have examined contemporaneous relations between sources of stress – especially major life events – and behavior problems (see Compas, 1987; Johnson, 1986). More recently, however, there has been a recognition of the need to assess longitudinal relations to determine if there are relatively long-term consequences of stress. An especially important methodological consideration is the need to partial out the effects of earlier behavior problems when determining longitudinal relations between stress and later behavior problems (e.g., Compas et al., 1989). Such a strategy assesses longitudinal effects of stress independent of previous levels of behavior problems; in other words, it helps to determine if stress is associated with changes in behavior problems over time.

The longitudinal design of the CAP enables us to assess relations between first-grade stress and behavior problems as reported by teachers two years later. This relationship will be examined after controlling for the previous level of behavior problems (as reported by teachers at the same time that first-grade stress was assessed).

Genes, environment, and first-grade stress

Traditionally, stress has been conceptualized in research with both children and adults as an environmental risk factor. For example, one of the most common etiological models for psychopathology is the diathesis–stress model, in which both a predisposition to psychopathology (diathesis) and precipitating events or stressors (stress) contribute to the expression of symptomatology. Although the diathesis–stress model does not necessarily specify how genes and environment contribute to psychopathology, it is often hypothesized that genetic factors represent the diathesis, and environmental factors are responsible for the stress (see Rende & Plomin, 1992a). However, recent behavioral genetic research has demonstrated that proto-typical measures of stress in adults – such as perceptions of life events – show as much genetic influence as do measures of personality (Plomin, Lichtenstein, Pedersen, McClearn, & Nesselroade, 1990). Hence, rather than assuming that first-grade stress represents an environmental risk factor, the relative contributions of genes and environment should be assessed empirically.

We have used the sibling adoption component to the CAP to investigate genetic and environmental contributions both to the occurrence of events in the first grade, and the perception of the stressfulness of these events. In addition, the sibling adoption component has provided data on the etiology of behavior problems in school, as reported by teachers. These data, in conjunction with the data on first-grade stress, will be used to inform a diathesis-stress model of school-related behavior problems.

Method

Sample

The sample consisted of 206 children (108 males, 98 females) in the adoptive families, and 208 children (111 males, 97 females) in the nonadoptive families, participating in the year 7 assessments in the CAP. In addition, 177 children (93 males, 84 females) in the adoptive families and 169 children (79 males, 90 females) in the nonadoptive families participated again in the year 9 assessments.

In the adoptive families, complete data were available for 50 sibling pairs assessed at age 7, and 33 pairs assessed again at age 9. In the nonadoptive families, complete data were available for 54 sibling pairs assessed at age 7, and 47 pairs assessed again at age 9.

Measures and procedures

During the year 7 assessments in the CAP, children and parents were administered the first-grade stress interview which is described below. For each child, one teacher was asked to complete the teacher version of the Child Behavior Checklist (CBCL) which is also described below; teacher reports were acquired during the 7- and 9-year assessments.

First-grade stress

Children and parents were administered an inventory containing 18 items associated with first grade (see Rende & Plomin, 1992b); the items were chosen based on open-ended interviews with children, parents, teachers, and clinicians designed to generate a list of representative items associated with first grade. Areas represented include academic concerns (3 items), peer relations (6 items), teacher relations (3 items), parent relations (2 items), and general hassles (4 items) (see Appendix).

One parent of each child was given a form listing the 18 items and was instructed to (1) indicate if the item was experienced by the child during the previous year, and (2) for items which were experienced, rate the upsettingness of the item to the child on a 4-point scale (0 = not at all upsetting, 3 = very upsetting).

Table 11.1 Descriptive statistics for first-grade stress and behavior problems

	Adoptive probands				Nonadoptive probands			
	Males		Females		Males		Females	
	N	M(SD)	N	M(SD)	N	M(SD)	N	M(SD)
Stress								
Child report								
Total events	108	10.18(6.91)	98	11.06(6.37)	111	10.46(7.19)	97	9.35(6.47)
Total upsettingness	108	7.92(3.04)	98	8.05(2.96)	111	7.69(3.01)	97	7.29(2.50)
Parent report								
Total events	108	9.18(4.72)	98	7.18(4.83)	111	9.46(4.52)	97	7.83(4.70)
Total upsettingness	108	6.14(2.45)	98	5.37(2.86)	111	6.25(2.45)	97	5.51(2.62)
Behavior problems								
Internalizing								
Age 7	108	5.16(5.95)	98	5.19(4.93)	111	4.90(7.88)	97	4.95(5.60)
Age 9	93	7.07(6.15)	84	6.59(6.72)	79	5.53(5.69)	90	5.58(6.85)
Externalizing								
Age 7	108	9.17(10.22)	98	5.24(8.35)	111	6.82(10.94)	97	3.70(6.02)
Age 9	93	8.82(9.29)	84	4.46(4.74)	79	6.35(8.08)	90	3.77(5.99)

Each child was interviewed separately concerning the occurrence and upsetting-ness of each item. The interview was semi-structured, with the interviewer asking the child if s/he experienced each item during the past year, and then asking the child to discuss his or her feelings concerning each experienced item (with the interviewer being blind to the parent's responses). In particular, the interviewer asked the child "How did [the item] make you feel? How upset were you?", and then asked the child to rate the upsettingness on the 4-point scale. Pilot work ($N = 30$) indicated that 1-week test-retest reliability of the ratings was .83.

Two composite scores were computed for children and parents. First, the total number of items reported by each child and parent as experienced was determined by a simple count of events. In addition, a total rating of upsettingness was calculated by summing the ratings for experienced items. The creation of composite scores is consistent with current approaches in the study of childhood stress (e.g., Compas, 1987).

Behavior problems

Teachers completed the teacher form of the Child Behavior Checklist (CBCL: Achenbach & Edelbrock, 1986). The CBCL consists of 113 items of behavioral symptomatology that yield two second-order factors, Internalizing and Externaliz-ing, as well as a total score. In the present sample, the externalizing dimension was highly correlated with the total score ($r = .95$) but less so with the internalizing dimension ($r = .36$); hence, only the internalizing and externalizing scales were included in analyses.

Results

Descriptive statistics: First-grade stress and behavior problems

Table 11.1 presents descriptive statistics for the two composite scores – total number of events and total upsettingness score – as provided by both child and parent report. 2 × 2 MANOVAs (gender, adoptive status, gender × adoptive status) revealed no significant effects of adoptive status for the variables listed in Table 11.1. Gender effects were found for parental report of both total number of events (overall $F_{3,391} = 3.06$, $p < .03$) and total upsettingness score (overall $F_{3,391} = 5.53$, $p < .01$); in both cases boys were given higher stress scores than girls.

The relation between child and parent report was assessed using correlations. Child and parent report were not interchangeable; child and parent correlated .26 ($p < .01$) for number of events and .29 ($p < .01$) for total upsettingness rating in nonadoptive families, and .25 ($p < .01$) for number of events and .24 ($p < .01$) for total upsettingness rating in adoptive families.

Descriptive statistics for internalizing and externalizing behavior problems as reported by teachers in the year 7 and year 9 assessments are also shown in Table 11.1.

2×2 MANOVAs revealed main effects of both gender and adoptive status for externalizing problems at the year 7 (overall $F_{3,391} = 6.28$, $p < .01$) and year 9 (overall $F_{3,332} = 8.02$, $p < .01$) assessments; boys had higher externalizing scores than girls, and adoptive children had higher scores than nonadoptive children. In each case, however, the effects of adoptive status and gender were small; these main effects accounted for 4% of the variance in externalizing problems at 7, and 7% of the variance in externalizing problems at 9. No significant effects of gender and adoptive status were found for internalizing problems at the year 7 and year 9 assessments.

Relations between first-grade stress and behavior problems

Child report of first-grade stress   Table 11.2 presents both contemporaneous and longitudinal correlations between the composite scores of first-grade stress based on children's self-reports, and behavior problems in school as reported by teachers. In the nonadoptive sample, all first-grade stress measures correlated significantly with both internalizing and externalizing problems, as assessed both contemporaneously and two years later. The adoptive sample showed a somewhat different pattern, however. Although there were also contemporaneous relations between first-grade stress and behavior problems, no significant longitudinal associations were found.

Table 11.2 Contemporaneous and longitudinal correlations between child report of first-grade stress, and behavior problems

Behavior problems	Adoptive probands		Nonadoptive probands	
	Total events	Total rating	Total events	Total rating
Internalizing				
Age 7	.08	.18*	.18*	.25**
Age 9	.03	.05	.21*	.33**
Externalizing				
Age 7	.17*	.16*	.19*	.33**
Age 9	.09	.06	.22*	.40**

*$p < .05$, **$p < .01$.

The next step was to determine the relation between first-grade stress and behavior problems two years later after controlling for the initial level of behavior problems assessed at the first-grade visit. Multiple regression models were fitted (separately for internalizing and externalizing problems) in which behavior problems at age 9 were predicted from (1) behavior problems at age 7 and (2) total first-grade upsettingness rating (since this variable was more strongly related to behavior problems than the total number of events).

In the nonadoptive sample, a model predicting internalizing problems at age 9 was significant ($F_{2,127} = 14.79$, $p < .01$) and explained 19% of the variance; both internalizing problems at age 7 ($t = 3.59, p < .01$) and total upsettingness ($t = 3.38$,

$p < .01$) made independent contributions to the total variance explained. Similarly, a model predicting externalizing problems at 9 was significant ($F_{2,127} = 40.35$, $p < .01$) and explained 28% of the variance; both externalizing problems at 7 ($t = 4.92$, $p < .01$) and total upsettingness ($t = 3.03$, $p < .01$) made independent contributions to the total variance explained.

In the adoptive sample, however, there were no significant effects of total upsettingness in predicting behavior problems at age 9; such a finding is consistent with the low correlations presented in Table 11.2.

Parent report of first-grade stress   Table 11.3 presents contemporaneous and longitudinal correlations between the composite scores of first-grade stress based on parental report, and behavior problems in school as reported by teachers.

Table 11.3 Contemporaneous and longitudinal correlations between parental report of first-grade stress, and behavior problems

Behavior problems	Adoptive probands		Nonadoptive probands	
	Total events	Total rating	Total events	Total rating
Internalizing				
Age 7	.20*	.23**	.12*	.18*
Age 9	.00	.09	.17*	.18*
Externalizing				
Age 7	.22*	.29**	.19*	.25**
Age 9	.00	.16*	.18*	.22*

*$p < .05$, **$p < .01$.

In the nonadoptive sample, the correlations mirrored those found with child as respondent, with the primary difference being that the correlations were lower when parents provided information on the child's stress. In the adoptive sample, parental report also yielded an overall pattern similar to the child report, although the correlations were higher when the parent was respondent. As was the case with the child report, there was less evidence of longitudinal relations between first-grade stress and behavior problems at age 9, although the total upsettingness rating was significantly related to externalizing problems (see Table 11.3).

As with the child report of first-grade stress, the relation between first-grade stress and behavior problems two years later was assessed after controlling for the initial level of behavior problems. In the nonadoptive sample, both externalizing problems at age 7 ($t = 5.89$, $p < .01$) and the parental rating of total upsettingness ($t = 2.13$, $p < .04$) made an independent contribution in a model predicting externalizing problems at age 9 (overall $F_{2,127} = 22.88$, $p < .01$). Parental rating of upsettingness did not significantly predict internalizing problems at 9 after controlling for internalizing problems at 7.

In the adoptive sample, there were no significant relations between total upsettingness as rated by the parent and behavior problems at 9 after controlling for initial level of problems at 7.

Genes, environment, and first-grade stress and behavior problems

Table 11.4 presents correlations for the nonadoptive and adoptive sibling pairs in the CAP for the total number of events and the total upsettingness rating (after partialling out gender), for both child and parent report. The low correlations for both nonadoptive and adoptive pairs suggest little influence of genes and shared environment on child reports of event occurrence and total upsettingness. There is a suggestion of shared environmental effects on parental reports based on the correlations for adoptive siblings (and the similar correlations for nonadoptive siblings).

Table 11.4   Adoptive and nonadoptive sibling correlations for first-grade stress and behavior problems

	Adoptive	*Nonadoptive*
First-grade stress		
Child report		
Total events	.06	.05
Upsettingness	.04	.06
Parent report		
Total events	.29	.26
Upsettingness	.26	.20
Behavior problems		
Internalizing		
Age 7	.09	.04
Age 9	.26	.37
Externalizing		
Age 7	.00	.17
Age 9	.05	.48

The multiple regression method developed by DeFries and Fulker (1985) was used to estimate the effects of genes, shared environment, and nonshared environment; results are shown in Table 11.5. As expected based on the correlations shown in Table 11.4, there were no significant effects of genes or shared environment on three of the four measures of first-grade stress. Most of the variance was due to nonshared effects, or environmental factors which operate to make siblings different (which also include error of measurement). The one exception was a significant shared environmental effect for the parental rating of total number of events, as shown in Table 11.5.

Table 11.4 also presents the adoptive and nonadoptive sibling correlations for teacher ratings of internalizing and externalizing behavior problems at 7 and 9. The higher correlations for nonadoptive sibling pairs for externalizing problems at both ages is suggestive of genetic influence. For internalizing problems, the correlations suggest possible shared environmental effects at age 9.

Table 11.5 presents results of model-fitting using the multiple regression method. Although the heritability estimate for externalizing problems at age 9 is significant, the estimate at age 7 is not. The shared-environment parameter at age 9 is also not significant.

Table 11.5 Estimates of heritability and shared environment for first-grade stress and behavior problems

	Heritability	Shared environment
Stress		
Child report		
Total events	.00	.01
Upsettingness	.04	.04
Parent report		
Total events	.00	.29*
Upsettingness	.00	.26
Behavior problems		
Internalizing		
Age 7	.00	.09
Age 9	.18	.26
Externalizing		
Age 7	.34	.00
Age 9	.86*	.05

*$p < .05$.

Discussion

Assessing the stress of first grade

One interesting finding from the CAP data set is that child and parent reports of the stress of first grade correlate only moderately. This result is consistent with recent research on childhood stress which has focused especially on the assessment of major life events (e.g., Brown & Cowen, 1988; Colton, 1985; Compas et al., 1989; Rende & Plomin, 1991; Sandler et al., 1986). As it is not possible to determine from the present data which informant is better, the low correlations between child and parent reports suggest that both respondents should be considered when assessing first-grade stress. However, it must be emphasized that there is special interest in the child's self-report, because perceptions of the children who experience stressful events are theoretically the most direct information on subjective experiences (Compas, 1987).

Relations between stress and behavior problems

There were contemporaneous correlations between measures of first-grade stress and behavior problems in school as reported by teachers. It is notable that these relations were found across respondents, as one confound in earlier research on childhood stress was having one informant provide the information on both stress and outcome (Compas, 1987; Johnson, 1986). Especially interesting was the finding that the child's self-report of first-grade stress was significantly correlated with teacher report of behavior problems. Because theories of stress hypothesize that stressful experiences should be linked with adverse outcomes such as behavior problems in childhood (Compas, 1987; Johnson, 1986), this finding may be taken as an index of the predictive validity of children's self-reports of their experiences in first grade.

Both promising and confusing findings emerged from the assessment of longitudinal relations between first-grade stress and behavior problems two years later. The promising finding was that first-grade stress was not only significantly associated with later externalizing and internalizing problems, but also made an independent contribution to later behavior problems after partialling out of the earlier level of problems at age 7. This finding is important because it suggests that first-grade stress may have long-term consequences which are independent of "baseline" levels of problem behavior. Such a finding carries many implications for developmentalists; most salient are the need to identify children who are reporting difficulties in adjusting to first grade, and the hope of targeting interventions for these children to prevent the development of later problems in school.

The confusing finding was that the results discussed above apply only to the nonadoptive sample. There was little evidence of longitudinal associations between first-grade stress and behavior problems in the adoptive sample. The basis for this discrepancy is not clear. No mean or variance differences were found for the first-grade stress measures based on adoptive status. There were also only small effects of adoptive status on behavior problems. Hence, there does not seem to be any obvious reason why the predictive validity of the first-grade stress measure was established only with the nonadoptive sample. Future studies assessing first-grade stress and its consequences in other samples may more clearly establish the predictive validity of the measures used in the CAP.

Genes, environment, and first-grade stress

The major finding from the sibling adoption component to the CAP was that there is little evidence for genetic influence on first-grade stress. This result is not consistent with recent research on adults which has revealed genetic influence on perceptions of life events (e.g., Plomin et al., 1990). However, to date there has been virtually no research on genetic influences on stress in childhood, so the results from the CAP should be interpreted as an initial study of genetic influence on childhood stress. In addition, the results are consistent with the traditional view that some forms of stress represent an environmental risk factor, and are not influenced by genetic differences between individuals.

The most striking finding was that siblings – both adoptive and nonadoptive – apparently had very different experiences in the first grade. The only finding that poses an exception to this claim is the shared environmental influence detected for parental report of the total number of events experienced. However, it is possible that this finding reflects the effect of having a single rater – a parent – provide the information on both siblings. Even when taking this result into account, most of the variance on the first-grade stress measures is not shared by siblings.

The finding of nonshared experiences in first grade is consistent with the emerging literature on the notable differences between siblings as they grow up (Dunn & Plomin, 1990). The results of the CAP analyses suggest that first-grade stress may represent an environmental risk factor for children, but a risk factor which is not common to children growing up in the same family. It should be noted that nonshared environment includes error of measurement along with systematic differences between siblings. However, even in the nonadoptive sample – in which the predictive validity of the first-grade stress measure was established – siblings had essentially uncorrelated experiences, suggesting that some reliable systematic differences between the siblings are operating in the experiences in first grade.

Analyses of behavior problems in school suggested a genetic contribution to externalizing, but not internalizing, problems, especially at age 9. The CAP will offer a unique opportunity to track developmental changes in behavior problems through adolescence; this database will be highly informative for determining etiological influences on psychopathology as children develop from middle childhood through adolescence. With respect to the goals of this chapter, the overall pattern of results is suggestive of a diathesis-stress model in which a genetic predisposition to externalizing problems and the environmental risk factor of first-grade stress both contribute to problems as rated by teachers. For internalizing problems, it appears that genetic factors are not yet influential; however, given that adolescence is the period of greatest risk for internalizing problems such as depression, it may be premature to make definitive conclusions about the most appropriate etiological model.

Appendix: Items relating to first-grade stress

Academic concerns
   School work too hard
   Academic pressure
   School work too easy

Peer relations
   Picked on by bully
   Teased/scared by older kids
   Teased/scared by other first-graders
   Unpopular with peers
   Hard to make friends
   Popular with peers

Teacher relations
   Not get along with teacher
   Scolded by teacher
   Teacher's pet

12 Height, Weight, and Obesity

Lon R. Cardon

Although the primary focus of the Colorado Adoption Project (CAP) is on traditional psychological concerns such as personality and cognitive functioning, the project also obtains data on health-related measures. In particular, the CAP is the only adoption study that has obtained longitudinal measurements of obesity-related variables in childhood. Weight and height data have been collected at each measurement occasion in the CAP. Such data facilitate analyses of genetic and environmental influences on childhood fatness which may lead to the onset of adult obesity.

Height and Weight in Early Childhood

Measures of height and weight have been extensively analyzed in the CAP sample for infancy and early childhood (years 1 through 4; Plomin, DeFries, & Fulker, 1988). Comparisons of nonadoptive and adoptive sibling correlations indicate substantial heritabilities for both height and weight (approximately .50–.80), with negligible impact of the shared sibling environment. In addition, height and weight measures appear quite stable throughout early childhood and to some extent into adulthood; that is, age-to-age correlations for height and weight are substantial. Comparisons of correlations between nonadoptive and adoptive siblings and between biological, adoptive, and nonadoptive parents and their respective offspring suggest that the phenotypic stability for height and weight is mediated largely by genetic factors. These results strongly indicate a genetic component to general body size which has a pervasive effect throughout early childhood.

Development of the Body Mass Index from Birth to Age 9

Recent research in the CAP has turned from height and weight as measures of general body size to a related measure, the Body Mass Index (BMI). The BMI, calculated as weight over height squared in the metric of kilograms and meters, is used as an index of body fat which is largely independent of height. It is essentially equivalent to the deviation of the observed weight from that expected based upon the regression of weight on height, and is often used as a measure of fatness, adiposity, or obesity.

Several family and twin studies of adult BMI have recently been reported, all suggesting a strong hereditary component (Fabsitz, Carmelli, & Hewitt, 1992; Price, Cadoret, & Stunkard 1987; Selby et al., 1990; Stunkard, Foch, & Hrubec, 1986; Stunkard, Harris, Pedersen, & McClearn, 1990). There is also some evidence for a genetic etiology of body fat and BMI during the teenage years (Brook, Huntley, & Slack, 1975; Price et al., 1990), but little is known about the etiology and development of BMI throughout early and middle childhood. In this chapter we summarize results from several analyses of sibling and parent–offspring data aimed at characterizing the pattern of influences for genetic and environmental effects on the development of BMI from birth to 9 years.

Measures and Models

Height and weight measurements defining the BMI are obtained annually in the CAP children from birth to age 9 with the single exception of age 8. At most ages the children are measured in the home or laboratory, but measurements at 5, 6, and 9 years of age are obtained by parental report. Assessments of adults were obtained in the laboratory during the initial testing sessions. The numbers of siblings and parents available at the time of analysis are presented in Table 12.1, where it may be seen that data from nearly 100 nonadoptive and 100 adoptive sibling pairs are available at the early years; however, the sample sizes diminish for middle childhood because many of the younger siblings in the CAP sample have not yet reached the later testing ages.

Mean BMI values for the CAP children at each age are presented in Figure 12.1. Differences in means as a function of gender or adoption status are negligible and non-significant at these ages. Descriptive statistics for height and weight in the parents have been presented by Plomin and DeFries (1985); adult BMI statistics may be found in Cardon and Fulker (in press).

Given the phenotypic stability of weight and height in childhood, the BMI would be expected to reveal some consistency during this period of development. Phenotypic correlations of BMI for ages 0 to 9 illustrate the expected continuity, as shown in Table 12.2. We employ the simplex and common factor models described in Chapter 3 to examine the genetic and environmental sources of this continuity in

Table 12.1 Body Mass Index: Sample sizes for children and parents

Age	AC	NC	US	NS	AM	AF	BM	BF	CM	CF
Birth	239	221	92	74	231	227	232	48	219	219
1	213	209	77	80	207	203	205	42	206	206
2	203	180	86	76	196	193	195	37	179	179
3	190	181	83	79	196	193	183	42	179	179
4	179	187	70	82	174	172	173	35	186	185
5	172	177	60	64	168	166	169	34	177	177
6	160	151	50	51	155	153	156	34	149	150
7	161	157	37	45	162	162	163	31	141	141
9	126	95	14	17	120	117	122	29	94	93

Note: Parent Ns refer to number of parent–offspring measurements.
AC = adopted child; NC = nonadopted child; US = unrelated sibling; NS = natural sibling; AM = adoptive mother; AF = adoptive father; BM = biological mother; BF = biological father; CM = control mother; CF = control father.

siblings and to compare alternative mechanisms which may foster this developmental stability. The common factor model includes a single set of genetic or environmental influences which contribute to developmental patterns of consistently high or low BMI observations at all childhood ages. Deviations may occur, but they are not expected to have any lasting effects on subsequent measurements. In contrast, the simplex model incorporates direct transmission of genetic or environmental influences

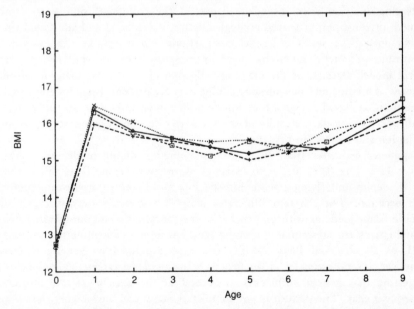

Figure 12.1 Body Mass Index (BMI) means for adopted and nonadopted children in the Colorado Adoption Project. Symbols are as follows: Adopted male (◇), nonadopted male (□), adopted female (×), nonadopted female (+).

from occasion to occasion. For BMI, such transmission implies that fatness perturbations at a specific age could have important consequences for obesity levels later in life.

Table 12.2　Phenotypic correlations of Body Mass Index from birth to age 9

Age	Age								
	0	1	2	3	4	5	6	7	9
0	1.00								
1	0.09	1.00							
2	0.14	0.52	1.00						
3	0.11	0.46	0.52	1.00					
4	0.13	0.39	0.45	0.56	1.00				
5	0.10	0.33	0.42	0.47	0.44	1.00			
6	0.11	0.28	0.37	0.44	0.42	0.49	1.00		
7	0.11	0.34	0.32	0.47	0.45	0.48	0.47	1.00	
9	0.06	0.25	0.28	0.34	0.32	0.34	0.47	0.56	1.00

Model-Fitting Results

Analyses of BMI in the CAP have examined both sibling relationships and parent –offspring resemblances. The sibling analyses were directed toward evaluation of genetic and environmental mediation of childhood continuity and change; the parent –offspring analyses were intended to extend the childhood sibling assessments through investigation of shared etiologies between childhood and adulthood. In the sibling analysis, a series of model comparisons was undertaken to examine the significance of genetic and environmental parameters in the common factor/simplex hybrid model. Because of the ongoing collection of CAP data, many longitudinal patterns of missing and non-missing sibling data are present; thus, we employed the maximum-likelihood pedigree function described in Chapter 3 to take full advantage of all non-missing data. Results of the full series of significance tests are reported in Cardon and Fulker (in press).

Parameter estimates from the most parsimonious sibling model are shown in Figure 12.2. The P_i, $i = 0, \ldots, 9$, variables represent observed BMI values and F_{G_i} and F_{S_i} depict underlying genetic and specific, nonshared environmental factors at each occasion. The diagram illustrates simplex regression estimates and time-specific factor loadings with no genetic or environmental common factor. Environmental effects are substantial at specific ages, but do not contribute to continuity in BMI, as all observed BMI stability could be explained by genetic mediation. Moreover, several estimates of the genetic transmission parameters are close to 1.0, suggesting that genetic influences at several of the ages persist in entirety to subsequent ages. The pattern is not completely static and unchanging, however, as the persistent genetic effects are augmented by new genetic variation at each age. Thus, the overall trend is one of cumulative genetic influence throughout early and middle childhood.

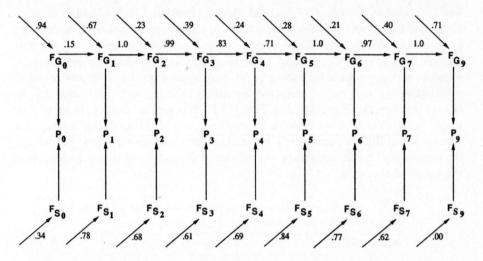

Figure 12.2 Final developmental model of body fat in siblings. Body Mass Index measurements are shown as phenotypes P, with genetic F_G and within-family environmental (F_S) influences.

Genetic correlations among the different ages, presented in Table 12.3, further illustrate the continuity of BMI during this developmental period. It is interesting that the persistent pattern does not begin at birth, but at age 1. Genetic influences on BMI at birth, although substantial in magnitude, do not show continuity with the later ages. This remarkable transition during the first year of age suggests that an important developmental genetic mechanism fosters BMI stability from infancy to middle childhood.

Table 12.3 Genetic correlations and variance components for Body Mass Index from birth to age 9

Age	Age								
	0	1	2	3	4	5	6	7	9
0	1.00								
1	.20	1.00							
2	.19	.95	1.00						
3	.17	.83	.88	1.00					
4	.16	.78	.82	.94	1.00				
5	.14	.68	.72	.82	.87	1.00			
6	.13	.64	.68	.77	.82	.94	1.00		
7	.11	.53	.56	.64	.68	.78	.83	1.00	
9	.08	.38	.40	.46	.49	.56	.59	.71	1.00
h^2	.88	.43	.53	.64	.52	.32	.39	.58	1.00
e^2	.12	.57	.47	.36	.48	.68	.61	.42	.00

The parent–offspring design of the CAP was used to assess genetic and environmental BMI stability from infancy to adulthood. The model selected for analysis was developed by Fulker, DeFries, and Plomin (1988) and is discussed in Chapter 3. As in the analysis of sibling data, a series of model comparisons was undertaken to assess the significance of parameters. Selective placement and cultural transmission for body fat were shown to be non-significant, and no differences were apparent between assortative mating parameters for biological and adoptive parents.

Estimates of the parent–offspring heritabilities (h^2), and assortative mating ($p = q$) parameters are presented in Table 12.4. Heritability estimates indicate little shared genetic resemblance between parents and their offspring during infancy, but increased resemblance during early and middle childhood. From 3 to 9 years of age, the heritability estimates stabilize somewhat and are similar to those obtained from the sibling analysis.

Table 12.4 Genetic parameter estimates from reduced model of parent–offspring resemblance

Parameter	Age								
	0	1	2	3	4	5	6	7	9
h^2_{po}	.09	.01	.09	.37	.52	.38	.55	.38	.57
$p = q$	.14	.12	.16	.13	.11	.12	.13	.13	.18

As discussed in Chapter 2, h^2 estimates obtained from evaluation of parents measured as adults and their offspring tested as children actually reflect the product of childhood (h_C) and adult (h_A) genetic effects and the genetic correlation between them (r_G). If we assume an estimate of adult heritability of 0.60 as indicated by twin and family studies (Price et al., 1987) and use the sibling estimates of child heritabilities, the genetic correlation at each age may be calculated as $r_G = h^2_{po}/\sqrt{.60h^2_C}$. Resulting estimates of these genetic correlations are .11, .02, .14, .52, .81, .75, .99, .56, and .64 between each of the childhood ages (0–9) and adulthood. The moderate to high correlations at age 3 (.52) and later clearly suggest substantial genetic continuity from early and middle childhood to adulthood.

Adiposity Rebound

From an epidemiological perspective, the observed continuity between early childhood and adult BMI suggests an important role for early prediction of obesity or obesity correlates. Diet and exercise are often considered to be paramount in this regard, but predictive utility has also been shown with the growth trend in the BMI itself. Rolland-Cachera et al. (1984, 1987, 1991) have used the term "adiposity rebound" to refer to the onset of a rapid growth in body fat that occurs at about 6 years of age. Adiposity rebound is based on the normal childhood development of adipose tissue: rapid increase in adiposity and adipocyte growth during the first year of life, followed by gradual decrement of adiposity and a pattern of height increase rather than weight growth until about age 6, with additional increases in adiposity eventually leading to adult levels (Knittle, Timmers, Ginsberg-Fellner, Brown & Katz, 1979; Tanner, Hughes, & Whitehouse, 1981). The growth trend may be observed in the CAP BMI means in Figure 12.1, which illustrate the decreasing, then increasing trend between ages 2 and 9. For an individual, the point of change between decreasing and increasing adiposity is his/her adiposity rebound.

The predictive function of adiposity rebound lies in the age at which the rebound phenomenon is experienced. A classification of three groups has been suggested (Rolland-Cachera et al., 1984): advanced or early rebound which occurs before age 5.5, average rebound occurring between ages 5.5 and 6.5, and late rebound occurring after age 6.5. Individuals who experience early rebound have been shown to be at greater risk for adult obesity than those in the average or late categories (Rolland-Cachera et al., 1984, 1987).

The standard method of determining the age of adiposity rebound for a child is to examine the BMI values at different ages and choose the minimum point as the rebound age. However, fluctuations in the BMI patterns are commonplace, leading to difficulties in determination of the minimum point. Cardon, DeFries, and Fulker (1993) have developed an objective procedure for rebound age assignment based on the regression equation

$$BMI_i = \alpha_i + \beta_{1_i} Age_i + \beta_{2_i} Age_i^2 \tag{1}$$

where BMI_i represents the BMI scores from individual i over time, and Age_i are the corresponding ages for the BMI values. The minimum, or rebound age, for each individual is then the point at which the differential equation $\dfrac{\delta BMI}{\delta Age}$ equals 0.0. Solving this equation for Age yields the following objective measure:

$$\text{Rebound age} = -1/2 \frac{\beta_1}{\beta_2} \tag{2}$$

Figure 12.3 illustrates mean BMI values for the CAP sample as grouped into the three adiposity rebound categories after objective determination of rebound age. The means illustrate large differences among groups, particularly between the advanced age group and the other groups. This trend is fully consistent with the previous finding indicating greater risk for later obesity by advanced rebound individuals.

An important question relating to the adiposity rebound prediction is whether the rebound age is influenced by heritable effects as is the BMI. Results from application of the DF model (see Chapter 3) indicate only a modest heritability, .26, and negligible c^2 effects, −.06. Thus, unique environmental influences are apparently highly important in determining the timing of changes in body fat growth during middle childhood. The predictive nature of this measure suggests that the frequently noted factors of diet and activity (Selby et al., 1990) may warrant close attention in young children. These results support the hypothesis that the propensity toward advanced rebound is genetically transmitted, but the actual occurrence is environmentally mediated (Rolland-Cachera et al., 1987).

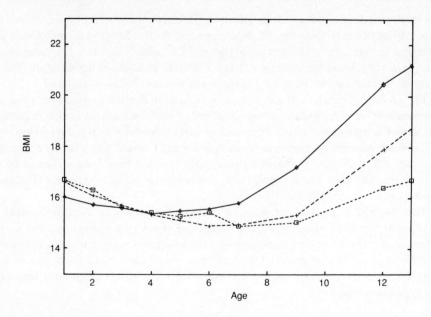

Figure 12.3 Adiposity rebound in the Colorado Adoption Project. Body Mass Index (BMI) means are plotted for three categories of adiposity rebound: early (◇), average (+), and late (□).

Conclusions

The CAP is an important source of information for behavioral health and anthropometry, as well as the more traditional psychological measures. The longitudinal aspect of the project provides a unique data set for evaluation of the development and etiology of body fat from infancy through middle childhood. Results from sibling and parent–offspring analyses of the Body Mass Index suggest a strong and pervasive role of genetic influences from early childhood and perhaps into adulthood. Additional analysis of the adiposity rebound, focused more specifically on obesity rather than general body fat, replicate findings from other longitudinal studies and provide the first evidence regarding the heritable nature of this predictive event. Additional longitudinal data from the CAP children should further elucidate the timing and etiology of the onset and maintenance of adult obesity.

13 Motor Development

Keith E. Whitfield
Stacey S. Cherny

Although motor development was of central concern to developmentalists in the 1920s and 1930s (see review by Dewey, 1935), interest dwindled in the 1950s and 1960s. The early work was largely descriptive and normative, that is, describing species-wide motor milestones as a function of age. It was assumed that motor development is largely biologically driven, an assumption embodied in the influential maturational theory of Arnold Gesell (1954). Individual variations on these normative themes emanated from this work and led to tests of motor development by Gesell, whose work was systematized in the widely used Motor Scale of the Bayley Scales of Infant Development (Bayley, 1969). It was assumed that, like species-wide developmental milestones, individual differences in rates of motor development are also biological in origin. This view clashed with the widespread environmentalism following the Second World War and may have contributed to the decline in interest in motor development.

One purpose of the present chapter is to examine the genetic and environmental origins of individual differences in motor development during infancy and early childhood. There is surprisingly little genetic research on individual differences in rates of motor development. The only report of which we are aware is from the Louisville Twin Study (Wilson & Harpring, 1972), in which at 6, 9, 12, and 18 months, the average identical and fraternal twin correlations for motor development were .79 and .69, respectively. Contrary to the widespread assumption that individual differences in motor development are highly heritable, these results suggest only slight genetic influence and an overwhelming importance of shared environmental effects.

A sibling study of children from ages 6 through 12 years that included measures of static strength and gross motor ability reported that 22 to 58% of the variance in performance was attributable to genetic variation (Malina & Mueller, 1981). However, because sibling studies without adoptees or twins cannot disentangle genetic from possible shared environmental effects, the results of that study should be interpreted only as evidence for familiality, i.e., genetic and/or shared environmental variance.

In a preliminary report from the Colorado Adoption Project (CAP), parent–offspring analyses also yielded little evidence of genetic influence when parental athletic competence was compared with infant motor development (Plomin & DeFries, 1985). Although results of analyses comparing nonadoptive and adoptive sibling pairs suggested genetic influence at 1 year of age, little genetic influence was found at 2 years (Plomin, DeFries, & Fulker, 1988).

A major objective of this chapter is to present results from a multivariate analysis of motor development, employing the CAP sibling adoption design at 1, 2, 3, and 7 years of age. The longitudinal nature of the CAP permits analyses of age-to-age change and continuity, in addition to analyses of the genetic and environmental etiologies of individual differences in motor ability at each of those ages.

Another objective of this chapter relates to renewed interest in motor development as viewed from process-oriented perspectives such as information-processing and dynamical systems approaches (e.g., Keogh, 1977; Payne & Isaacs, 1991). Links between motor development and mental development in infants have been frequently noted. For example, in the CAP, the Bayley Psychomotor Development Index (PDI) correlates .40 at 1 year of age with the Bayley Mental Development Index (MDI). The correlation between the PDI and MDI drops to .26 at 2 years. This developmental change in the relationship between mental and motor development is particularly interesting in light of the tentative CAP finding that genetic influence on motor ability is greater at 1 year than at 2 years. To what extent do the same genetic and environmental factors influence the expression of these two domains? The CAP design and data permit multivariate genetic analyses of motor and mental development to address this issue. Earlier analyses of this issue suggested that genetic influence on motor development at 1 year overlaps substantially with genetic influence on mental development (Plomin et al., 1988). By 2 years of age, however, mental development and motor development appear to be largely independent genetically.

In summary, the present chapter will address three major issues pertaining to motor development: (1) the extent to which genetic and environmental influences contribute to individual differences in motor development; (2) the extent to which these influences contribute to change and continuity from infancy to middle childhood; and (3) the nature of the relationship between mental and motor development. These issues will be addressed by subjecting data from CAP adoptive and nonadoptive sibling pairs to structural equation model-fitting analyses.

Method

Motor development was measured at 1 and 2 years of age using the Motor Scale of the Bayley Scales of Infant Development (Bayley, 1969). At age 3 years, an extension of this measure was used, which was designed to assess both fine and gross motor ability. The fine motor scale consisted of four tasks: (1) drawing, (2) stacking blocks, (3) stringing beads, and (4) throwing and catching. The gross motor scale also consisted of four tasks: (1) kicking a ball, (2) jumps, (3) leg coordination, and (4) balance beam.

Motor ability at 7 years of age was measured using the Bruininks-Oseretsky Test of Motor Proficiency (BRO) (Bruininks, 1978). The BRO is scaled to measure motor development from 4 1/2 to 14 1/2 years of age. The short form of the test was employed, which took 15 to 20 minutes to administer. This version provides a single gross motor-ability score based upon tester ratings of 14 items.

A factor analysis of the five measures of motor ability was conducted to evaluate whether these measures represent a single underlying construct. Two factors were indicated, based on both an inspection of the scree plot of the eigenvalues and the fact that only two eigenvalues were greater than unity. The factor pattern matrix obtained following oblique rotation is presented in Table 13.1. Clearly, the fine motor measure given at age 3 is not strongly related to the other four measures. Consequently, that fine motor skill measure was omitted from subsequent analyses.

Table 13.1 Factor pattern matrix for measures of motor ability

Measure	Age	Factor 1	Factor 2
Bayley	1	.41	
Bayley	2	.62	
Fine	3		.34
Gross	3	.56	
Bruininks	7	.55	

Note: Values less than .30 have been omitted for clarity.

The intellectual measures included the Mental Scale from the Bayley (1969) at 1 and 2 years of age, the Stanford-Binet (Terman & Merrill, 1973) at 3 years of age, and the Wechsler Intelligence Scale for Children–Revised (Wechsler, 1974) at 7 years of age.

Results

Descriptive statistics

In Table 13.2, means of the motor measures, standard deviations, and sample sizes, by sex and adoptive status, are presented for ages 1, 2, 3, and 7 years. A $2 \times 2 \times 4$ MANOVA, with sex and adoptive status (adopted proband versus nonadopted proband) as between-subjects factors and age as a within-subjects factor, was performed. Of the between-subjects effects, only adoptive status was significant ($F_{1,308} = 13.07$, $p < .001$). However, adoptive status accounted for less than 5% of the total variance. The main effect of age was also significant, but should not be interpreted since different scales are used at different ages. The status-by-age

Table 13.2 Means and standard deviations for motor measures from ages 1 through 7 years

Measure	Age	Adopted probands			Unrelated sibs			Nonadopted probands			Biological sibs		
		Mean	SD	N	Mean	SD	N	Mean	SD	N	Mean	SD	N
Males													
Bayley	1	91.40	15.59	130	93.13	14.96	46	95.90	14.04	133	93.33	16.34	58
Bayley	2	99.37	13.32	109	105.05	12.22	41	103.54	13.69	115	108.78	15.28	45
Gross	3	−1.00	3.17	99	0.58	2.56	31	0.66	2.96	95	1.31	3.12	44
Bruininks	7	59.50	9.31	103	64.38	7.30	29	62.91	8.64	116	64.82	8.88	44
Females													
Bayley	1	95.13	14.12	111	91.38	18.22	45	94.24	14.05	113	95.26	15.48	43
Bayley	2	99.12	12.11	99	103.53	11.91	45	104.10	12.95	105	109.73	17.38	41
Gross	3	0.27	3.11	84	0.51	2.70	46	0.56	2.94	94	2.03	2.85	39
Bruininks	7	60.08	9.04	91	60.80	11.16	35	60.40	9.23	99	61.00	8.96	26

interaction was also significant ($F_{3,306} = 3.10$, $p < .05$), but, again, the proportion of variance that this effect explains is very small. Since these significant effects account for only a very small proportion of the variance, they are unlikely to have any substantial effect on second-degree statistics such as correlations and covariances, upon which our subsequent model-fitting procedures are based.

Longitudinal correlations at 1, 2, 3, and 7 years of age are shown in Table 13.3. From these correlations, there appears to be modest stability from infancy to middle childhood in motor development.

Table 13.3 Phenotypic correlation matrix among motor measures

Age	1	2	3	7
1	229.26	594	523	532
2	.35	188.57	506	490
3	.28	.37	9.59	444
7	.19	.27	.24	85.19

Note: *r*s below diagonal, *N*s above diagonal, and variances on diagonal. $p < .01$ for all *r*s.

Nonadoptive and adoptive sibling correlations are presented in Table 13.4. The larger nonadoptive than adoptive sibling correlations at 1 and 2 years of age suggest some genetic influence. The correlations for years 3 and 7 suggest some shared environmental variance and little or no genetic variance.

Table 13.4 Sibling correlations for motor measures

Age	Adoptive		Nonadoptive	
	r	*N*	*r*	*N*
1	.07	88	.38**	101
2	−.09	76	.26*	80
3	.28*	66	.11	77
7	.25	61	.27*	68

*$p < .05$, **$p < .01$.

Model-fitting analyses

Analysis of motor measures   In order to estimate genetic, shared environmental, and unique environmental components of variance and covariance from these data, and to allow for specific tests of those components, model-fitting analyses were performed. A full-rank Cholesky decomposition model was fitted to the data at the four ages (1, 2, 3, and 7) using a maximum-likelihood pedigree approach. The data were standardized across individuals, within each age, before analysis. This standardization effectively eliminates variance differences across ages, which are most likely a result of using different tests at different ages, while preserving adoptive versus control and sibling variance differences. The model-fitting procedures are described in detail in Chapter 3.

Estimates of h^2, c^2 and e^2, obtained from fitting this full model, are presented along the diagonals in Tables 13.5, 13.6, and 13.7, respectively. None of the genetic or shared environmental variance components was statistically significant, although the h^2 estimates are substantial at both 1 and 2 years of age. At years 3 and 4, there is some shared environmental variance and little genetic variance. Finally, although neither the overall test of genetic nor shared environmental variance was statistically significant, a combined test of the two components was ($\chi^2_{20} = 37.43, p < .02$), indicating that familial resemblance for motor ability is significant at these ages.

Table 13.5 Genetic correlations for motor measures

Age	Age			
	1	2	3	7
1	.55			
2	.77	.28		
3	.95	.92	.09	
7	-.08	.08	-.09	.14

Note: h^2s appear on the diagonal.

Table 13.6 Shared environmental correlations for motor measures

Age	Age			
	1	2	3	7
1	.06			
2	.48	.06		
3	-.18	.78	.18	
7	.61	.99	.67	.22

Note: c^2s appear on the diagonal.

Table 13.7 Unique environmental correlations for motor measures

Age	Age			
	1	2	3	7
1	.38			
2	.02	.66		
3	.14	.14	.73	
7	.27	.18	.12	.64

Note: e^2s appear on the diagonal.

The genetic correlation matrix among these measures is presented in Table 13.5. Genetic correlations among the motor measures at 1, 2, and 3 years of age are high, but those with the measure at 7 years are low. All but one of the shared environmental correlations are relatively high, suggesting that shared environmental influences on motor ability persist over time (see Table 13.6). Finally, the unique environmental correlations among the four ages are all relatively small, indicating that unique environmental influences are only transient (see Table 13.7).

Covariance between mental and motor ability In order to investigate the relationship between motor and mental ability, a Cholesky decomposition was fitted to motor ability at age 1 and IQ at ages 1, 2, 3, and 7 years. Only motor ability at age 1 was included because results of previous CAP analyses suggested that motor and mental development are highly correlated genetically at only 1 year of age (Plomin et al., 1988). For this Cholesky model, the motor ability measure was entered first, followed by the four IQ measures. This allows parameter estimates for the relationship between motor ability at year 1 and IQ at each of the four ages. Tests of the various components of the full model were performed to determine which parameters are essential for explaining these data and to arrive at the most parsimonious model (see Table 13.8).

Table 13.8 Model comparisons for covariance between mental and motor ability

Model	Form	$-LL$[a]	$NPAR$[b]	χ^2	df	p
1	Full model	1232.287	45			
2	Model 1, drop unique environmental covariances	1236.513	35	8.452	10	>.50
3	Model 2, drop shared environmental parameters	1245.212	20	17.398	15	>.25
4	Model 3, drop genetic motor–mental covariances	1314.333	16	138.242	4	<.001

[a] Log-likelihood of the data (without the addition of the constant).
[b] Number of free parameters.

First, results of a test of the unique environmental covariances indicated that the environmental influences specific to the individual are not responsible for the relationships between motor ability and mental ability, or those among mental ability measures across time (Model 2). Next, the shared environmental parameters could also be dropped from the model without a significant change in model fit (Model 3). Finally, the loadings of the first genetic factor (motor at 1 year of age) on IQ at the four time points were dropped as an overall test of the genetic covariance between motor ability at 1 year and IQ at 1, 2, 3, and 7 years of age. This resulted in a highly significant change in model fit ($p < .001$), suggesting significant genetic covariance between motor development at age 1 and mental ability (Model 4).

The final model of the relationship between motor ability at age 1 and IQ at ages 1, 2, 3, and 7 is presented in Figure 13.1. All loadings are standardized to a phenotypic variance of unity. There is clearly a strong genetic correlation between motor ability at 1 year of age and IQ during infancy, but this relationship does not persist into middle childhood.

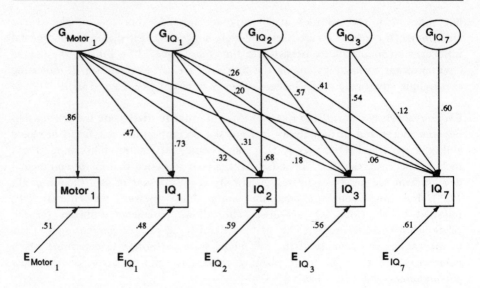

Figure 13.1   Genetic (*G*) and unique environmental (*E*) influences on motor ability at age 1 and IQ at ages 1, 2, 3, and 7.

Discussion

Although caution is warranted because so little is known about the etiology of individual differences in motor development, the results are intriguing in that they are counterintuitive. Motor development scores show considerable genetic influence at age 1, declining influence at age 2, and negligible genetic influence at ages 3 and 7. In contrast, shared environmental effects show increasing influence from infancy to middle childhood, although shared environmental influences later in development covary with shared environmental influences in infancy.

The puzzling finding that the greatest genetic influence in motor development is at age 1 is elucidated by the multivariate analysis which indicates that this genetic influence is somewhat due to an overlap with mental development scores. Results of unpublished analyses suggest that genetic correlations between mental and motor development drop sharply at age 2 and remain low at ages 3 and 7.

That individual differences in motor development show little genetic influence at later ages may also seem somewhat counterintuitive in that common wisdom suggests that motor development should be heritable. Such intuitions are a poor guide to the realities of genetic influence. It should be pointed out that these results do not contradict those from the sibling analysis in middle childhood of Malina and Mueller (1981). Although their sibling correlations provide evidence for the familiality of motor development, the results of the present analysis indicate that this familial resemblance is largely due to shared environmental influences.

In conclusion, this first genetic analysis of individual differences in motor development from infancy to middle childhood suggests that there may be some surprising discoveries when this uncharted territory is further explored.

14 Sex Differences in Genetic and Environmental Influences for Cognitive Abilities

Laura A. Baker
Hsiu-zu Ho
Chandra Reynolds

Questions about nature and nurture have dominated gender research in the social sciences. The social and biological bases of differences between males and females, as well as the extent to which within-sex variation may be differentially affected by heredity and environment, continue to be important questions considered by gender researchers. Yet, by and large, these questions have rarely been addressed within a behavioral genetic framework. Instead, studies of biological and social processes related to gender have employed methods requiring strong assumptions, which are rarely tested formally. For example, environment–behavior relationships have traditionally been inferred on the basis of mother–child interactions, in spite of the known genetic confound which exists in such relationships.

In this chapter we examine several questions of importance in the study of sex differences in cognitive abilities.* The primary focus is on whether or not males and

---

* The term "sex differences" is employed throughout this chapter in reference to any differences, real or hypothesized, which may exist between males and females for whatever reasons, social or biological. Since our objective is to compare the relative effects of both biological and cultural transmission for males and females, we have opted not to use the currently popular term "gender" here, as it often implies that culture is the primary source of differences between the sexes (see Jacklin, 1992).

females are differentially affected by heredity and environment. We address this question in two different ways, first via the analysis of parent–offspring resemblance using model-fitting, and second by examining relationships among cognitive abilities and measures of the child's home environment. In the model-fitting analyses of parent–offspring resemblance, we also consider sex differences in parental influences, that is, whether or not maternal and paternal effects on children's cognitive abilities are comparable.

We first review previous research which has investigated differences in hereditary and environmental influences for boys and girls. For the most part, this research has focused on relationships among cognitive abilities and measures of the home environment, based primarily on mother–child dyads from intact nuclear families. While a few investigators have examined gender differences in heritability in studies of personality variables (Eaves, Eysenck, & Martin, 1989), criminal behavior (Baker, Moffitt, Mack, & Mednick, 1989), and educational attainment (Heath et al., 1985), there have been very few systematic studies of whether or not relative genetic variability (i.e., heritability) differs between the sexes within the cognitive domain. Thus, our analyses provide one of the first comprehensive examinations of sex differences in heritability in cognitive development.

Several longitudinal studies in child development have reported differential responses by boys and girls to the early environment and suggested that boys are more susceptible to early environmental influences. In the Berkeley Growth Study, Bayley and Shaefer (1964) found relationships between maternal behaviors and children's intellectual outcomes to be stronger for boys than for girls. Moreover, Bayley and Schaefer suggested a "genetic sex difference in resistance to or resilience in recovery from environmental influences" and stated that "such a sex difference would be in accord with a number of observed physical sex differences." Further support of differential responses of girls and boys to aspects of the early environment came from Bradley and Caldwell (1980) who reported stronger relationships for boys than for girls between early availability of appropriate toys and materials and later cognitive abilities. In a study of the relationship between "inanimate stimulation" in infancy and later cognitive development, Wachs (1979) reported consistently stronger relationships for boys than for girls. Wachs has termed such associations the "gender specificity" of environment–outcome relationships. Of the differences found by Bee, Mitchell, Barnard, Eyres, and Hammond (1984), stronger predictions for boys of IQ or language skill were reported from measures of the mother's developmental expectations, the extent of the father's involvement in the infant's early care, the provision of appropriate play materials, and the extent of parental life change. For girls, however, stronger predictions of IQ or language skill were found for a measure of restriction and punishment.

Several studies on the relationship between environmental quality and children's language competence have shown sex differences in favor of girls. For example, Elardo, Bradley, and Caldwell (1977) found aspects of the home environment to be more "frequently and strongly related to aspects of language performance for females than for males." Similarly, Moore (1968) found vocabulary scores to be more highly correlated with home environment for girls than for boys.

A critical problem in the interpretations drawn from these studies of "environmental" influences is that the confound of genetic influences has been ignored. Nearly all of the studies that have examined the impact of early home environment on cognitive development have used intact nuclear families. Unfortunately, in such families, factors that appear to be environmental in nature may be associated with heritable parental characteristics and therefore be transmitted genetically to the child. For example, it is likely that associations between maternal responsivity and the child's cognitive ability are partly due to genetic influences. That is, mothers who are more responsive to their child may also be more "intelligent," and thus contribute genes which influence greater cognitive development in the child. Hence, the hypothesis that boys are more influenced than girls by the early environment (or that girls are more influenced genetically) cannot be adequately tested by examining families where parents and their children are genetically related.

It is acknowledged that the adoption method is the most powerful behavioral genetic method for separating genetic from environmental influences. In their report of the Colorado Adoption Project derived from analyses conducted on infants at 1 and 2 years of age, Plomin and DeFries (1985) examined sex differences utilizing univariate and bivariate analyses and found little support for the hypothesis that boys are more susceptible to environmental influences and girls are more influenced genetically.

In the present chapter, we will extend the investigation of Plomin and DeFries (1985) on the issue of differential responses of girls and boys to genetic and environmental influences to: (1) developmental measures examined later, at 3, 4, and 7 years of age, (2) the application of multivariate methods of data analysis in examining sex differences and environment–behavior relationships, and (3) the utilization of model-fitting procedures in examining parent–offspring resemblance.

Method

Subjects

Analyses presented in this chapter are based on data available for adopted and nonadopted control children's cognitive abilities measured at 1, 2, 3, 4, and 7 years of age, as well as cognitive abilities in the biological, adoptive, and control parents of these children. For model-fitting analyses of parent–child resemblance for general cognitive ability, data were available for at least one family member from 131 families of adopted boys, 132 families of control boys, 114 families of adopted girls, and 113 families of control girls. Multivariate analyses of the specific cognitive abilities in the children (see preliminary analyses to model-fitting described below) were based on the following numbers: year 3, 112 adopted boys, 91 adopted girls, 112 control boys, 101 control girls; year 4, 109 adopted boys, 87 adopted girls, 113 control boys, 100 control girls; and year 7, 108 adopted boys, 92 adopted girls, 114 control boys, 101 control girls. Data from a subset of these children (75 adopted boys, 64 adopted girls, 89 control boys, 79 control girls) at all ages were also subjected to multivariate analyses of variance.

Measures

Both global and specific measures of cognitive abilities were examined at each age. Global measures for the children at various ages were the Bayley Mental Index at 1 and 2 years of age, and the first principal component from a test battery of specific cognitive abilities at ages 3, 4, and 7 years. Correspondingly, the first principal component from a battery of 13 tests of specific cognitive abilities (corrected for sex and age effects) was used as an indicator of global ability in the biological and adoptive parents (DeFries, Plomin, Vandenberg, & Kuse, 1981). Global ability measures were examined at all five ages in all types of analyses, including mean differences, model-fitting of parent–child resemblance, and environment–behavior relationships.

Specific abilities examined at each age were based upon subscales derived from factor analyses of the Bayley Mental items at year 1 (Means–End, Imitation, Verbal Skill) and at year 2 (Lexical, Spatial, Verbal–Symbolic), and factor scores from the cognitive abilities test batteries at years 3, 4, and 7 (Verbal, Spatial, Perceptual Speed, Memory) (Plomin & DeFries, 1985; Singer, Corley, Guiffrida, & Plomin, 1984). Mean differences between girls and boys were examined for specific abilities at all five ages, as were sex differences in environment–behavior relationships. However, model-fitting analyses of parent–offspring resemblance were applied only to years 3, 4, and 7 where comparable measures of verbal and spatial abilities, memory, and perceptual speed were available in both parent and child generations. Due to the obvious diversity of constructs between specific abilities for adults and children at ages 1 and 2 years, model-fitting analyses were not performed at these ages.

Environmental measures were based upon the Caldwell Home Observation for the Measurement of the Environment (HOME: Caldwell & Bradley, 1978), measured during home visits at ages 1, 2, 3, and 4 years. Both global scores and subscales were examined in our analyses. Global measures of the early environment were based on first principal component scores of quantitatively scored items in the HOME inventory (Plomin, DeFries, & Fulker, 1988). Subscales of the HOME at each year of age have been previously derived through factor analyses (Plomin et al., 1988). For years 1 and 2 these subscales were: Toys, Maternal Involvement, Encouraging Developmental Advancement, and Restriction–Punishment. Subscales of the HOME at year 3 are similar to those obtained in infancy, except that a Restriction–Punishment factor did not emerge in the analyses. At year 4, only two factors emerged: Toys and Maternal Involvement.

Results

Mean differences

Recognizing the intercorrelations among the various longitudinal measures of cognitive development and among the environmental measures, multivariate analyses of variance (MANOVA) procedures were conducted to assess mean differences between girls and boys. Separate MANOVAs were conducted for adopted and control children in order to check for consistency of results across the two samples. Adhering to the protection-levels approach recommended by Cliff (1987), univariate analyses of specific subscales were only conducted following a significant omnibus test (i.e., the multivariate F ratio), in order to maintain $\alpha = .05$ in our analyses of mean differences. (Note that comparisons of adopted and control children were also made, both with respect to means, variances, and correlations among cognitive measures. These are described later as preliminary analyses to model-fitting procedures.)

Cognitive measures   Tables 14.1 and 14.2 present descriptive statistics with respect to global cognitive measures for adopted and nonadopted children respectively. Multivariate analyses of variance yielded no significant differences between girls and boys on these measures for either adopted or control children. However, when MANOVA analyses were conducted for specific cognitive measures, a few significant sex differences were found for both adopted and control samples. Tables 14.3 and 14.4 present descriptive statistics for the specific cognitive measures and results of subsequent univariate analyses. It should be noted that while only the sex difference in Verbal–Symbolic ability at year 2 replicated across adopted and control samples, mean differences for three of the four variables with significant sex differences in either sample were consistently in the same direction for both samples. However, the sex effect explains little of the variance (ranging from 3.0 to 9.45% of the variance) in the variables with significant differences between girls and boys.

Table 14.1   Means and standard deviations for global cognitive measures for adopted probands

Age	Boys		Girls	
	Mean	*SD*	*Mean*	*SD*
1	106.91	11.27	107.38	11.49
2	106.84	14.95	109.30	13.44
3	.07	.98	−.01	.83
4	−.07	1.06	.05	.82
7	−.14	.97	−.27	.90

Note: Multivariate test yielded no sex difference ($F_{5,133} = .73$, $p > .05$); $N = 75$ boys and 64 girls.

Table 14.2   Means and standard deviations for global cognitive measures for nonadopted probands

Age	Boys		Girls	
	Mean	*SD*	*Mean*	*SD*
1	110.79	12.64	109.63	12.28
2	108.36	16.29	112.09	15.43
3	.11	1.09	.29	.99

Table 14.2 (Cont.)

Age	Boys		Girls	
	Mean	SD	Mean	SD
4	−.09	1.12	.27	1.03
7	.16	1.04	.10	1.08

Note: Multivariate test yielded no significant sex difference ($F_{5,162} = 1.68$, $p > .05$); $N = 89$ boys and 79 girls.

Table 14.3 Means and standard deviations for measures of specific cognitive abilities of adopted probands

Cognitive variables	Boys		Girls	
	Mean	SD	Mean	SD
Year 1				
Means–End	1.87	1.11	1.83	1.09
Imitation*	2.24	1.71	1.71	1.26
Verbal Skill	1.55	1.12	1.81	1.22
Year 2				
Lexical	8.36	3.34	8.92	2.66
Spatial	1.12	1.97	.81	2.09
Verbal–Symbolic*	7.94	1.68	8.54	1.45
Imitation	2.58	1.03	2.51	1.24
Year 3				
Verbal	.17	.71	−.05	.69
Spatial	.10	.84	.01	.71
Perceptual Speed	.06	.75	.06	.67
Memory	−.02	.96	.00	.94
Year 4				
Verbal	.01	.69	.08	.62
Spatial	.15	.99	−.12	.73
Perceptual Speed*	−.17	.95	.19	.87
Memory	.06	.93	−.11	.94
Year 7				
Verbal	−.01	.83	−.24	.81
Spatial	−.06	.87	−.08	.60
Perceptual Speed	.01	.83	−.17	.84
Memory	−.04	.91	−.14	.98

Note: Multivariate test by sex yielded a significant sex difference ($F_{19,110} = 2.03$, $p < .05$); $N = 67$ boys and 63 girls.
*Indicates a significant sex difference for univariate test ($p < .05$)

Environmental measures Tables 14.5 and 14.6 present descriptive statistics for HOME scales for adoptive and control families. MANOVA analyses yielded significant sex differences with respect to the quantitatively scored HOME factors for only adoptive families. Subsequent univariate tests yielded differences for 3 of the 13 scores (Developmental Advancement at year 1, Maternal Involvement at year 2 and Responsivity of Mother at year 3). Because these differences were not replicated in non-adoptive families, they may, in fact, be spurious.

Table 14.4 Means and standard deviations for measures of specific cognitive abilities of nonadopted probands

Cognitive variables	Boys		Girls	
	Mean	*SD*	*Mean*	*SD*
Year 1				
Means-End	2.05	1.28	1.73	1.19
Imitation	2.64	1.82	2.21	1.69
Verbal Skill*	1.58	1.09	2.03	1.02
Year 2				
Lexical	8.56	3.13	9.26	2.37
Spatial	1.03	1.95	1.17	2.10
Verbal–Symbolic*	7.93	1.83	8.50	1.43
Imitation	2.50	1.24	2.56	1.06
Year 3				
Verbal	.11	.79	.24	.81
Spatial	.12	.73	.23	.84
Perceptual Speed	.05	.80	.13	.75
Memory	.36	1.06	.23	1.02
Year 4				
Verbal	.06	.63	.18	.71
Spatial	.15	.89	−.01	.76
Perceptual Speed	−.21	.93	.08	.92
Memory*	−.10	.83	.41	.90
Year 7				
Verbal	.12	.92	.17	.78
Spatial	.23	.87	.18	.82
Perceptual Speed	.07	.86	−.01	.97
Memory	.08	.79	.01	.90

Note: Multivariate test by sex yielded a significant sex difference ($F_{19,130} = 2.18$, $p < .05$); $N = 80$ boys and 70 girls.
*Indicates a significant sex difference for univariate test ($p < .05$).

Table 14.5 Means and standard deviations for Caldwell HOME scales for adoptive families

HOME scores	Boys		Girls	
	Mean	*SD*	*Mean*	*SD*
Year 1				
Toys	−1.00	4.93	.57	4.99
Maternal Involvement	−.18	3.00	.82	2.93
Encouraging Developmental				
Advancement*	−.94	3.96	1.12	4.49
Restriction–Punishment	.41	2.07	−.70	1.38

Table 14.5 (Cont.)

HOME scores	Boys		Girls	
	Mean	SD	Mean	SD
Year 2				
Toys	−.09	4.87	.49	4.31
Maternal Involvement*	−.19	3.02	1.24	3.20
Encouraging Developmental Advancement	.55	4.19	.68	4.72
Restriction–Punishment	.38	2.03	−.29	1.85
Year 3				
Toys	−1.33	4.88	−.57	4.78
Encouraging Developmental Advancement	.90	3.23	.08	2.56
Responsivity of Mother*	−.35	2.93	1.20	2.63
Year 4				
Toys	−1.30	8.04	.50	8.01
Responsivity of Mother	.19	4.30	.76	4.54

Note: Multivariate test by sex yielded a significant sex difference ($F_{13,101} = 2.85$, $p < .05$); $N = 68$ boys and 47 girls.
*Indicates a significant sex difference for univariate test ($p < .05$)

Table 14.6 Means and standard deviations for Caldwell HOME scales for nonadoptive families

HOME scores	Boys		Girls	
	Mean	SD	Mean	SD
Year 1				
Toys	2.25	5.47	.59	5.23
Maternal Involvement	−.72	3.03	−.16	3.20
Encouraging Developmental Advancement	.78	4.44	.57	4.28
Restriction–Punishment	−.54	1.72	−.45	1.75
Year 2				
Toys	1.36	4.89	1.19	5.43
Maternal Involvement	−.61	3.08	.02	2.75
Encouraging Developmental Advancement	.58	4.50	−.16	3.75
Restriction–Punishment	−.48	1.58	−.22	1.84
Year 3				
Toys	−.31	4.22	−.37	4.79
Encouraging Developmental Advancement	.29	3.22	−.11	3.20
Responsivity of Mother	−.46	2.88	−.20	2.63
Year 4				
Toys	−1.03	7.21	.17	6.49
Responsivity of Mother	−.48	4.06	.23	4.52

Note: Multivariate test by sex yielded a non-significant mean difference between sexes ($F_{13,126} = .78$, $p < .05$); $N = 77$ boys and 63 girls.

In summary, for cognitive abilities in infancy and early childhood, no sex differences were found in global ability measures. However, when examining more specific tests of ability, sex differences do emerge, albeit only consistently across adopted and control children for Verbal–Symbolic ability at year 2. It should be noted that the finding of mean differences between males and females is not a necessary condition to finding differential gender responses to genetic or environmental influences. Measures with no mean sex differences may be differentially heritable in boys and girls, or vice versa. Similarly, environment–behavior relationships may differ as a function of gender even when mean differences are absent.

Model-fitting analyses of parent–child resemblance

Univariate model-fitting analyses were conducted for global cognitive ability at ages 1, 2, 3, 4, and 7 years, employing the full adoption model described by Fulker and DeFries (1983). For specific cognitive abilities at years 3, 4, and 7, multivariate models of genetic and environmental influence (Fulker, 1988) were fit separately for boys and girls. These models included the effects of additive genetic variance (h), environmental variance (e), maternal (m) and paternal (f) cultural transmission, genotype–environment correlations (s), and (in multivariate analyses) genetic (R_G) and environmental (R_E) correlations. Also in multivariate analyses, assortative mating and cross-assortative mating were modeled using the co-path method described by Carey (1986), such that the matrix of phenotypic marital correlations (M) is expressed as a function of a matrix of conditional paths (D) and the within-person phenotypic correlations among the traits (R_P):

$$M = R_P D R_{P'}$$

(1)

(see also Fulker, 1988).

Both univariate and multivariate model-fitting were performed using the pedigree-based approach described by Lange, Westlake, and Spence (1976) which uses all available data from each family. The estimation procedure also allows the simultaneous modeling of phenotypic means of the traits, as well as their variance–covariance structure. All model-fitting analyses were accomplished through Mx (Neale, 1991), a special-purpose optimization program designed for the genetical analysis of covariance structures.

Preliminary analyses: Genotype–environment correlation, selective placement, and assortative mating Several preliminary analyses were conducted prior to fitting models to parent–child resemblance. These were performed to simplify the models, particularly with respect to selective placement, assortative mating, and genotype–environment correlation. First, variance–covariance matrices among measures of children's cognitive abilities were compared for adopted and control girls and boys. Within each age of measurement, a full model was fitted which allowed estimates of means, variances, and intercorrelations among the four specific cognitive abilities separately for female and male adoptees and controls and their parents. This full model was then compared to a second, more restricted, model which constrained both correlations and variances to be equal across the four groups. These models were not significantly different at year 4 ($\chi^2 = 37.17$, $df = 30$, $p < .20$) or at year 7 ($\chi^2 = 29.57$, $df = 30$, $p > .50$), and were significantly different only marginally at year 3 ($\chi^2 = 43.05$, $df = 30$, $p < .10$). The absence of any difference in variances or inter-test correlations between adoptees and controls suggests that passive gene–environment correlations are negligible (Baker, 1986; Fulker, 1988; Fulker & DeFries, 1983).

The effects of selective placement and assortative mating, and especially how these might differ between families of boys and girls, were next evaluated. Previous analyses of these data have shown that selective placement for cognitive abilities (i.e., correlations between adoptive and biological parents) is negligible (Fulker & DeFries, 1983; Rice, Fulker, & DeFries, 1986), although these results were based on the combined sample of boys and girls. Since it is conceivable that differential methods of placement may have occured for male and female adoptees, these effects were examined separately for parents of girls and boys in our analyses.

Correlation matrices (16×16) among the four specific factors of cognitive abilities for biological and adoptive mothers and fathers, along with 16 variances, were estimated separately for parents of adopted boys ($N = 131$ families) and adopted girls ($N = 114$ families) using the Mx program. Constraining selective placement (i.e., correlations between adoptive and biological parents) to be zero led to a non-significant difference in fit compared to the unconstrained correlations both for parents of girls ($\chi^2 = 64.49$, $df = 64$, $p > .05$) and parents of boys ($\chi^2 = 61.77$, $df = 64$, $p > .05$). Given the negligible and non-significant selective placement correlations for specific cognitive abilities in these data, model-fitting was performed without inclusion of these parameters.

The magnitude and nature of assortative mating were explored further in another set of analyses, where correlations between mothers and fathers were compared across adoptive, biological, and control parents of boys and girls. First, multivariate models of assortative mating were fitted separately for parents of girls and parents of boys, and then compared to a model where assortative mating was constrained to be equal across parents of girls and parents of boys. Constraining the estimates of assortative mating to be equal for parents of girls and boys was not significantly different from the models allowing for separate estimates ($\chi^2(24) = 14,096.89 - (7521.41 + 6551.77) = 23.71$, $p > .05$). For parents of girls and parents of boys combined, several additional models of assortative mating were evaluated. A model constraining equal assortative mating effects across biological, adoptive, and control parents did not significantly increase the fit over the full model which allowed separate assortative mating across each type of parent ($\chi^2(32) = 34.66$, $p > .05$). Thus, differential assortative mating was not found among control, adoptive, and biological couples. Constraining the D matrix to be diagonal (i.e., eliminating the effects of cross-trait assortment among the four specific abilities) fit significantly worse than a model including the off-diagonal elements of D ($\chi^2(12) = 26.07$, $p < .05$). Thus, in subsequent model-fitting of parent–child resemblance the D matrix in assortative mating was constrained to be equal across biological, adoptive, and control parents, and was retained as a full matrix which included the effects of cross-trait assortment.

Global measures of cognitive ability Results of univariate analyses of first-principal-component scores for boys and girls at years 1, 2, 3, 4, and 7 are shown in Tables 14.7 and 14.8. Comparison of models fitted separately for boys and girls and those where biometrical effects were constrained to be equal across sex are provided in Table 14.7, where it can be seen that no significant differences were found between the sexes at any age. Parameter estimates from the full model for each sex are presented in Table 14.8. At all ages of testing, maternal and paternal cultural transmission (m and f, respectively) as well as genotype–environment correlation (s) parameters are negligible for both girls and boys. These effects could, in fact, be dropped from each model without significant loss of fit. Contrary to the hypothesis that boys are more influenced by environment than girls, the cultural transmission parameters are not greater in boys. With respect to estimates of genetic effects, the h parameter is similar between the two sexes at each age. The largest difference occurs in year 3, where heritability estimates (h^2) are .10 for girls and .20 for boys. Moreover, estimates of h are greater in absolute value in boys for all except year 2. Thus, heritable effects are not greater for girls than boys, at least for general cognitive abilities between years 1 and 7.

Table 14.7 Univariate analyses of first principal component for years 1, 2, 3, 4, and 7

Model	Year 1	Year 2	Year 3	Year 4	Year 7
1 Boys = Girls: − 2ln(L)	7,310.34	7,200.46	4,615.29	4,601.92	4,633.56
2 Boys ≠ Girls: − 2ln(L)	7,306.02	7,190.55	4,608.71	4,590.63	4,627.64
Difference[a]	4.32	9.91	6.58	11.29	5.92

[a] Difference between −2ln(L) from Models 1 and 2 is distributed approximately as chi-square variates with $df = 8$, and provides a test of equality of the full model across girls and boys.

Table 14.8 Maximum-likelihood parameter estimates for first principal component from (univariate) full models, separately for boys and girls

Parameter	Year 1		Year 2		Year 3		Year 4		Year 7	
	Boys	Girls	Boys	Girls	Boys	Girls	Boys	Girls	Boys	Girls
h	.36	.24	.34	.40	.45	.31	.49	.41	.60	.49
s	.02	.01	.03	.02	.03	.05	.01	.04	.05	.04
μ	.24	.32	.24	.32	.24	.33	.24	.32	.23	.32
f	.05	.03	.05	.03	.11	.13	.04	−.03	.17	.08
m	.03	.00	.10	.05	.00	.07	−.03	.18	−.05	.04

Specific factors of cognitive ability Log-likelihood values were compared for models fitted with and without equality constraints across sexes (see Table 14.9). (Note, however, that due to findings of mean sex differences described earlier, means were still allowed to vary across sex in the constrained models. All other parameters were equated across families of boys and girls, including those contained in h, m, f, s, R_G, R_C, as well as phenotypic variances and covariances, and parental means.) Contrary to univariate model-fitting analyses of global cognitive abilities, where no noteworthy sex differences were found, multivariate analyses of specific cognitive abilities did show some sex differences. At year 3 in particular, constraining the full model to be equivalent for boys and girls yielded a significant worsening of fit compared to a model with different effects across sex of offspring. Although the full model could be constrained to be equal across families of girls and boys at years 4 and 7, inspection of parameter estimates for the full model within each sex revealed enough differences that further comparisons seemed warranted.

Table 14.9 Multivariate analyses of specific cognitive abilities for years 3, 4, and 7

Model	Year 3	Year 4	Year 7
1 Boys = Girls[a]: −2ln(L)	17,875.84	17,960.39	18,111.65
2 Boys ≠ Girls: −2ln(L)	17,761.84	17,883.90	18,033.53
Difference[b]	114.00*	77.48	78.12

[a] All parameters in full model are constrained to be equal across sex, except means of boys' and girls' specific cognitive abilities.

[b] Difference between −2ln(L) from Models 1 and 2 is distributed approximately as chi-square with $df = 70$, and provides a test of equality of the full model across boys and girls.

*$p < .05$.

Given the low power available to detect sex differences in biometrical effects using these data, we deemed it necessary to look more closely at results for girls and boys, and to search for consistent patterns of differences. Fitting the full model within each sex yielded values of −2lnL as follows: at year 3, 9530.41 for boys and 9612.86 for girls; at year 4, 9612.86 for boys and 8271.04 for girls; and at year 7, 9670.58 for boys and 8362.95 for girls. Several reduced models were then fitted within each year and sex and compared to the full model in each case (see Table 14.10). For boys at year 3, all submodels gave a significant worsening of fit; thus the full model appears most parsimonious. For girls at year 3, equating maternal and paternal cultural transmission ($m = f$) resulted in a non-significant increase in chi-square. Dropping correlated genetic influences ($R_G = 0$) also resulted in a non-significant change in fit. However, all other constraints produced significant worsening in model fit.

For boys at year 4, while we could not drop cultural transmission and G − E correlation altogether ($m = f = s = 0$), the fit of the model did not significantly worsen when constraining these effects to be diagonal (i.e., specific to each measure). We could also equate maternal and paternal cultural transmission ($m = f$) without loss of fit. Dropping genetic and correlated genetic influences ($h = R_G = 0$) did not result in a significant reduction in fit. Additionally, one could drop either R_G or R_C (but not both) without causing a significant worsening in model fit. These results do not provide a clear indication of whether genetic or cultural transmission is relatively more important in explaining parent–child resemblance in boys at 4 years of age. For girls at year 4, we could drop cultural transmission and G–E correlations ($m = f = s = 0$) without significant worsening of model fit. Additionally, one could drop both R_G and R_C without a significant increase in chi-square.

Table 14.10 Model comparisons for multivariate analyses of cognitive abilities at years 3, 4 and 7

Constraint		Year 3		Year 4		Year 7	
		Boys	Girls	Boys	Girls	Boys	Girls
	df	χ^2	χ^2	χ^2	χ^2	χ^2	χ^2
s = 0	32	87.95*	52.46*	48.26*	37.76	44.44	50.23*
m, f, s diag.	24	74.35*	44.47*	33.99	32.02	33.13	42.93*
m = f	16	46.99*	17.79	20.53	—	—	19.37
R_G = h = s = 0	10	32.17*	22.96*	15.33	23.92*	29.04*	25.07*
R_G = 0	6	13.14*	3.29	8.78	8.11	10.91	4.25
R_C = 0	6	12.99*	22.00*	4.38	8.07	5.33	7.42
R_G = R_C = 0	12	—	—	21.56*	20.74	21.84*	18.48
orthogonal, full **D**	36	—	77.92*	63.96*	57.90*	57.93*	66.02*

Note: The difference between $-2\ln(L)$ for each constrained model and the corresponding full model for each sex and year are approximately distributed as chi-square variates with degrees of freedom noted.
*$p < .05$.

Results for boys at year 7 indicated that cultural transmission and G–E correlations could be dropped (m = f = s = 0) without a significant increase in chi-square. While either R_G or R_C could be dropped without loss of fit, dropping both resulted in a significant increase in chi-square. For girls at year 7, cultural transmission and G–E correlations (s) could not be dropped though maternal and paternal cultural transmission could be equated (m = f). However, we could drop both R_G and R_C without a significant loss of fit.

It is noteworthy that, for models which require non-zero cultural transmission effects (i.e., where setting m = f = s = 0 provides a significant chi-square change), equating maternal and paternal effects (m = f) does not provide any significant change in fit for girls or boys at any age, with the exception of boys in year 3. Thus, for the most part, there does not appear to be differential environmental influence from mothers and fathers to their children's specific cognitive abilities.

In spite of the non-significant differences between girls and boys for the full model as a whole at years 4 and 7, the model comparisons within each sex do not yield similar patterns of results for boys and girls. This prompted us to inspect parameter estimates for the full model within each sex at each age.

The most notable sex differences are apparent in matrices of genetic correlations (R_G) at each age (see Tables 14.11–14.13). The patterns of correlated genetic influences are markedly different between girls and boys, and do not show evidence in either sex for the single genetic factor which was suggested by previous multivariate analyses of a combined sample of girls and boys (see Rice et al., 1986).

Contrary to predictions, comparison of heritability estimates (h) across the two sexes does not reveal a pattern of greater genetic effects in girls. In fact, the only specific cognitive ability which shows a consistent direction of difference between the sexes across the three years is Memory, where heritability is greater for *boys* at years 3 ($h^2 = .35$), 4 ($h^2 = .07$), and 7 ($h^2 = .32$) than for girls ($h^2 = .04, .04,$ and .003

at years 3, 4, and 7, respectively). The sex differences in heritability of Memory are larger at years 3 and 7 than for any other specific cognitive ability during this age range. It is noteworthy, however, that the next largest differences in absolute value of h are for Spatial ability at years 4 and 7, where heritabilities do appear greater for girls ($h^2 = .55$ at year 4 and $h^2 = .61$ at year 7) compared to boys ($h^2 = .17$ at year 4 and $h^2 = .23$ at year 7). Otherwise, sex differences in heritable variation are unimpressive for Verbal ability and Perceptual Speed measures.

Table 14.11   Maximum-likelihood estimates of heritability (**h**), genetic correlations ($\mathbf{R_G}$), and environmental correlations ($\mathbf{R_C}$) from full models of specific cognitive abilities, for boys and girls at year 3

	Boys				Girls			
	Spatial	Verbal	Perceptual Speed	Memory	Spatial	Verbal	Perceptual Speed	Memory
Parameter matrix	(S)	(V)	(P)	(M)	(S)	(V)	(P)	(M)
h								
S	0.42	0	0	0	0.31	0	0	0
V	0	0.50	0	0	0	0.46	0	0
P	0	0	0.25	0	0	0	0.35	0
M	0	0	0	0.59	0	0	0	0.21
$\mathbf{R_G}$								
S	1.00				1.00			
V	.62	1.00			−.10	1.00		
P	1.00	−.12	1.00		.40	.76	1.00	
M	1.00	.69	−.09	1.00	−.32	.52	1.00	1.00
$\mathbf{R_C}$								
S	1.00				1.00			
V	.07	1.00			.18	1.00		
P	.17	.25	1.00		.19	.13	1.00	
M	.03	−.05	.20	1.00	.21	.24	.32	1.00

Table 14.12   Maximum-likelihood estimates of heritability (**h**), genetic correlations ($\mathbf{R_G}$), and environmental correlations ($\mathbf{R_G}$) from full models of specific cognitive abilities, for boys and girls at year 4

	Boys				Girls			
	Spatial	Verbal	Perceptual Speed	Memory	Spatial	Verbal	Perceptual Speed	Memory
Parameter matrix	(S)	(V)	(P)	(M)	(S)	(V)	(P)	(M)
h								
S	0.41	0	0	0	0.74	0	0	0
V	0	0.33	0	0	0	0.27	0	0
P	0	0	0.45	0	0	0	0.41	0
M	0	0	0	0.26	0	0	0	0.19
$\mathbf{R_G}$								
S	1.00				1.00			

Parameter matrix	Boys				Girls			
	Spatial	Verbal	Perceptual Speed	Memory	Spatial	Verbal	Perceptual Speed	Memory
	(S)	(V)	(P)	(M)	(S)	(V)	(P)	(M)
V	.32	1.00			−.12	1.00		
P	1.00	−.01	1.00		1.00	−.27	1.00	
M	1.00	1.00	−.86	1.00	−.10	.22	1.00	1.00
R_C								
S	1.00				1.00			
V	.10	1.00			.26	1.00		
P	.01	.14	1.00		−.03	.06	1.00	
M	.06	−.06	.10	1.00	.28	.07	.04	1.00

Table 14.13 Maximum-likelihood estimates of heritability (**h**), genetic correlations (**R**$_G$), and environmental correlations (**R**$_C$) from full models of specific cognitive abilities, for boys and girls at year 7

Parameter matrix	Boys				Girls			
	Spatial	Verbal	Perceptual Speed	Memory	Spatial	Verbal	Perceptual Speed	Memory
	(S)	(V)	(P)	(M)	(S)	(V)	(P)	(M)
h								
S	0.48	0	0	0	0.78	0	0	0
V	0	0.32	0	0	0	0.50	0	0
P	0	0	0.49	0	0	0	0.25	0
M	0	0	0	0.57	0	0	0	0.06
R$_G$								
S	1.00				1.00			
V	1.00	1.00			.26	1.00		
P	.99	.15	1.00		1.00	.88	1.00	
M	−.22	.75	.00	1.00	1.00	1.00	−.09	1.00
R$_C$								
S	1.00				1.00			
V	−.07	1.00			−.11	1.00		
P	−.05	.06	1.00		.18	.04	1.00	
M	.06	.03	.22	1.00	.17	.14	.09	1.00

Examination of cultural transmission effects for mothers (**m**) and for fathers (**f**), as well as genotype–environment correlations (**s**) (see Tables 14.14–14.16), does not reveal any clear pattern of sex differences for specific cognitive abilities at any age of testing. In spite of the fact that these effects appeared significant as a whole in all cases except girls at year 4 and boys at year 7, the magnitude of these parameters is unimpressive, suggesting only small or even negligible effects of cultural transmission and gene–environment correlation between 3 and 7 years of age. It is clearly not the case that environmental effects, as reflected in cultural transmission parameters, are greater for boys than for girls.

Table 14.14 Maximum-likelihood estimates of paternal (f) and maternal (m) cultural transmission and genotype–environment correlation (s) from full models of specific cognitive abilities, for boys and girls at year 3

Parameter matrix	Boys				Girls			
	Spatial	Verbal	Perceptual Speed	Memory	Spatial	Verbal	Perceptual Speed	Memory
	(S)	(V)	(P)	(M)	(S)	(V)	(P)	(M)
f								
S	.11	−.08	−.01	−.04	−.04	−.05	.09	.12
V	.04	.12	.03	−.34	−.04	.06	.09	.16
P	.04	.14	−.01	.06	−.08	−.09	.11	.01
M	.03	−.06	.01	−.20	.08	.00	.20	−.08
m								
S	−.25	.08	.01	−.06	.21	−.11	.00	.04
V	.06	−.03	.22	−.06	.04	−.07	−.04	.12
P	.08	−.09	−.04	−.26	−.01	−.06	−.02	.09
M	−.26	.03	.09	.05	.13	.05	−.02	.08
s								
S	−.04	−.28	−.52	−.51	−.01	−.06	−.38	−.37
V	−.03	−.08	−.38	−.11	−.07	−.04	−.40	−.40
P	−.01	.14	.05	−.30	−.06	−.08	−.05	−.88
M	−.05	−.08	−.04	−.09	.01	.06	.01	.05

Table 14.15 Maximum-likelihood estimates of paternal (f) and maternal (m) cultural transmission and genotype–environment correlation (s) from full models for boys and girls at year 4

Parameter matrix	Boys				Girls			
	Spatial	Verbal	Perceptual Speed	Memory	Spatial	Verbal	Perceptual Speed	Memory
	(S)	(V)	(P)	(M)	(S)	(V)	(P)	(M)
f								
S	.15	−.04	−.05	.01	−.15	.15	−.03	.15
V	.07	.11	.04	−.16	−.03	.05	.05	.04
P	−.02	−.01	.05	.00	−.17	.00	−.06	−.13
M	−.08	−.04	−.04	−.12	−.03	.01	.06	−.01
m								
S	−.12	−.01	.09	−.01	−.02	.13	.14	.10
V	.07	.12	.02	.02	−.03	.08	−.05	.08
P	−.08	−.10	−.07	.02	−.07	−.19	.06	.11
M	−.17	−.12	.19	.01	.03	.09	−.04	.17
s								
S	.01	−.16	−.44	−.25	−.10	−.18	−.25	−.21
V	−.01	.03	−.35	.08	.03	.01	.08	−.20
P	.02	.07	−.04	.16	−.03	−.01	−.10	−.11
M	−.01	.07	−.05	−.13	.08	.03	−.03	.04

Table 14.16 Maximum-likelihood estimates of paternal (**f**) and maternal (**m**) cultural transmission and genotype–environment correlation (**s**) from full models for boys and girls at year 7

Parameter matrix	Boys				Girls			
	Spatial	Verbal	Perceptual Speed	Memory	Spatial	Verbal	Perceptual Speed	Memory
	(S)	(V)	(P)	(M)	(S)	(V)	(P)	(M)
f								
S	.15	−.06	.07	.11	.02	−.04	.05	.13
V	.07	.10	.01	.00	−.09	.05	.16	−.09
P	−.14	.07	−.05	.01	−.15	−.09	.07	.08
M	.05	.00	.00	−.06	.02	.01	.12	.10
m								
S	−.12	.07	−.03	.19	−.08	−.11	.11	−.12
V	−.14	.10	.09	.04	−.05	−.06	.00	.15
P	−.11	−.13	.09	−.01	.03	−.19	−.03	−.15
M	−.02	−.16	.09	−.01	.11	.02	.05	.21
s								
S	.00	−.15	−.28	.02	−.05	.19	−.30	−.16
V	.00	.03	−.07	−.59	−.04	−.01	−.17	−.22
P	−.03	.01	−.06	−.37	−.05	−.04	−.12	−.05
M	.10	.06	−.01	−.05	−.07	−.06	−.14	.08

Although the lack of any clear sex difference in cultural transmission parameters fails to support the hypothesis that boys are more influenced by environment than girls, this does not unequivocally rule out the possibility of such a hypothesis. The cultural transmission effects in Fulker's (1988) model reflect primarily the extent to which factors related to these four specific abilities in the parents are correlated with abilities in the children. There may be other aspects of parents' behavior apart from their performance on the cognitive test battery which represent an important part of their children's environment, which in turn relates to children's intellectual outcomes. Thus, we next explored this possibility by examining correlations among children's abilities with aspects of the home environment as measured in the Caldwell HOME Inventory.

Environment–behavior relationships

Whether relationships between aspects of the early environment and children's intellectual outcomes are stronger for boys or for girls can only be appropriately addressed utilizing data from adoptive families where no genetic confound exists. Multiple regression analyses were conducted to predict cognitive developmental measures at each year of age. The predictors included in each set of regression analyses were the first principal component for quantitatively-scored HOME items assessed at 1, 2, 3, and 4 years of age. Only those HOME scores relevant to the age of the cognitive measure being predicted were included as predictors. For example, in predicting the intellectual performance at age 4, HOME scores for years 1, 2, 3 and 4 were included as predictors, whereas HOME scores for only years 1 and 2 were included when predicting intellectual performance at year 2.

Table 14.17 gives the standardized regression coefficients for each of the global HOME environmental variables when predicting global cognitive measures. Of the 10 regression analyses conducted, four HOME scores showed significant partial regression coefficients when predicting general cognitive performance – two for boys and two for girls. Additional regression analyses including interaction terms (see Ho, 1987) to test for significant sex differences between regression coefficients yielded only one significant difference, the prediction of cognitive performance at year 7 from HOME scores at year 1, in which the environment–cognitive performance regression is significantly higher for girls than for boys.

Table 14.17 Regression of global cognitive measures on Caldwell HOME scores

	Boys					Girls				
Global cognitive measures	Home predictors (at years 1–4)					Home predictors (at years 1–4)				
	E_1	E_2	E_3	E_4	R^2	E_1	E_2	E_3	E_4	R^2
Year 1	.14				.02	.01				.00
Year 2	.04	.29*			.10	−.12	.11			.01
Year 3	−.04	.23	.00		.04	−.05	.24	−.17		.06
Year 4	−.01	.18	−.05	−.10	.02	.25	−.02	−.39*	.29	.10
Year 7	−.01	.06	.18	.38*	.09	.59*	−.33	.02	−.01	.23

Note: Sample sizes for the 10 regressions ranged from 61–77 for boys and 49–69 for girls. Note that these analyses were conducted on only the adoptive families.
*$p < .05$

Examination of the proportions of variance in general cognitive ability (R^2) for girls and boys does not indicate any greater effect of environment on boys. In fact, for three of the five global ability measures, the absolute value of R^2 in girls exceeds that in boys.

Table 14.18 gives the standardized regression coefficients for each of the global HOME environmental variables when predicting specific cognitive measures for each age. Of the 38 regression analyses conducted, 14 HOME scores showed significant partial regression coefficients when predicting specific cognitive performance – 4 for boys and 10 for girls. Subsequent regression analyses that utilized interaction terms to test for significant sex differences yielded five pairs of regression coefficients that are significantly different for boys and girls. Three of the significant differences were HOME environmental measures predicting cognitive factors at year 7: HOME environmental measure at year 2 predicting Verbal performance; HOME assessed at year 1 predicting Spatial performance; and HOME assessed at year 4 predicting Perceptual Speed. The remaining two significant sex differences are predictive of Spatial performance at years 2 and 3 (for year 2, the significant predictor was HOME assessed at the same age; for year 3, the significant predictor was HOME assessed during year 1).

Table 14.18 Regression of specific cognitive measures on Caldwell HOME scores

| | Boys | | | | | Girls | | | | |
| | Home predictors (at years 1–4) | | | | | Home predictors (at years 1–4) | | | | |
	E_1	E_2	E_3	E_4	R^2	E_1	E_2	E_3	E_4	R^2
Year 1										
Means–End	−.06				.00	−.14				.02
Imitation	−.04				.00	.13				.02
Verbal Skill	.15				.02	.25*				.06
Year 2										
Lexical	−.08	.22			.03	−.27	.47*			.14
Spatial	.06	.37*			.17	−.11	−.11			.04
Verbal–										
Symbolic	−.03	.07			.00	−.07	.19			.02
Imitation	.13	−.04			.01	.16	.02			.03
Year 3										
Verbal	−.09	.34*	−.08		.07	.01	.00	−.23		.05
Spatial	.17	−.10	−.05		.02	−.42*	.33	.19		.11
Perceptual										
Speed	−.10	.21	.03		.03	.14	.11	−.24		.06
Memory	.13	.01	.00		.02	.03	.19	−.10		.04
Year 4										
Verbal	.02	−.08	.18	.29	.14	.17	.12	−.20	.32	.16
Spatial	−.13	.26	.06	−.18	.05	.04	.08	.10	−.09	.02
Perceptual										
Speed	.18	−.06	−.12	−.13	.05	.21	−.33*	−.46*	.33*	.18
Memory	−.13	.03	−.23	.16	.06	.15	.14	−.23	.06	.07
Year 7										
Verbal	−.27	.45*	−.03	−.16	.11	−.08	.38*	.25	−.01	.19
Spatial	.02	−.25	.04	.02	.05	.58*	−.44*	−.08	−.23	.28
Perceptual										
Speed	.04	−.03	.06	−.20	.03	.40*	−.33	.11	−.03	.15
Memory	−.09	.07	.24	−.44*	.12	.36	−.18	−.22	.25	.10

Note: Sample sizes for the 38 regressions ranged from 60–87 for boys and 49–69 for girls. Note that these analyses were conducted on only the adoptive families.
*$p < .05$.

As in the previous analyses of global measures of cognitive ability, proportions of variance in specific abilities explained by HOME scores are not greater for boys than for girls. For 15 of the 19 variables examined, absolute values of R^2 in girls exceeded that in boys. Thus, neither for global nor for specific cognitive abilities does it appear that environmental effects are greater in boys.

Conclusions

These data provide little or no support for the hypothesis that boys and girls are differentially affected by heredity and environment. For measures of global cognitive ability between 1 and 7 years of age, there are no apparent differences between girls and boys in mean performance, or in the relative effects of heredity and environment on variation within each sex. Model-fitting analyses of parent–child resemblance showed remarkably similar patterns of genetic and environmental influence within samples of girls and boys across ages 1, 2, 3, 4, and 7 years. Multiple regression analyses of environment–behavior relationships, furthermore, showed no evidence that boys were more susceptible to environmental influences.

While similar analyses of specific cognitive abilities did show some suggestions of sex differences, both in mean performance and in the relative effects of heredity and environment within each sex, the patterns of the few differences that appeared did not replicate across ages. Mean differences were robust across adopted and control samples for only a single measure, Verbal–Symbolic skill at 2 years of age. Cultural transmission, when present, does not appear to differ by sex of parent. To the extent that parental abilities relate to environments important to their children's intellectual outcomes, maternal and paternal influences appear to be similar.

In general, heritabilities of specific cognitive abilities did not significantly differ between the two sexes, although heritable variation for Memory was consistently greater in boys at 3, 4, and 7 years of age. There was also some suggestion of a sex difference in genetic correlations among the four specific cognitive abilities. This may indicate some sex-limited effects, whereby genetic factors important in one sex may not be isomorphic with those important in the other sex. Models of sex-limitation (see Eaves, Last, Young, & Martin, 1978) might provide one direction for future research in this area.

Future research should also focus on more specific abilities and tasks, rather than merely looking at global performance measures. However, large samples will be needed to detect sex differences in genetic and environmental variation, because such differences are likely to be small.

siblings differ for environmental reasons on a trait has been referred to by a range of terms: nonshared, within-family, specific, individual, unique, and E1 environmental influences (Plomin, 1986). In theory, nonshared environment is defined as the proportion of reliable variance not explained by genetic and shared environmental influences (Plomin, 1986; Plomin & Daniels, 1987). In practice, however, error variance is rarely excluded from estimates of nonshared environment.

Evidence for the importance of nonshared environmental influence has been reviewed (see Dunn & Plomin, 1990; Plomin, Chipuer, & Neiderhiser, 1993; Plomin & Daniels, 1987). Sibling analyses from the CAP support the conclusion that nearly all *environmental* variance is of the nonshared variety. For example, a direct test of the importance of shared environment comes from the correlation for adoptive siblings; adoptive sibling correlations in the CAP for parental reports of children's temperament were near zero in infancy and early childhood (Plomin, Coon, Carey, DeFries, & Fulker, 1991). Although genetic influences were significant, heritability estimates rarely accounted for as much as half of the variance of the temperament measures. Children's intelligence, on the other hand, has shown significant shared environmental influence (Cardon, Fulker, DeFries, & Plomin, 1992; Plomin, DeFries & Fulker, 1988), but research on adolescents has found little sibling similarity for intelligence (Loehlin, Horn, & Willerman, 1989; Scarr & Weinberg, 1977). This suggests that nonshared environmental influences may become more important as the CAP children grow older.

While behavioral genetic analyses have documented the importance of nonshared environment as an "anonymous" component of variance, such analyses do not identify specific aspects of siblings' experiences that make them so different despite growing up in the same family. It is important to move beyond the indirect evidence and investigate directly those environmental influences which siblings do not share, and to relate these experiences to outcomes (Rowe & Plomin, 1981). Several researchers have warned that nonshared experiences may be too idiosyncratic to be of help in understanding development (see responses to Plomin & Daniels, 1987). That is, events such as meeting the President of the United States or being in a car accident may be very important in a person's life, but not systematic enough to aid us in understanding individual differences in development across the population.

Studies of children's differential experiences, which we review in the next section, suggest that such pessimism may be unwarranted. During the last decade, researchers in the fields of child development and family studies have been examining the nature of children's different experiences within the family (Dunn & Plomin, 1990). The majority of these studies have focused on siblings' experiences with their mothers on the same measurement occasion. That is, mothers and children are interviewed and observed at the same time. When assessed in this manner, the siblings' nonshared environment reflect children's day-to-day experiences. It includes the extent to which siblings experience distinct environments, attend to different aspects of their surroundings, and perceive situations uniquely.

In this chapter, we approach siblings' nonshared environment from a different perspective. Instead of focusing on the differences in the siblings' experiences of family life on a day-to-day basis when they were different ages, we consider the possibility that differences in their experiences when they were the same age may be important in their development. Were the experiences of one child as a 7-year-old different from the experiences of her sibling as a 7-year-old? If so, do these differences relate to the children's outcome? If their experiences were similar – if, for example, there is consistency in parental behavior to the siblings when they are the same age – this would imply that such experiences do not contribute to the differences in their development. In contrast, the impact of witnessing and perceiving different experiences from those of your sibling may be important. Analyzing the differential experiences of siblings when they were the same age can, then, begin to clarify the significance of differences in these "direct" experiences.

To address this issue, the mothers were interviewed about their relations with each child on different occasions when the siblings were the same age. That is, each mother discussed her first child when he or she was 7 years old and was interviewed again when the second child became 7 years old. Comparisons of mothers' self-reports of their treatment of the siblings when the children were the same age indicates how the children were treated. They do not include the children's perceptions or observations of maternal treatment (although they of course reflect the mothers' perceptions of their own behavior rather than those of independent observers). In a way, this approach is similar to the between-family approach in which children are compared across families; however, we are comparing two children within a family.

Maternal Differential Treatment and Children's Outcomes

In this section, we review the literature documenting maternal treatment when the children are different ages and the same age, and consider the research which has examined links between such experiences and children's outcomes.

Siblings at different ages

Studies of young siblings have shown that from the beginning children are very sensitive to their mothers' attention toward their siblings. After the birth of a sibling, firstborns rarely ignore interactions between their mothers and siblings, but frequently respond to mother–sibling interactions with protests or demands for attention (Dunn & Kendrick, 1982). During their second year, secondborns are in turn very attentive to older sibling–mother interactions (Dunn & Munn, 1985). Koch (1960) interviewed children in early childhood about the parent and sibling relationship. She found that children – particularly firstborns – were very concerned about inequality in parental attention. Interviews with older children and adults reveal that differential treatment is still an issue for adolescents (Daniels, 1986; Daniels & Plomin, 1985) and adults (Baker & Daniels, 1990). Moreover, the discrepancies reported in maternal behavior are relatively large. In a study of the CAP children, 32% of the mothers reported large discrepancies in their affectionate behavior and 29% in their controlling behavior toward their two children, suggesting that differential treatment is also an important issue for parents (Dunn et al., 1991).

Siblings at the same age

Previously, CAP data have been reported concerning mothers' behavior toward two siblings at 1, 2, and 3 years of age (Dunn, Plomin, 1986). These studies found that mothers were relatively consistent in their behavior to their two children when each child was the same age. The mothers, however, did not behave consistently to the same child across ages – at least in the circumstances in which the data were collected. These results suggest that during infancy and the preschool period, the developmental stage of the children was a particularly powerful influence on mothers' behavior. To put it concretely, some mothers appear to be particularly interested and affectionate toward their children when they are babies; some are more interested in them as they become preschoolers; and so on. We now examine the question of whether children's family experiences during middle childhood are similar or different to those of their siblings. Do some children, for example, experience a different family environment as 7-year-olds than their siblings did as 7-year-olds? And are these differences in experience related to their behavioral outcomes when they reach 9 years old?

Links to outcomes

The demonstration that differences in parental treatment existed and that children were aware of such differences in their parent–child experiences was just the initial step in examining the role of such experiences as influences on development. Several related, but relatively unexplored, issues include the links between such differential experiences and children's temperament, sibling relationships, and individual outcomes. In addition, few studies of siblings' differential experiences have investigated children across early childhood into adolescence. It was this gap in the literature which led to the development of the Colorado Sibling Study (CSS), a longitudinal study of sibling relationships and nonshared environmental experiences and a subproject of the Colorado Adoption Project (Dunn, Stocker, & Plomin, 1991; Stocker, Dunn, & Plomin, 1989). In addition to the reasonably well-documented effects of maternal differential treatment on children's sibling relationships and temperament (Boer, 1990; Brody, Stoneman, & Burke, 1987; Bryant & Crockenberg, 1980; Hetherington, 1988; Stocker et al., 1989), nonshared environment has also been found to be linked to outcomes such as behavior problems, anxiety, and low self-esteem (Daniels, Dunn, Furstenberg, & Plomin, 1985; Dunn & Plomin, 1990; Dunn et al., 1991; McHale & Gamble, 1989; McHale & Pawletko, 1992). These studies found that when mothers were differentially attentive, affectionate, or controlling to their various children, the siblings were more likely to be hostile toward one another and the less favored child was more likely to have behavior problems and lower self-esteem. These results appeared in studies focusing on early childhood to adolescence which used diverse methods such as observations, data collected over the telephone, interviews, and questionnaires.

The studies linking outcomes to differential treatment examined differences in siblings' experiences with their mothers when the two children were at different ages and developmental stages. The research has shown that at a given point in time mothers do treat their children differently and that such nonshared experiences are linked to children's outcomes and relationships with others. Since the siblings were different ages, it is perhaps not surprising to find that their interactions with their mothers were different. As Sandra Scarr puts it, "Can you imagine speaking to a 1-year-old as you would to a 2-year-old, even if you were the most insensitive person in the world?" (Scarr, 1987). However, we should also consider the possibility that parents' behavior differed toward their children when the siblings were the same age (i.e., at different measurement occasions). Did each child have similar experiences of love, attention, and control from their parents during childhood? If so, do such differences contribute to individual differences in outcome?

Method

Sample

For this study we examined the CAP children when they were 7, 9, and 10 years old. We used measures which had been completed by the mothers and both siblings across the three time points. Since the CAP is an ongoing longitudinal study, the sample sizes decreased across time, and therefore, results presented here must be interpreted with caution and will require replication. There were 100 (49 adoptive and 51 nonadoptive) pairs at 7 years, 77 (32 adoptive and 45 nonadoptive) pairs at 9 years, and 54 (22 adoptive and 32 nonadoptive) pairs at 10 years.

Measures

At all three time points, mothers completed Dibble and Cohen's (1974) Parent Report in which they rated their behavior toward the child using a 7-point scale (0 = never to 6 = always). The 48-item measure included eight positive and negative subscales which were factor analyzed. The results revealed three factors: Acceptance or Warmth (Acceptance of Child as a person, Child Centeredness, Sensitivity to Feelings, Positive Involvement, and Shared Decision Making), Inconsistency (Inconsistent and Lax Enforcement of Discipline, and Detachment), and Negative Control (Control through Guilt, Control through Hostility, Control through Anxiety, and Withdrawal of Relationship). Only the Acceptance and Negative Control subscales were used in our analyses because they most closely parallel other measures found in the literature.

Three sets of outcome variables were examined: psychopathology, competence and self-worth, and intelligence. Children's internalizing and externalizing problems were assessed by mothers and teachers at 7, 9, and 10 using the Child Behavior Checklist (CBCL) (Achenbach 1991a, b). Social competence (Confidence, Leadership, Popularity, and Problem Behavior) was assessed by teachers at all three time points using the CAP Social Competence Scale (CSCS) (see Chapter 10 for details). In addition, the children reported their feelings of behavioral, scholastic and social competence, and self-worth using Harter's (1982) Self-Perception Profile (SPP) at ages 9 and 10. Each child's IQ was assessed using the Wechsler Intelligence Scale for Children–Revised (WISC-R) at age 7 (see Chapter 4).

We also investigated individual differences in temperament as potential correlates of sibling environmental differences. Mothers and teachers rated each child's level of emotionality, activity, and sociability using the Colorado Childhood Temperament Inventory (CCTI) at ages 7, 9, and 10 (see Chapter 9).

Results

In this section that follows, differences in maternal treatment of siblings are related to individual differences in the outcome and temperament variables across time. For a more detailed explanation of the advantages and disadvantages of analyses using difference scores, see Rovine (1993).

Maternal behavior

As with the results from the study of mother–child interaction at 1, 2, and 3 years, the sibling correlations for parental Acceptance and Negative Control were moderate to high across the three time points (see Table 15.1). At each of the ages, mothers' reports of their behavior to their two children were similar. The magnitude of the correlations was greater for nonadoptive pairs than adoptive pairs, suggesting genetic influence on the Dibble and Cohen parenting measure. These analyses are presented and discussed in greater detail in Chapter 17; however, the results remind us that siblings may differ for genetic as well as environmental reasons. In addition, correlations across time showed considerable stability in the degree of maternal warmth (*r*s ranged from .65 to .70) and control (*r*s ranged from .49 to .68) directed toward the two children.

We created difference scores by subtracting the scores for mothers' reports of their behavior toward the younger siblings from the scores for mothers' reports of their behavior toward the older siblings; these differences, even though small in scale, might be significantly related to differences in the children's outcomes. Correlations among the difference scores showed little stability across time for the two scales. As with the previous analyses of sibling data (Dunn et al., 1986; Dunn et al., 1985; Dunn & Plomin, 1986), this result suggests that differences in maternal behavior are not stable and may be influenced by the age or developmental stage of the children. Some mothers are consistent toward their two children as 7-year-olds, but are not toward those same siblings at 11 years of age.

Table 15.1 Sibling correlations and sample sizes for mother-reported maternal parenting measures at 7, 9, and 10 years.

Measure	Year		
	7	9	10
Maternal Acceptance	.61*	.70*	.58*
(N)	(100)	(79)	(53)
Maternal Negative Control	.71*	.60*	.32*
(N)	(99)	(79)	(54)
Maternal Inconsistency	.64*	.47*	.63*
(N)	(100)	(78)	(54)

$*p < .01.$

Links with outcomes

In general, the number of significant correlations between maternal differential treatment and mother-rated competence was not significantly above chance. We did find correlations between differences in maternal treatment and older and younger siblings' psychopathology (Tables 15.2 and 15.3). Children who received more maternal warmth than their sibling scored lower on externalizing and internalizing problems. There were a few significant associations between maternal negative control and both siblings' externalizing problems. Children receiving more discipline than their sibling scored higher on externalizing problems. Maternal behaviors were related to teachers' ratings of externalizing behavior for the older siblings (see Table 15.2) and to mothers' reports of internalizing and externalizing problems for younger siblings (see Table 15.3).

Table 15.2 Correlations between differences in mothers' reports of maternal Acceptance and Negative Control toward the two siblings and older siblings' scores on the Child Behavior Checklist at 7, 9, and 10 years.

Outcome	Maternal Acceptance			Maternal Negative Control		
	7	9	10	7	9	10
Year 7						
Internalizing (M)	.09	.25**	.22	.07	−.02	−.26*
Externalizing (M)	−.01	.13	.19	.11	.14	−.19
Internalizing (T)	−.08	−.15	.14	−.14	.13	−.16
Externalizing (T)	−.31**	−.37**	−.25*	−.02	.42**	.32**
Year 9						
Internalizing (M)	−.12	−.16	−.01	.06	.09	−.20
Externalizing (M)	−.01	.08	−.03	.05	.29**	−.05
Internalizing (T)	.08	−.30**	.07	.02	.01	.06
Externalizing (T)	−.08	−.43**	−.01	.04	.20*	.17

Table 15.2 (Cont.)

Outcome	Maternal Acceptance			Maternal Negative Control		
	7	9	10	7	9	10
Year 10						
Internalizing (M)	.05	.23	.21	.11	.10	−.12
Externalizing (M)	.07	.16	.13	.12	.08	−.01
Internalizing (T)	−.05	.01	.28*	.18	.21*	−.27*
Externalizing (T)	−.02	−.11	.14	.04	.16	−.12

Note: M = mother report; T = teacher report.
*$p < .10$, two-tailed, **$p < .05$, two-tailed.

We do not, then, have evidence from these data that differences in how mothers treat their children when they are the same age explain why children differ in competence. The low number of significant correlations, however, may be due to the lack of stability over time in the difference scores used to assess maternal differential treatment; such instability suggests that the scores may reflect mostly error. It should also be noted that these measures of maternal behavior are solely self-report assessments; clearly it would be preferable also to have other assessments of maternal behavior, such as observational measures or children's reports.

Table 15.3 Correlations between sibling differences in mother's reports of maternal Acceptance and Negative Control and younger siblings' scores on the Child Behavior Checklist at 7, 9, and 10 years.

Outcome	Maternal Acceptance			Maternal Negative Control		
	7	9	10	7	9	10
Year 7						
Internalizing (M)	.27**	.33**	.12	−.12	−.07	−.06
Externalizing (M)	.28**	.16	.27*	−.12	−.05	−.08
Internalizing (T)	−.22**	−.01	−.17	−.16	.14	.11
Externalizing (T)	.02	−.08	−.15	−.09	.11	.01
Year 9						
Internalizing (M)	.24*	.39**	.31**	.09	−.09	−.20
Externalizing (M)	.23*	.30**	.30**	.04	−.20*	−.27*
Internalizing (T)	−.15	−.02	.24*	.24*	.05	−.19
Externalizing (T)	−.07	.01	.12	.29**	.03	−.17
Year 10						
Internalizing (M)	.05	.30**	.22*	.08	.16	−.18
Externalizing (M)	.12	.32**	.32**	.09	.03	−.12
Internalizing (T)	.16	.10	−.10	−.16	.06	−.32**
Externalizing (T)	.47**	.29*	−.02	−.25	−.18	−.12

Note: M = mother report; T = teacher report.
*$p < .10$, two-tailed, **$p < .05$, two-tailed.

Links with temperament

Some of the older siblings' temperament characteristics were related to differences in maternal warmth and negative control (see Table 15.4). We also examined links between differences in maternal behavior and the younger siblings' temperament characteristics but found few significant correlations. For older siblings, higher emotionality and activity were related to less maternal warmth. Most significant correlations occured between temperament at 9 years and later differences in maternal treatment at 10 years. In general, maternal differential behavior was not related to later sibling temperament.

Table 15.4 Correlations between differences in mother's reports of maternal Acceptance and Negative Control toward the two siblings and older siblings' scores on the CCTI at 7, 9, and 10 Years

Temperament differences	Maternal Acceptance			Maternal Negative Control		
	7	9	10	7	9	10
Year 7						
Emotionality (M)	−.05	.01	−.14	.03	−.01	.05
Activity (M)	−.14	.03	.12	.02	.04	.03
Sociability (M)	−.02	−.08	.10	.01	.05	.08
Attention (M)	.06	.22*	.18	−.04	−.22*	.03
Emotionality (T)	−.01	−.10	.12	−.10	.16	.01
Activity (T)	−.08	−.13	−.24*	.10	−.01	.02
Sociability (T)	−.13	.02	−.13	.16	.16	.12
Shyness (T)	.16	−.07	.07	−.14	−.01	−.23
Year 9						
Emotionality (M)	−.01	−.05	−.18	−.10	.04	−.05
Activity (M)	−.08	.07	−.31**	−.07	−.08	.15
Sociability (M)	.01	−.06	.11	.11	−.01	.05
Attention (M)	.10	.21+	.28**	−.03	−.14	−.15
Emotionality (T)	−.24**	−.30**	.12	.06	−.17	−.03
Activity (T)	.06	−.07	−.34**	.02	−.09	−.07
Sociability (T)	.06	−.11	−.35**	.13	.02	−.03
Shyness (T)	.09	.05	.21	−.06	−.14	−.13
Year 10						
Emotionality (M)	−.05	.11	−.07	.18*	.05	−.16
Activity (M)	−.09	−.05	−.32**	−.21**	−.13	.26*
Sociability (M)	.08	.12	.20	.06	−.03	−.15
Attention (M)	.10	.10	.23*	−.14	−.16	−.30**
Emotionality (T)	−.03	−.08	.25*	.09	.14	−.04
Activity (T)	−.13	−.12	.02	.08	−.07	.11
Sociability (T)	−.14	−.08	−.06	−.01	.01	.14
Shyness (T)	−.08	.22**	.08	.22**	.01	.11

Note: M = mother report; T = teacher report.
*p < .10, two-tailed, **p < .05, two-tailed.

The relationships between maternal behavior and children's outcome and personality across time highlight the issue of direction of effects: we clearly cannot assume that maternal differential behavior is a cause and not a consequence of children's behavior problems or personality differences. Indeed, recent studies of children's conduct-disorders have suggested that children's behavior plays a significant role in both the parenting process and their own adjustment and social development (Anderson, Lytton, & Romney, 1986; Lytton, 1990; Patterson, 1986).

In order to address the issue of the direction of effects, the older siblings' temperament data were subjected to regression analyses. Differential maternal affection when the older sibling was 10 years old was not correlated with earlier temperament (as established from both mother and teacher ratings). There were, however, links between older siblings' temperament at 9 years and differences in maternal warmth at 10 years. We examined the association between temperament at 9 years and later maternal warmth after controlling for differences in maternal warmth at 9 years and temperament at 10 years to determine if the longitudinal associations were a function of the contemporaneous correlations between the measures at 9 and 10 years or the stability of mothers' reports of their own behavior. Since the sample sizes were small, we used only the temperament characteristics significantly correlated with differences in maternal warmth: mothers' reports of attention and activity and teachers' reports of activity and sociability (see Table 15.4).

Using hierarchical regression, maternal reports of temperament at 9 years predicted an additional 15% of the variance in differential maternal warmth at 10 years after controlling for previous maternal warmth (total $R^2 = .27$). In a separate analysis, teachers' reports of children's temperament predicted an additional 19% in differences in maternal warmth after controlling for mothers' behavior at 9 years (total $R^2 = .31$). In addition, maternal behavior at 9 years was not correlated with the temperament variables at 9 years (see Table 15.4). These results suggest that the links between older siblings' temperament at 9 years and later differential maternal affection was not just a function of the relationship between these two variables at 9 years or the stability over time in maternal behavior.

In a second set of regression analyses, we examined whether or not temperament at 9 years predicted differences in maternal affection at 10 years after controlling for temperament at 10 years. We wanted to know whether or not the relationship between temperament at 9 years and later maternal behavior reflected the link between the two variables at 10 years. We found that mothers' reports of temperament at 9 years did not predict a significant amount of the variance (2%) in later differences in maternal behavior at 10 years after controlling for temperament at 10 years (total $R^2 = .15$). Teachers' reports of the older siblings' temperament at 9 years, however, did predict an additional 34% of the variance in differences in maternal warmth when the siblings were 10 years after controlling for earlier temperament (total $R^2 = .35$). It appears that differences in mothers' warmth toward the siblings were related to past but not present teacher ratings of the children's temperament and to both past and present maternal ratings of temperament. Perhaps the mothers' behavior toward the older siblings was affected by information they received from the teachers' concerning their children's behavior and they were not aware of the teachers' present views. We should stress, however, that the sample sizes in these analyses were small and that the links between earlier temperament and later maternal behavior were not replicated across all the measurement occasions.

Conclusions

In this chapter, we have considered the extent to which siblings experienced different environments during middle childhood when they were the same age. In general, the findings indicate that mothers treat their children rather similarly when they are at the same age. This is an interesting contrast to the evidence that at any one point in "real time," when children are different ages, their mothers behave quite differently toward them. These results echo those found for the toddler and preschool period (Dunn et al., 1985; Dunn et al., 1986; Dunn & Plomin, 1986). The results do not support arguments that ordinal position plays a significant role in children's nonshared experiences (Hoffman, 1991). In fact, strong birth-order effects have not been found in large studies of children's personality (Ernst & Angst, 1983).

Secondly, the results indicate that the small differences in maternal differential treatment when the children were the same age were not related to the children's later competence. However, there were correlations between differences in maternal warmth and both older and younger siblings' behavior problems and the older siblings' temperament characteristics. It is possible that temperament and adjustment measures assess behaviors that are more relevant for understanding family relationships, whereas the competence measures assess more school-related behaviors. That is, whether or not a child is very emotional, active, or rebellious may be more important as an influence on the mother's behavior than a child's popularity or confidence with peers.

There is, however, growing evidence that differential experiences in siblings' everyday lives (that is, at the same time when the children are different ages) are related to children's relationships and adjustment (Dunn & Plomin, 1990). It appears that this second type of differential experience, that which siblings encounter everyday, may be more important in relation to children's development. The willingness of mothers to report such disparity in their treatment of their children points to how common the situation is for parents (Dunn et al., 1991). In addition, it may be that children's perceptions of differential treatment, compared to their 'objective' experiences, are especially significant for children's adjustment. We know from other research that from very early in development children are extremely sensitive to such differences (Dunn, 1988; Koch, 1960).

In general, our results showed that two siblings have similar experiences with their mother at the same age. This finding suggests that such experiences do not explain why the siblings are so different and, thus, other factors must be contributing to individual differences in children's development. As we have noted, witnessing *differential* behavior to self and to other children may be more important than similar experiences of direct interaction with the parents. Seeing your mother's evident affection for your sibling may override the impact of the affection you in fact received at the same age. This is an idea directly at odds with the usual view of what matters in parenting; an idea that collides with current psychological theory, but is definitely worth pursuing (Dunn & Plomin, 1990).

There are two issues concerning the longitudinal nature of these experiences which should be noted. First, the connections over time remind us that a key issue for future research will be to clarify the question of the direction of effects in these links between individual differences and differential experiences. Parents may be reacting to temperamental and behavioral differences in their children which have developed over time as the children react to parental treatment. Second, longitudinal data may also help us to understand how children's developmental level is linked to their differential experiences within the family.

In this chapter, we have examined only one example of nonshared environment: maternal treatment. Many other dimensions remain to be explored – father–child relations, sibling relationship experiences, extra-familial experiences. These aspects of nonshared environment may be of greater importance for understanding children's development (Rowe & Plomin, 1981). Studies have shown that children in middle childhood report very different experiences of their relationships with each other (Boer, 1990; McGuire & McHale, 1993; Stocker & McHale, 1992), and that these relationship experiences are linked to children's adjustment (Daniels et al., 1985). A recent study of paternal differential behavior points to the importance of understanding the role of fathers here (Brody, Stoneman, & McCoy, 1992). In the Colorado Sibling Study, we found that siblings have very different relationships with their friends and teachers (see Dunn & McGuire, 1993). In addition, differences in children's extra-familial experiences were related to differences in the siblings' temperamental characteristics (McGuire, Dunn, & Plomin, 1991). It is likely that differences in extra-familial relations may become more important as the CAP children become older and spend more time outside the family (Dunn & Plomin, 1990).

Another important set of differential experiences are major life events and illnesses. A study of 40 sibling pairs showed that a number of events such as school problems, accidents, or illnesses were not directly experienced by both siblings at the same time, although some of these events did have an impact on the other sibling (Beardsall & Dunn, 1992). Moreover, 63% of the major life events that were apparently shared by siblings (e.g., death or illness of a family member) had a different impact on the two children. It may be that certain events are too idiosyncratic to be important in development, like visits to the doctor. Temporary illnesses such as winter colds may influence a child's life very little. On the other hand, chronic hay fever or asthma may have large effects by restricting a child's activities. A recent study by McHale and Pawletko (1992) showed that the links between nonshared experiences and children's well-being were different for children with disabled siblings, than for those with nondisabled siblings. Being treated better than one's sibling was associated with higher anxiety for those with disabled siblings; however, for those with nondisabled siblings, preferential treatment resulted in lower anxiety. This suggests that families with chronically ill or disabled children may be viewed as a different context for other dimensions of sibling differential experiences, and that illness and disabilities should be studied as forms of nonshared environment.

Understanding the nature of nonshared experiences over time will not be an easy task for researchers. Studies are already moving beyond the focus on maternal behavior that until recently has been the most frequently investigated source of nonshared environment (e.g., Brody et al., 1992). However, longitudinal studies such as the CAP are needed to unravel the direction and strength of these associations. Such studies may help us to better grasp the complex nature of siblings' different lives inside the family.

16 Sibling Relationships in Childhood and Adolescence

Clare Stocker
Judy Dunn

In this chapter we explore three aspects of sibling relationships: (1) stability and change in relationship characteristics as younger siblings make the transition from early childhood to middle childhood and as older siblings move from middle childhood to early adolescence; (2) the connections between sibling relationships and children's relationships with parents; and (3) genetic influence on children's sibling relationships. Because the Colorado Adoption Project is one of the first studies to investigate sibling relationships longitudinally from childhood to adolescence and is the only study to include both adoptive and nonadoptive siblings, it offers a unique opportunity to study these aspects of sibling relationships that have not previously been investigated.

Stability and Change in Sibling Relationships

The longitudinal design of the Colorado Adoption Project (CAP) allows us to examine stability and change in individual sibling relationships, as well as mean differences in relationships of siblings of varying ages. Previous research has shown that there are marked differences between pairs of siblings in characteristics of their relationships (Dunn, 1983). Some siblings have relationships characterized by high levels of cooperation and affection whereas other siblings are competitive and conflictual. In this chapter, we examine whether differences such as these are stable across a four-year interval.

To date, there has been little research on the stability or change of sibling relationship characteristics in middle childhood. There are several reasons to expect stability in sibling relationships. If parents' behavior to children is consistent over time, and parents' behavior is related to siblings' behavior, there is likely to be consistency in sibling relationship characteristics over time. Also, consistency in sibling relationships might be expected if sibling relationship qualities are influenced in part by children's stable temperamental characteristics. Some longitudinal studies have been completed in early childhood and suggest that individual differences in older siblings' aggression and friendliness to younger siblings persist across the transition from toddlerhood to early childhood. There appears to be less consistency in younger siblings' behavior to older siblings (Stillwell & Dunn, 1985).

There are also reasons to expect change in individual sibling relationships. Major life events such as parental divorce or moving may be associated with changes in sibling relationships. There is some evidence that divorce changes the characteristics of children's relationships with their siblings – although the pattern of change is not simple and varies as a function of children's gender and the quality of parents' relationships with their children (Hetherington, 1988). Structural characteristics such as gender and age spacing between two siblings may be related to individual differences in how sibling relationships change. Previous research on associations between sibling structure variables and sibling relationship quality have produced mixed findings; some studies have found influences of these variables (Buhrmester & Furman, 1990) while others have not (Dunn, 1988; Stocker, Dunn, & Plomin, 1989). Therefore it is unclear if sibling structure is associated with change in sibling relationships.

Children's temperamental characteristics may also be related to change in their sibling relationships. Previous research has shown that children with difficult temperaments experience more conflict with their siblings than other children do (Brody, Stoneman, & Burke, 1987; Stocker et al., 1989). It is possible that the sibling relationships of children with difficult temperaments become more negative as children enter adolescence and that the relationships of children with easy temperaments become more positive or remain the same across this transition. Other factors that could affect change in sibling relationships are characteristics of siblings' earlier relationships with parents and with each other. In families in which children have been treated inequitably by parents, sibling relationships may become more negative over time. If sibling relationships are unusually close or hostile in childhood, they may become more intimate or more conflictual across development than will more neutral sibling relationships.

A separate issue from the degree of stability or change in individual sibling relationships is whether there are mean differences in characteristics of sibling relationships at different developmental stages. Changes in children's social, cognitive, and emotional development across the transition from childhood to adolescence may be related to differences in their sibling relationships over time. During the transition from early childhood to middle childhood, younger siblings become more cognitively sophisticated, and thus may be more skilled at communicating, negotiating, and perhaps also manipulating their siblings. Younger siblings' increased social competence may diminish the power differences between siblings and result in more egalitarian sibling relationships. This developmental change may also reduce the amount of caretaking behavior older siblings direct toward younger siblings.

Previous findings on developmental changes in sibling relationships have been based on cross-sectional studies and have produced mixed results. Several studies have found that sibling relationships become more egalitarian as children develop (Buhrmester & Furman, 1990; Vandell, Minett, & Santrock, 1987). In one study, levels of both cooperation and conflict increased between ages 8 and 11 (Vandell et al., 1987). In another study, siblings spent less time with each other and reported lower levels of both affection and conflict in adolescence than in childhood (Buhrmester & Furman, 1990). Buhrmester (1992) argues that these changes in sibling relationships may parallel decreases in intimacy found in parent–child relationships during the transition to adolescence. The Colorado Adoption Project offers one of the first opportunities to study changes in sibling relationships using longitudinal data rather than cross-sectional data. The longitudinal design also allows us to examine whether correlates of individual differences in sibling relationships change across development. In the following sections we will discuss parent–child relationships and genetic similarity as correlates of individual differences in sibling relationships.

Links Between Sibling Relationships and Parent–Child Relationships

There are a number of theoretical reasons to expect that children's relationships with their parents should be linked to individual differences in their sibling relationships. If children model their parents' behavior, as a social learning perspective would suggest, it is likely that there will be consistencies between sibling relationships and parent–child relationships. Attachment theorists propose that, based on their early relationships with parents, children develop "internal working models" of relationships (Bowlby, 1969; Bretherton, 1985). These models are expected to influence the quality of relationships children develop with others, including, presumably, their siblings (Teti & Ablard, 1989). Connections between sibling relationships and parent–child relationships could also arise if children's stable temperamental characteristics lead them to behave similarly with, and elicit similar responses from, both siblings and parents. It is also possible that the connections between parent–child relationships and sibling relationships are compensatory in nature. If children do not receive support and affection from their parents, they may develop particularly intimate and dependent relationships with their siblings.

Several studies have shown that the quality of children's relationships with their parents is associated with the quality of their sibling relationships and that these connections appear to be congruous rather than compensatory. Most of this research has focused on mothers; less is known about links between father–child relationships and sibling relationships. Maternal affection has been associated with more positive sibling relationships (Bryant & Crockenberg, 1980; Hetherington, 1988;

Stocker et al., 1989) and maternal rejection and control with more negative sibling relationships (Brody et al., 1987; Patterson, 1986; Stocker et al., 1989). But the connections between parent–child relationships and sibling relationships are not always simple. Research on infants and preschool-aged siblings has shown that first born daughters who had particularly close relationships with their mothers before the birth of their second child behaved with relatively high levels of negative behavior toward their new siblings (Dunn & Kendrick, 1982). Research on young children has also found that the way mothers talked to their older children about the new infant was related to the quality of the relationship that later developed between the children. Siblings developed closer relationships in families in which mothers talked with the older sibling about the infant sibling's feelings (Dunn & Kendrick, 1982; Howe & Ross, 1990). Clearly there are a number of processes that may operate to link parent–child relationships and sibling relationships, and these processes may vary for different relationship dimensions and at different developmental stages (Dunn, 1988).

One of the most important findings to come out of recent research on sibling relationships is that in addition to the quality of each sibling's relationship with his or her parent, differences in parents' behavior to siblings are associated with variations in the quality of sibling relationships (Dunn & Plomin, 1990). Results from a number of studies form a consensus that in families in which parents are differentially affectionate or controlling to their children, siblings have more conflictual relationships than in families in which they are treated similarly (Brody et al., 1987; Bryant & Crockenberg, 1980; Hetherington, 1988; McHale & Pawletko, 1992; Stocker et al. 1989). Researchers have yet to determine the process that links differential parental treatment to increased sibling conflict. Children may observe inequities in their parents' behavior toward them and their siblings, feel jealous, and in turn behave aggressively to their brothers and sisters. It is also possible that in families in which one sibling is particularly hostile to the other, parents respond by being more punitive or less affectionate to that child than to his or her sibling.

The issue of whether connections between parent–child relationships and sibling relationships change over children's development has not been previously studied. As children enter middle childhood and adolescence their focus may shift to relationships outside the family and the connections between parent–child relationships and sibling relationships may weaken, while the links between sibling and peer relationships may become stronger. On the other hand, given the early importance of parent–child relationships for sibling relationships, the links between these relationships may persist across development. Changes in parents' own development also could be related to changes in children's family relationships, including their sibling relationships.

Genetic Influence on Sibling Relationships

In addition to children's developmental levels and their relationships with parents, the degree to which siblings are genetically related may be linked to characteristics of their relationships. If children's behavior to their siblings is influenced in part by traits that are heritable, nonadoptive siblings, who share approximately 50% of their heritable material, should behave more similarly to each other than adoptive siblings, who are genetically unrelated. Additionally, if genetically influenced characteristics elicit particular behaviors from parents, nonadoptive parents would be likely to treat their children more similarly than adoptive parents. Finally, some researchers working in the sociobiological tradition suggest that family members who are genetically related will be more altruistic and less hostile toward each other than family members who are not genetically related (see Smith, 1987). If this hypothesis is true, there should be group differences in sibling warmth and hostility for adoptive and nonadoptive siblings. To investigate this, in addition to comparing the correlations of adoptive and nonadoptive sibling and maternal behavior, we examined mean differences in sibling behavior in adoptive and nonadoptive families.

We have already investigated genetic influence on siblings' and parents' behavior in early childhood (Rende, Slomkowski, Stocker, Fulker, & Plomin, 1992). In this chapter, we extend these analyses to examine genetic influence when older siblings have entered adolescence and younger siblings have reached middle childhood.

To reiterate and amplify the three aspects of sibling relationships on which we are focusing. First, we consider the stability and change in characteristics of siblings' relationships. We examine the impact of sibling structure variables, children's temperamental characteristics, earlier parent–child relationships and sibling relationships, and major life events on changes in sibling relationships. We also discuss *mean differences* in sibling relationship characteristics across the transitions from early childhood to middle childhood for younger siblings and from middle childhood to early adolescence for older siblings. Second, we explore links between sibling relationships and parent–child relationships and pay particular attention to differences in parents' relationships with siblings. We investigate whether these connections among family relationships change across development. Finally, we consider possible genetic influence on children's behavior to their siblings and on parents' behavior to siblings.

Method

Sample

The sample includes 118 families participating in the Colorado Sibling Study (CSS), a subproject of the Colorado Adoption Project. Families were visited in their homes on two occasions. At the first time of measurement (Time 1) older siblings were $7\frac{1}{2}$ years old and younger siblings were $4\frac{1}{2}$ years old on average. At the second data collection point (Time 2) older siblings were 11 years old and younger siblings were 8 years old on average. There were 28 brother–brother sibling pairs, 23 sister–sister pairs, 30 older brother–younger sister pairs, and 37 older sister–younger brother pairs. Seventy-three of the sibling pairs were from two-child families, 35 were from three child families, and 10 were from families with four or more children. Fifty-two of the sibling pairs were in adoptive families and genetically unrelated to each other; 66 of the sibling pairs were in nonadoptive families and were genetically related. It should be noted that in the analyses that follow, sample sizes vary because data were not always available on the full sample.

Procedure

One of the strengths of the Colorado Sibling Study is that we were able to collect multiple measures of each relationship, including both observational data and mothers' and children's interview responses. At each home visit, mothers and siblings were interviewed about their relationships. During the first home visit, siblings and mothers were videotaped while they participated in structured activities for 30 minutes, and siblings were observed while they interacted for 30 minutes in an unstructured setting. Children participating in the CSS also completed regular yearly CAP assessments. In this chapter we use a measure of life events that older siblings and mothers completed during the summer after older siblings had finished first grade (in most cases the CAP assessment occurred within a year of the CSS assessment).

Measures

Sibling relationships At Time 1 and Time 2, mothers were interviewed about their children's sibling relationships, using the Maternal Interview of the Sibling Relationship (Stocker et al., 1989). Mothers were asked 17 open-ended questions about the frequency of particular behaviors their children directed toward each other. Interviewers rated mothers' responses on 6-point Likert scales ranging from 0 = almost never/rarely to 6 = regularly/just about every day. The interview yielded two scales: Warmth and Hostility/Rivalry. The internal consistency alphas of the Warmth and Hostility scales were .44 and .73 respectively for older siblings and .52 and .66 for younger siblings at Time 1, and .59 and .64 respectively for older siblings and .64 and .71 for younger siblings at Time 2. For some analyses, older siblings' and younger siblings' scores were summed to create dyadic Warmth and Hostility/Rivalry scores.

During the Time 1 home visit, siblings were videotaped while they played six games together, with their mothers present. The games were designed to elicit particular behaviors such as cooperation and competition (see Stocker et al., 1989 for a full description). Each sibling's behavior to the other was rated on four 5-point Likert scales: Conflict, Cooperation, Control, and Competition. Dyadic scores were created by summing older siblings' and younger siblings' scores on each of the four scales. During the Time 1 home visit, siblings were also observed while they interacted in a 30-minute unstructured setting. Children's behavior was coded by the observer during the session and their conversation was audiotaped and coded later. Coded behaviors and conversational turns were summed to create a positive scale and a negative scale for each sibling. Siblings' scores on each scale were summed to create dyadic positive and dyadic negative scores.

During the Time 2 visit, siblings were interviewed about the sibling relationship using semi-structured interviews. The interviews were transcribed and coded for Warmth and Hostility on 5-point Likert scales ranging from low to high. To establish interrater reliability, two coders independently rated 40 sibling interviews. Finn's R indicated high interrater reliabilities for the two dimensions ($r = .79$ for Warmth and $r = .85$ for Hostility).

To recapitulate, measures of sibling relationships included at Time 1: maternal reports of sibling warmth and hostility, measures of sibling cooperation, conflict, control, and competition from structured observations, and measures of positive and negative sibling behavior from the unstructured observations. At Time 2, measures included both mothers' and children's reports of sibling warmth and hostility.

Change in sibling relationships   Changes in children's behavior to their siblings were measured by subtracting children's scores on Warmth and Hostility to their siblings at Time 2 (based on mothers' reports) from their scores on Warmth and Hostility at Time 1. Thus, positive scores indicate that children became less warm or hostile over time and negative scores indicate that they became more warm or hostile.

Parental behavior direct to each sibling   At Time 1, mothers' behavior to each sibling during the videotaped observations was rated on four 5-point scales: Control, Affection, Attention, and Responsiveness (see Stocker et al., 1989, for a description of the measures). Mothers' direct behavior to each sibling was not assessed at Time 2.

Differential parental behavior   Measures of differences in mothers behavior toward the two siblings were collected from the videotaped observations and from interviews with mothers. At Time 1, differences in mothers' Control, Affection, Attention, and Responsiveness to the two siblings during the videotaped observation were measured by subtracting mothers' scores for younger siblings from their scores for older siblings. Positive scores indicate that mothers directed more of a behavior to older siblings than to younger siblings and negative scores mean that mothers directed more of a behavior to younger siblings than to older siblings.

At Time 1 and Time 2, mothers were interviewed about differences in their behavior to the two siblings. The interviews yielded scores for differential affection and differential control. Scores ranged from 1 to 5, with 1 indicating that the mother directed the behavior to the younger sibling much more often than to the older sibling, 2 = a bit more often to the younger sibling, 3 = the same amount to both children, 4 = a bit more often to the older sibling, and 5 = a lot more often to the older sibling.

During the Time 2 visit, older siblings and younger siblings were interviewed about differences in their mothers' and fathers' behavior toward them and their sibling. In this chapter we use children's reports of their satisfaction with mothers' and fathers' differential treatment. Children's reports of satisfaction were rated on 5-point Likert scales ranging from very unsatisfied to very satisfied.

To recapitulate, measures of parental behavior included at Time 1: assessments of mothers' control, affection, attention, and responsiveness to each sibling during the videotaped observations, differences in maternal control, affection, attention, and responsiveness to siblings, and mothers' interview reports about differences in their control and affection to siblings. Measures of parental behavior at Time 2 included: mothers' interview responses about differential control and affection, and children's interview reports about satisfaction with differences in their mothers' and fathers' behavior to them and their siblings.

Temperament During the Time 1 maternal interview, mothers reported on children's temperamental characteristics using an interview based on Buss & Plomin's 1975 temperament dimensions. Mothers' responses to questions about each sibling's Activity, Fear, Anger, Sociability, Shyness, and Emotionality were rated by the interviewer using 5-point, low–high Likert scales.

Life events Mothers and older siblings completed the Social Readjustment Rating Scale (Coddington, 1972) as part of their annual CAP assessment during the summer after older siblings had finished first grade. Mothers indicated which of 33 stressful events occurred in the child's life during the past year. Items included events such as death of a significant other, hospitalization or health problems, school and family problems, change in the family's financial status, and the birth of a sibling. Mothers and children then used 0–3 low–high scales to rate the degree of upset each event caused the child. Total scores for the number of events that occurred and the degree of upset were created from mothers' reports. A total score for degree of upset was created from older siblings' responses.

Results and Discussion

Stability and change in sibling relationships

Our first set of analyses examines the stability and change in individual differences in sibling relationships. We then examine mean differences in siblings' behavior to each other across the four year transition from Time 1 to Time 2. In this section we consider mothers' reports about sibling relationships because these measures were the same at the two time periods.

Individual differences Correlational analyses revealed that there was some stability in children's behavior to their siblings across the four-year period. Correlations for mothers' reports of children's warmth to their siblings were $r = .51$ for older siblings and $r = .41$ for younger siblings (both significant at $p < .05$). Mothers' reports of children's hostility to their siblings were also significantly associated across the two time periods: $r = .33$ for older siblings and $r = .37$ for younger siblings. Children's hostile behavior was slightly less stable than their warmth, although the difference between these correlations was not significant.

Table 16.1 Variance in sibling relationship measures at Time 2, explained by Time 1 measures

Sibling measure Time 2	Sibling measure Time 1	R^2	F	Variance unique to other Time 1 measure	R^2	F
Older siblings						
Warmth	OS Warmth	$R^2 = .26^*$	$F_{1,80} = 28.13$	YS Activity	$R^2 = .05^*$	$F_{1,78} = 6.12$
				YS Fear	$R^2 = .05^*$	$F_{1,78} = 5.09$
				YS Hostility	$R^2 = .06^*$	$F_{1,77} = 7.09$
Hostility	OS Hostility	$R^2 = .11^*$	$F_{1,80} = 9.90^*$	OS Activity	$R^2 = .07^*$	$F_{1,78} = 6.78$
				Differential maternal control	$R^2 = .07^*$	$F_{1,76} = 6.68$
				YS Hostility	$R^2 = .05^*$	$F_{1,77} = 4.45$
Younger siblings						
Warmth	YS Warmth	$R^2 = .17^*$	$F_{1,80} = 16.03$	OS Anger	$R^2 = .05^*$	$F_{1,78} = 5.09$
				OS Fear	$R^2 = .05^*$	$F_{1,77} = 5.13$
				YS Fear	$R^2 = .05^*$	$F_{1,78} = 5.45$
				OS Warmth	$R^2 = .05^*$	$F_{1,78} = 5.10$
				YS Hostility	$R^2 = .06^*$	$F_{1,77} = 6.00$
Hostility	YS Hostility	$R^2 = .14^*$	$F_{1,79} = 12.79$	OS Emotionality	$R^2 = .05^*$	$F_{1,77} = 4.48$
				Differential maternal affection	$R^2 = .06^*$	$F_{1,72} = 5.47$

Note: Only Time 1 variables that accounted for significant variance in Time 2 sibling measures are included in the table.
$*p < .05$ OS = older sibling; YS = younger sibling.

These correlations are very high given the developmental changes children typically experience across this four-year interval. There are several factors that could be related to the stability of children's behavior to their siblings. Continuities in parents' relationships with each sibling could contribute to the stability of siblings' behavior across the two time periods. Children's stable temperamental characteristics could be partially responsible for these high correlations if they influence children to behave similarly to their siblings over time. Additionally, siblings may develop patterns of interaction that are stable despite the developmental changes each child experiences across the four-year interval. Finally, as these data are based on maternal reports at Time 1 and Time 2, it is possible that the significant correlations may be due in part to consistencies in mothers' perceptions across time. Although the correlations between siblings' behavior at the two time periods are significant, siblings' behavior at Time 1 explained only between 11% and 26% of the variance in siblings' behavior at Time 2. Thus these correlations can be interpreted to suggest that there is change as well as stability in children's behavior to their siblings across the four-year interval.

To investigate change in sibling relationships we adopted a two-stage data analysis strategy to explore two models of change. First, we examined associations between measures of family structure (children's gender, age, adoptive status, the gender composition of the sibling dyad, and the age difference between siblings), children's temperament, maternal behavior, and sibling behavior collected at Time 1 and measures of sibling relationships collected at Time 2. Because connections between measures collected at Time 1 and sibling behavior at Time 2 could be due to stability of sibling behavior over time, we conducted a series of hierarchical multiple regression analyses to examine how much variance in sibling relationship measures at Time 2 was explained by these variables after measures of sibling behavior at Time 1 had been entered into the equations.

In the second stage of data analysis, we correlated Time 1 variables that explained variance in sibling relationship measures at Time 2 with measures of change in sibling relationships. Positive change scores mean children become less warm or less hostile between Time 1 and Time 2 and negative scores indicate that siblings became more warm or more hostile across this transition. This two-stage procedure enabled us to examine two models of change; first, how much of the variance in Time 2 sibling measures was explained by particular Time 1 measures, and second, whether the Time 1 measure was associated with positive or negative changes in sibling behavior.

Table 16.1 describes the results of the first type of analysis. The first row shows the results of predicting older siblings' warmth at Time 2 from older siblings' warmth at Time 1 and from structure, temperament, maternal, and younger sibling behavior measures at Time 1. The last column of Table 16.1 indicates that three of these variables accounted for significant variance in older siblings' warmth at Time 2 after older siblings' warmth at Time 1 was taken into account. All of these predictors involved younger sibling characteristics at Time 1: activity, fear, and hostility.

Variance in older siblings' hostility at Time 2 was explained by measures of younger siblings' hostility, differences in maternal control, and older siblings' activity at Time 1. For younger siblings' warmth at Time 2, unique variance was explained by Time 1 measures of older siblings' anger and fear and younger siblings' fear. Younger siblings' hostility at Time 1 and older siblings' warmth at Time 1 also accounted for unique variance in younger siblings' warmth at Time 2 (see Table 16.1). Earlier measures of differences in maternal behavior and family structure variables were not associated with younger siblings' warmth to older siblings at Time 2. Results in Table 16.1 indicate that younger siblings' hostility at Time 2 was associated with older siblings' emotionality and differences in maternal affection at Time 1.

Table 16.2 presents data from the second type of analysis. Correlations between change in sibling relationships and Time 1 measures of sibling temperament, sibling behavior, and maternal behavior are included in the table. The first column in Table 16.2 presents correlations between changes in older siblings' warmth and Time 1 measures. Decreases in older siblings' warmth over time were associated with younger siblings' fear and activity and younger siblings' hostility at Time 1. The second column in Table 16.2 indicates that decreases in older siblings' hostility over time were associated with younger siblings' hostility and older siblings' activity at Time 1. Differences in maternal control were associated with increases in older siblings' hostility over time. Older siblings who received more maternal control than younger siblings at Time 1 became more hostile toward their siblings by the second time of measurement. Correlations in the third column reveal that Time 1 measures of temperament and older siblings' warmth and younger siblings' hostility were associated with decreases in younger siblings' warmth over time. The final column of Table 16.2 presents correlations between Time 1 measures and changes in younger siblings' hostility. Older siblings' emotionality was associated with decreases in younger siblings' hostile behavior over time.

Table 16.2　Correlations between sibling change scores and Time 1 measures

Time 1 measures	Sibling change scores			
	Older sibling		Younger sibling	
	Warmth	Hostility	Warmth	Hostility
Sibling relationship (N = 82)				
Older sib. Warmth			.40*	
Younger sib. Hostility	.35*	.43*	.31*	
Sibling temperament (N = 82)				
Older sib. Emotionality				.25*
Older sib. Anger			.26*	
Older sib. Fear			.20	
Younger sib. Fear	.26*		.27*	
Older sib. Activity		.13		
Younger sib. Activity	.33*			
Maternal differential behavior				
Affection (N = 77)				−.02
Control (N = 80)		−.19		

Notes: Positive change scores indicate that sibling warmth or hostility decreased between Time 1 and Time 2. Negative change scores mean that warmth or hostility increased between Time 1 and Time 2. Only Time 1 variables that accounted for significant variance in Time 2 sibling measures are included in the table.
*p < .05.

Taken together, these findings extend previously reported results indicating little effect of family-structure variables on the quality of sibling relationships (Brody et al., 1987; Stocker et al., 1989) by showing that family-structure variables were also unrelated to change in children's sibling relationships. There were a number of associations between earlier measures of children's temperamental characteristics and change in their sibling relationships. It is interesting to note that change in children's behavior to their siblings was more likely to be associated with siblings' temperamental characteristics at Time 1 than with their own earlier temperamental characteristics. The associations between temperament and changes in sibling relationships differed for older siblings and younger siblings. This is not surprising given that older and younger siblings have different roles in sibling relationships (Dunn, 1983). The same temperamental characteristic can have quite a different meaning for a 7-year-old younger sibling and for an 11-year-old older sibling. Future research in this area should explore whether this pattern of results is related to children's age or birth order.

There were some connections between differences in maternal treatment at Time 1 and changes in children's behavior to their siblings over time. Older siblings became more hostile to their siblings over time if their mothers controlled them more than their sibling at Time 1. Although differences in maternal affection at Time 1 accounted for significant variance in younger siblings' hostility at Time 2, these differences were not significantly correlated with the direction of change in younger siblings' hostility. If younger siblings received less affection than their siblings they were more hostile than other children. This connection appeared to remain stable and was not associated with children becoming more hostile over time.

The level of children's warmth and hostility toward their siblings at Time 1 was associated with changes in these behaviors over time. Younger siblings who had high scores on hostility at Time 1 became less warm and had siblings who became both less warm and less hostile between Time 1 and Time 2. These findings suggest that high levels of sibling hostility may be related to distancing or disengagement in sibling relationships over time.

In an earlier study using these data, McGuire and Dunn (1993) examined the associations between stressful life events at Time 1 and dyadic measures of sibling relationships at Time 2. Levels of sibling hostility at Time 2 were associated with the number of life events ($r = .28, p < .05$) and the degree of upset the life events caused, reported by mothers ($r = .41, p < .05$) and by children ($r = .33, p < .05$). Mothers' reports about the degree of upset were also negatively correlated with warmth in the sibling relationship at Time 2 ($r = -.28, p < .05$). Regression analyses indicated that major life events accounted for a significant portion of the variance in Time 2 sibling relationship measures after children's gender, adoptive status, age spacing between siblings, and measures of sibling relationships at Time 1 had been entered into the multiple regression equations. These analyses indicate that the longitudinal connections between earlier life events and sibling relationship characteristics were not simply due to life events being associated with the initial level of sibling relationships at Time 1. Although the life events measure was collected prior to the Time 2 sibling relationship measures, because these analyses are correlational we cannot be sure that the life events caused the later characteristics of children's behavior to their siblings. It is also possible that early sibling behavior influenced the life events: sibling hostility could lead to marital discord, for example, which could in turn influence children's behavior to their siblings.

In summary, longitudinal analyses showed that there was both stability and change in sibling relationships over time. Individual differences in changes in siblings' warmth and hostility were associated with earlier measures of children's temperamental characteristics, sibling behavior, maternal differential behavior, and stressful live events.

Mean differences We next examined whether there were mean differences in children's sibling relationships from Time 1 to Time 2. A repeated measures ANOVA indicated that older siblings were reported by mothers to be significantly less positive to their younger siblings at Time 2 than at Time 1 ($F_{1,81} = 40.45$, $p < .05$) (see Table 16.3 for means). They were also significantly less negative toward their younger siblings at the second time point than at Time 1 ($F_{1,81} = 15.76$, $p < .05$). Like older siblings, younger siblings became less positive and less negative to their siblings across the two time points. The differences in younger siblings' behavior, however, were not significant. These findings suggest that as they enter adolescence, children may become more indifferent toward their siblings and both the positive and negative behavior they direct toward them may decrease. When they reach adolescence, older siblings may place more emphasis on their peers and spend less time with their younger siblings. Younger siblings as a group, on the other hand, do not appear to change the warmth and hostility they direct to their older siblings as much when they move from early childhood to middle childhood. There were no significant group differences in younger siblings' behavior to their brothers and sisters at Time 1 and Time 2.

Table 16.3 Means of sibling relationship measures at Time 1 and Time 2

Sibling relationship measure: Mother report	Time 1 Mean	Time 2 Mean
Older sibling (*N* = 82)		
Warmth*	18.17	15.26
Hostility*	8.50	6.51
Younger sibling (*N* = 82)		
Warmth	16.89	15.86
Hostility	8.17	7.68

* Significantly different between Time 1 and Time 2, $p < .05$.

Links between sibling relationships and parent–child relationships

The next question we addressed was whether there were links between parent–child relationships and sibling relationships and if these associations changed from Time 1 to Time 2. At the first time of measurement, when on average older siblings were 7 years old and younger siblings were 4 years old, there were many connections between mothers' behavior to children and dyadic measures of children's sibling relationships (father–child relationships were not measured at this time point). Table 16.4 shows that mothers' controlling behavior and attention to children, assessed during the videotaped observations, was associated with more competitive and controlling behavior between siblings. These findings suggest that the connections between parent–child relationships and sibling relationships are congruous rather than compensatory. Siblings did not develop closer relationships to compensate for conflict in their relationships with mothers. It is unclear if mothers' controlling behavior caused the siblings to behave more negatively to each other or if mothers responded to siblings who were competitive and controlling by directing more negative behavior to them. Whatever the causal direction of these associations, they fit with results from other studies showing increased levels of sibling hostility in families with rejecting and controlling mothers (Brody et al., 1987; Patterson, 1986; Stocker & McHale, 1992).

Table 16.4   Correlations between sibling relationships and maternal behavior at Time 1

| Maternal behavior | Dyadic measures of sibling relationships | | | | |
| | Video observation | | | Maternal interview | |
	Competition	Cooperation	Control	Warmth	Hostility
Video observation (N = 96)					
Direct behavior to older sibling					
Control	.30*	−.04	.27*	.17	−.07
Attention	.18	.03	.06	.16	.08
Direct behavior to younger sibling					
Control	.39*	−.20	.30*	.06	−.01
Attention	.44*	−.17	.29*	.05	.09
Differential behavior					
Dif. control	.22*	−.21*	.12	−.16	.12
Dif. affection	.18	.10	.22*	−.02	.12
Dif. attention	.34*	−.20	.19	.14	.14
Dif. responsiveness	.24*	−.02	.34*	−.24*	−.11
Maternal interview (N = 118)					
Dif. affection	−.03	−.03	.00	.17	.10
Dif. control	.00	−.04	−.03	−.22*	−.07

Note: For the video and interview measures of differential treatment, high scores indicate that mothers directed more of the behavior to older siblings than to younger siblings.
*$p < .05$.

At both Time 1 and Time 2, there was more negative and less positive behavior between siblings in families in which mothers treated siblings differently on measures of control and affection than in families in which mothers treated siblings similarly (see Tables 16.4 and 16.5). At the second observation, children rated how satisfied they were with differences in their parents' behavior to them and their

Table 16.5 Correlations between sibling relationships and parental differential treatment at Time 2

	Sibling relationships							
	Maternal interview				Child interview			
	Older		Younger		Older		Younger	
Parental treatment	Warmth	Hostility	Warmth	Hostility	Warmth	Hostility	Warmth	Hostility
Maternal report (N = 118)								
Dif. control	-.07	.14	.07	-.26*	.01	.01	.03	.01
Dif. affection	.20*	.01	-.04	.00	-.01	.30*	-.00	.07
Older sibling report								
Satisfaction with maternal dif. (N = 118)	.05	-.10	-.01	-.16	.30*	-.26*	.16	.19
Satisfaction with paternal dif. (N = 117)	.21*	.02	.16	-.04	.06	-.05	.01	-.09
Younger sibling report								
Satisfaction with maternal dif. (N = 114)	.04	-.14	-.04	-.04	.20*	-.09	.21*	-.18
Satisfaction with paternal dif. (N = 110)	-.02	-.07	-.06	-.03	.16	-.06	.28*	-.25*

Note: High scores on mothers' reports indicate they directed more of behavior to older siblings than to younger siblings.
*$p < .05$.

siblings. Children who were more satisfied with how their mothers and fathers treated them compared to how they treated their siblings reported being more affectionate and less hostile to their siblings than other children (see Table 16.5).

The pattern of associations between sibling relationships and parent–child relationships was similar at Time 1 and at Time 2. At both time periods, differential parental behavior was associated with increased negativity in the sibling relationship (parents' direct behavior to each sibling was not assessed at Time 2). The importance of parents' differential behavior for sibling relationships appears to remain strong, rather than diminish, as children reach adolescence.

The process that links differences in parental behavior to increased levels of conflict in sibling relationships has not been established. Children may observe differences in their parents' behavior and respond to these inequities by directing negative behavior to their siblings. Alternatively, parents may respond to a child who is hostile to his or her sibling by controlling that child more than his brother or sister. It is also likely that these influences are bidirectional. These findings add to those already in the literature by showing that in early adolescence, as has been shown in childhood (Brody et al., 1987; Stocker et al., 1989), differences in parents' behavior to siblings were associated with increased levels of sibling hostility. These findings also show for the first time that children's satisfaction with fathers', as well as mothers', differential behavior was related to the quality of their sibling relationships.

Genetic influence on sibling relationships

In addition to children's developmental level, relationships with parents, temperament, and major life events, siblings' genetic similarity may be associated with characteristics of their relationships. If genetic factors influence children's behavior with their siblings, one would expect that genetically related siblings would treat each other more similarly than siblings who are not related genetically. The same logic holds for parental behavior.

We have examined this issue with data collected at the first time point of the CSS, when on average older siblings were 7 years old and younger siblings were 4 years old (Rende et al., 1992). Fisher Z tests were used to test the significance of differences between adoptive siblings' and nonadoptive siblings' correlations for sibling and parental behaviors. Table 16.6 shows that there was genetic influence on some behaviors children directed toward their siblings and on some maternal behaviors. Nonadoptive siblings were more similar to each other on maternal reports of warmth and hostility and on the positive measure from the unstructured observation than adoptive siblings. Mothers treated nonadoptive siblings more similarly on attention and control than they treated adoptive siblings.

At the second observation point, mothers' reports of sibling hostility showed significant genetic influence. Nonadoptive siblings were significantly more similar to each other than were adoptive siblings (Fisher $Z = 2.16, p < .05$). Unlike the results from Time 1, mothers' reports of sibling warmth were not genetically influenced at Time 2. Children's reports about warmth and hostility in the sibling relationship did not show genetic influence. Results for the Warmth dimension were in the direction of genetic influence but those for Hostility showed the opposite pattern (see Table 16.6).

Table 16.6   Adoptive siblings' and nonadoptive siblings' correlations for sibling behavior and maternal behavior at Time 1 and Time 2

Measure	Adoptive siblings	Nonadoptive siblings
Time 1	(N = 57)	(N = 67)
Maternal report of sibling behavior		
Warmth*	.62	.81
Hostility*	.65	.85
Videotaped observation of sibling behavior		
Cooperation	.81	.76
Competition	.46	.66
Conflict	.85	.91
Control	.28	.36
Videotaped observation of maternal behavior		
Control*	.65	.86
Affection	.91	.88
Attention*	.36	.71
Responsiveness	.94	.95
Unstructured observation of sibling behavior		
Positive*	.63	.85
Negative	.62	.79
Time 2	(N = 52)	(N = 66)
Maternal report of sibling behavior		
Warmth	.72	.77
Hostility*	.29	.61
Child report of sibling behavior	(N = 46)	(N = 63)
Warmth	.39	.51
Hostility	.49	.25

* Correlations significantly different for adoptive siblings and nonadoptive siblings, $p < .05$.

At the first time of data collection, there were no group differences between adoptive and nonadoptive families on any of the measures of maternal or sibling behavior. There were also no significant differences in the mean levels of adoptive and nonadoptive siblings' warmth or hostility at Time 2. Thus, sociobiological theory suggesting that genetically related family members will behave more positively and less negatively than genetically unrelated family members was not supported either when siblings were in early and middle childhood or when older siblings had entered adolescence.

Previous studies have found that children's and adults' perceptions of their family relationships are genetically influenced to some degree (Plomin, McClearn, Pederson, Nesselroade, & Bergeman, 1988; Rowe, 1983). Results from the CAP are especially important because they show genetic influence on mothers' and siblings' observed behavior, rather than simply on their perceptions. Taken at face value, results suggest less genetic influence on sibling relationships at Time 2 than at Time 1.

This conclusion, however, cannot be substantiated for two reasons. First, with the exception of mothers' interview responses, measures of sibling behavior were different at the two times of measurement. Second, the sample size is small for detecting genetic influence and inadequate for comparing genetic influence across time and measures.

Conclusions

Longitudinal analyses on the Colorado Adoption Project indicated that there was both stability and change in sibling relationships. Children's warmth and hostility to their siblings was significantly correlated from Time 1 to Time 2. In addition, we found individual differences in how much and in what direction children's behavior to their siblings changed over time. Structure variables such as adoptive status, gender, gender composition of dyads, and age spacing between siblings were not related to change in sibling relationships. Children's temperamental charac-teristics, siblings' behavior, differences in mothers' behavior to siblings, and major life events at Time 1 were related to individual differences in change in sibling relationship characteristics.

Analyses at the group level revealed that older siblings were both less warm and less hostile to their siblings in early adolescence than they had been in middle childhood. Younger siblings did not differ significantly in the amounts of warmth and hostility they directed to their siblings in early childhood and middle childhood. As children enter adolescence and the peer group takes on added importance in their lives they may spend less time with their siblings and distance themselves from their brothers and sisters.

The second issue we addressed was whether there were connections between the quality of children's sibling relationships and their relationships with parents. At the first time point, when data on mothers' direct behavior to each sibling were collected, maternal control and attention were associated with more conflictual sibling relationships. Because our analyses were correlational we could not deter-mine the direction of influence between mother–child and sibling relationships, but it is likely that these connections are bidirectional to some degree. These findings from the CAP fit with other research on associations between mother–child relationships and sibling relationships in early and middle childhood.

Differences in mothers' treatment of siblings were associated with higher levels of sibling hostility and lower levels of sibling warmth both at Time 1 when siblings were children and at Time 2 when older siblings were early adolescents and younger siblings were in middle childhood. These findings indicate that the importance of differences in parents' behavior for sibling relationships does not diminish as children enter adolescence. The results are important also because they show, for the first time, an association between fathers' differential behavior and sibling relationships. We cannot tell if differential parental behavior leads to higher levels of sibling conflict or if the opposite direction of influence operates. Our findings highlight the fact that parents do not treat all their children identically and that these differences in parents' behavior are related to individual differences in children's relationships with siblings.

The final issue we addressed in this chapter was whether sibling relationships are genetically influenced. Results indicated genetic influence for some, but not all, dimensions of children's behavior to siblings and mothers' behavior to children. Given the small sample size in the CSS, we were unable to determine if genetic influence on sibling relationships changed across children's development. Further research using larger samples is needed to clarify this issue. There were no group differences between adoptive siblings' and nonadoptive siblings' relationships at Time 1 or at Time 2. Thus, our findings do not support sociobiological theories suggesting that genetically related siblings will behave more altruistically toward one another than genetically unrelated siblings. These results should alleviate fears that some adoptive parents may have about their children's sibling relationships.

In summary, we found both stability and change in children's sibling relationships across transitions in childhood and adolescence. Children's genetic similarity and characteristics of their relationships with parents were related to individual differences in sibling relationship quality. These findings from the Colorado Adoption Project have provided the first steps toward clarifying the nature of developmental changes and genetic influence on sibling relationships and have demonstrated that the connections between sibling relationships and differential parental behavior remain strong as children grow from childhood to adolescence.

17 Genetic Influence on "Environmental" Measures

Julia M. Braungart

The purpose of this chapter is to use the sibling adoption design to examine whether measures that are typically used to assess children's family environments are influenced genetically. Measures of the environment include the HOME at ages 1–4 (Caldwell & Bradley, 1978), children's self-perception of their family environment at age 7 (FES: Moos & Moos, 1981), and mothers' reports of parenting style at ages 7 and 9 (Parent Report: Dibble & Cohen, 1974). Furthermore, it is possible that children's genetically influenced attributes, such as temperament, affect their environment. Thus, a secondary goal of this chapter is to explore the degree to which antecedent and contemporaneous measures of temperament using maternal reports (CCTI: Rowe & Plomin, 1977) and tester ratings (IBR: Bayley, 1969) are correlated with environmental measures, and whether such associations are mediated genetically. Environmental measures and developmental outcomes are presented in the next chapter, and parent–offspring analyses of genetic involvement in associations between an index of the environment and children's IQ are reported in Chapter 19.

A myriad of studies have provided correlational evidence demonstrating that children's cognitive, emotional, social, and dysfunctional development are associated with qualities in the family environment, such as parenting styles, family relationships, and physical attributes in the home (e.g. Ainsworth, Blehar, Waters, & Wall, 1978; Baumrind, 1971; Belsky, 1984; Caldwell & Bradley, 1978; Emery, 1982; Moos, 1976; Wachs, 1992; Wachs & Gruen, 1982). Presently, most developmentalists also accept the possibility that socialization is not a one-way process. That is, children may contribute to their own environment (Bell, 1968). However, few studies pay more than lip service to this possibility. Because individuals and their family members participate in the family environment, genetic as well as environmental factors may contribute to aspects of the home environment. The main purpose of the present investigation is to examine genetic and environmental influences on measures of infants' and children's family environments by using the sibling adoption design.

Recently, research using quantitative genetic methods has found genetic influences on measures of the environment when environmental measures are treated as phenotypes (Plomin & Bergeman, 1991). The first two studies in this area assessed adolescent twins' perceptions of their family environments (Rowe, 1981, 1983). Both studies found that identical twin correlations for self-report measures of parental warmth were greater than fraternal twin correlations, suggesting genetic influence. Similar results were found in a study of middle-aged adult twins' retrospective reports of their childhood rearing environments (using Moos and Moos's (1981) Family Environment Scale) in a study that included twins reared apart as well as twins reared together (Plomin, McClearn, Pedersen, Nesselroade, & Bergeman, 1988). Such findings suggest that genetic factors influence perceptions of family environments and/or that family members respond to genetically mediated characteristics of twins.

Studies employing observational methods to assess genetic mediation of family environments eliminate the possibility that effects are merely due to subjective perceptions. The results of a small twin study of parental behavior indicated that parents respond to genetically influenced characteristics of children (Lytton, 1977, 1980). Similarly, observer ratings of maternal behavior demonstrated that genetically related siblings at ages 1, 2, and 3 were treated more similarly by their mothers than were adoptive siblings (Dunn & Plomin, 1986; Dunn, Plomin, & Daniels, 1986). In addition, a previous analysis of sibling adoption data in the CAP on the widely used Home Observation for Measurement of the Environment (HOME: Caldwell & Bradley, 1978) indicated that the total HOME score was genetically influenced at ages 1 and 2 (Braungart, Fulker, & Plomin, 1992).

The present study extends the search for genetic influence on environmental measures in CAP. In addition to total HOME scores during infancy, HOME subscales and total scores will be examined during early childhood. Moreover, children's perceptions of their family environment at age 7 and parental report of their own behavior toward their children at ages 7 and 9 will be investigated. A secondary goal of this study is to explore the degree to which genetic influence on children's temperament can explain genetic influence on measures of the family environment.

Method

Sample

CAP sibling data at 1–4, 7, and 9 years of age were subjected to genetic analyses. However, data from the entire sample of CAP children were used to estimate longitudinal correlations as well as differences in means for status groups and gender. The samples are described in more detail in Chapter 2.

Environmental measures

Observational ratings   Caldwell and Bradley's Home Observation for the Measure of the Environment (HOME: 1978) was employed at ages 1–4 in the CAP. The HOME involves both direct observations of the family environment by the interviewer and reports given by the parent. The HOME versions used at ages 1 and 2 were identical, whereas a nonstandard extension of the HOME was used at 3 and 4 because the preschool version of the HOME was not available at the time of testing (Plomin & DeFries, 1985). Based on Caldwell and Bradley's traditional dichotomous scoring, six scales from the HOME at 1, 2, and 3 included: Responsivity of Mother, Avoidance of Restriction and Punishment, Organization of the Environment, Appropriate Play Materials, Maternal Involvement, and Variety in Daily Stimulation. In addition, a principal-component score was created, which consisted of the sum of 37 standardized items (see Plomin & DeFries, 1985). At age 4, some scales differed from those scales in the earlier versions of the HOME: Language Stimulation, Warmth, Avoidance of Punishment, Stimulation through Toys/Reading, Encouragement of Social Maturity, and Variety in Daily Stimulation. A principal-component score representing the sum of standardized items at age 4 was also used.

Self-report   A child's version of the Family Environment Scale (Moos & Moos, 1981), which assesses children's perceptions of their family life, was administered to each child when they were 7 years of age. Twenty items constitute five scales: Cohesion, Expressiveness, Conflict, Achievement, and Control.

Parental ratings   Parents rated their own behaviors when each sibling was 7 and 9 years of age using Dibble and Cohen's (1974) Parent Report. This measure consists of eight socially desirable parental behaviors, and eight socially undesirable parenting techniques. The eight positive scales include: Acceptance of Child as a Person, Child Centeredness, Sensitivity to Feelings, Positive Involvement, Acceptance of Autonomy, Shared Decision Making, Consistent Enforcement of Discipline, and Control through Positive Discipline. Eight negative scales are labeled Detachment, Intrusiveness, Lax Enforcement of Discipline, Inconsistent Enforcement of Discipline, Control through Anxiety, Control through Guilt, Control through Hostility, and Withdrawal of Relationship. In addition, factor analyses of these items suggested that three factors were present: Acceptance, Inconsistency, and Negative Control.

Temperament measures

Tester ratings   Bayley's (1969) Infant Behavior Record (IBR) was used to assess children's temperament during the administration of the Bayley Scales at ages 1 and 2, and the Stanford-Binet at 3 and 4. The three major factors included: Affect-Extraversion, Activity, and Task Orientation.

Maternal ratings   Mothers rated their child's temperament at each age using the Colorado Childhood Temperament Inventory (CCTI: Rowe & Plomin, 1977). Three scales identified as the EAS traits (Emotionality, Activity, and Sociability: Buss & Plomin, 1984) are included in this measure.

Adjustment of scores and data reduction

To assess group (adopted versus nonadopted) and gender differences for each of the environmental measures, five MANOVAs were tested with the following sets of dependent measures: HOME subscales (1–4), HOME principal-component scores (1–4), child's FES scales at 7, Parent Report subscales at 7 and 9, and Parent Report summary scales at 7 and 9. Table 17.1 presents the *F* tests for the main effects for adoptive status and gender. Comparisons of the means for environmental measures by adoptive status and gender indicate that the adopted children received higher scores than did nonadopted children on the HOME principal component at age 2, Cohesion and Achievement scores on the FES at age 7, and Warmth at age 9. Boys also received higher scores than girls for the Control scale on the parenting measure at 7 and 9. Variance due to gender was subsequently removed from each measure so that adjusted values represent equivalent scores for each gender. The effects for adoptive status will be irrelevant because all further analyses will involve standardization within each status group, which removes differences in means between groups.

Table 17.1   Multivariate analyses of variance testing for the effects of adoptive status and gender on measures of the environment

Measure	Group $F(df)$	Gender $F(df)$
HOME (years 1–4)		
Subscales	$F_{24,386} = 1.05$	$F_{24,386} = .71$
Total scores	$F_{4,660} = 4.14^{**}$	$F_{4,660} = 1.17$
Child's FES (year 7)	$F_{5,631} = 2.28^{*}$	$F_{5,631} = .80$
Parent Report (years 7 and 9)		
Subscales	$F_{32,327} = 1.45$	$F_{32,327} = 1.59^{*}$
Summary scores	$F_{6,434} = 2.44^{*}$	$F_{6,434} = 3.05^{**}$

$^{*}p < .05$, $^{**}p < .01$.

Table 17.2 presents year-to-year longitudinal correlations for repeated scales. On average, the HOME scales showed moderate stability from ages 1 to 2, whereas correlations were lower from 2 to 3. The lower magnitude of correlations from 2 to 3 years may reflect discrepancies between the actual HOME instruments. In contrast, correlations for parenting styles from 7 to 9 were moderately high.

To reduce the number of variables examined, equivalent HOME scales were averaged from 1 and 2 to reflect the home environment during infancy. Although the CAP HOME scales at 3 were similar to those measured at 1 and 2, the HOME scales at 4 are somewhat different, with the exception of Variety in Daily Stimulation and the principal-component scales. Thus, measures of the HOME environment during early childhood included the HOME scales assessed at age 3 and the average of the scores at 3 and 4 years for Variety in Daily Stimulation and the principal component.

Table 17.2 Longitudinal correlations for the HOME from 1–4 years of age and Parent Report from 7 to 9 years of age.

Measure	Longitudinal Correlation		
HOME (N = 544–704)	Ages 1–2	Ages 2–3	Ages 3–4
Responsivity of Mother	.42	.19	NA
Avoidance of Restriction and Punishment	.28	−.07[a]	NA
Organization of the Environment	.29	.10	NA
Appropriate Play Materials	.65	.24	NA
Maternal Involvement	.45	.23	NA
Variety in Daily Stimulation	.51	.35	.23
Principal component	.64	.46	.42
Parent Report (N = 356–372)	Ages 7–9		
Acceptance of Child	.48		
Child Centeredness	.60		
Sensitivity to Feelings	.48		
Positive Involvement	.65		
Acceptance of Autonomy	.51		
Shared Decision Making	.49		
Consistent Enforcement Discipline	.63		
Control through Positive Discipline	.51		
Detachment	.36		
Intrusiveness	.57		
Lax Enforcement of Discipline	.67		
Inconsistent Enforcement of Discipline	.56		
Control through Anxiety	.58		
Control through Guilt	.62		
Control through Hostility	.65		
Withdrawal of Relationship	.66		
Acceptance	.72		
Inconsistency	.61		
Negative Control	.60		

[a] Non-Significant correlations; all others are significant at $p < .05$.

Although the same measure for parenting style was used at ages 7 and 9, those scales were not averaged for two reasons: (1) the sample size was substantially smaller at age 9, which would result in some children's scores reflecting the average of parenting at 7 and 9, whereas other children's scores would only include the 7 year values; and (2) examining parenting during different stages of middle childhood might yield interesting developmental results. Temperament scales were also averaged from 1 to 2 and from 3 to 4 to reflect temperament during infancy and early childhood.

In sum, measures of the environment included seven HOME scales during infancy, seven HOME scales during early childhood, children's ratings of the FES at age 7 (five scales), 19 scales for parenting at 7, and 19 scales for parenting at age 9. Temperament measures included three CCTI scales and three IBR scales during infancy and three scales from each measure during early childhood.

Results and Discussion

Is there genetic influence on environmental measures?

The main goal of this study was to explore whether measures of the environment in the CAP show genetic influence using the sibling adoption design. To test genetic and environmental influences on such measures, intraclass correlations were computed for adoptive and nonadoptive sibling pairs. Genetic influence is implied when nonadoptive sibling correlations are greater than those for adoptive siblings. Shared environmental influence is suggested when adoptive correlations are substantial. In addition, model-fitting analyses of these data yielded estimates of the magnitude of genetic and environmental variances, as well as a chi-square, which tests the goodness of fit of each model. To test whether genetic and environmental parameter estimates were significant, chi-squares from reduced models were compared to chi-squares from full models; a significant change in chi-square indicates that the parameter excluded in the reduced model was significant.

Intraclass correlations for adopted and nonadopted sibling pairs on the HOME, FES, and Parent Report measures are presented in Table 17.3. During infancy, correlations were greater (although not significantly) for nonadoptive than for adoptive siblings on three HOME scales: Maternal Involvement, Variety in Daily Stimulation, and the principal-component score, which suggests genetic influence. Both adoptive and nonadoptive sibling correlations are substantial for Responsivity of Mothers, Avoidance of Restriction, and Organization of the Environment, which suggests shared environmental contributions. Figure 17.1 depicts the components of variance on the HOME and FES measures based on maximum-likelihood model-fitting analyses using LISREL VI (Jöreskog & Sörbom, 1985). Components of variance for Dibble and Cohen's Parent Report appear in Figure 17.2 It is noteworthy that all chi-square tests were non-significant ($p < .05$), which indicates a good fit.

Figure 17.1 demonstrates that 20 to 30 percent of the variance is attributed to genetic contributions for Maternal Involvement, Variety in Daily Stimulation, and the principal component on the HOME during infancy, which confirms the correlational findings. Although reduced models that excluded the genetic parameter

Table 17.3 Intraclass correlations for measures of the environment

Measure	Correlations			
	Infancy		Childhood	
	Adopted	Nonadopted	Adopted	Nonadopted
HOME	(N = 86–100)	(N = 91–108)	(N = 80–93)	(N = 75–103)
Responsivity of Mother	.30*	.32*	.04	.05
Avoidance of Restriction and Punishment	.21*	.18*	.32*	.38*
Organization of the Environment	.37*	.30*	.34*	.21*
Appropriate Play Materials	.04	.02	.06	.07
Maternal Involvement	.26*	.42*	.19*	.10
Variety in Daily Stimulation	.46*	.61*	.19*	.52*
Principal component	.37*	.48*	.47*	.41*
FES	Age 7			
	(N = 63)	(N = 70)		
Cohesion	−.11	−.11		
Expressivity	.04	.04		
Conflict	−.04	.16		
Achievement	.06	.23*		
Control	.06	.15		
Parent Report	Age 7		Age 9	
	(N = 46–49)	(N = 49–51)	(N = 31–33)	(N = 43–46)
Acceptance of Child	.12	.29*	.45*	.46*
Child Centeredness	.36*	.72*	.43*	.61*
Sensitivity to Feeling	.57*	.63*	.33*	.49*
Positive Involvement	.38*	.70*	.28*	.66*
Acceptance of Autonomy	.40*	.48*	.37*	.56*
Shared Decision Making	.41*	.39*	.40*	.59*
Consist. Enforce. Discipline	.60*	.65*	.48*	.65*
Control through Positive Discipline	.50*	.47*	.64*	.56*
Detachment	.06	.47*	.18	.36*
Intrusiveness	.43*	.28*	−.05	.65*
Lax Enforce. of Discipline	.75*	.64*	.62*	.55*
Inconsistent Enforce. Disc.	.63*	.66*	.54*	.65*
Control through Anxiety	.58*	.61*	.38*	.63*
Control through Guilt	.56*	.61*	.54*	.54*
Control through Hostility	.40*	.62*	.34*	.62*
Withdrawal of Relationship	.63*	.64*	.47*	.62*
Acceptance	.42*	.76*	.52*	.79*
Inconsistency	.68*	.56*	.33*	.57*
Negative Control	.63*	.74*	.56*	.48*

*p < .05.

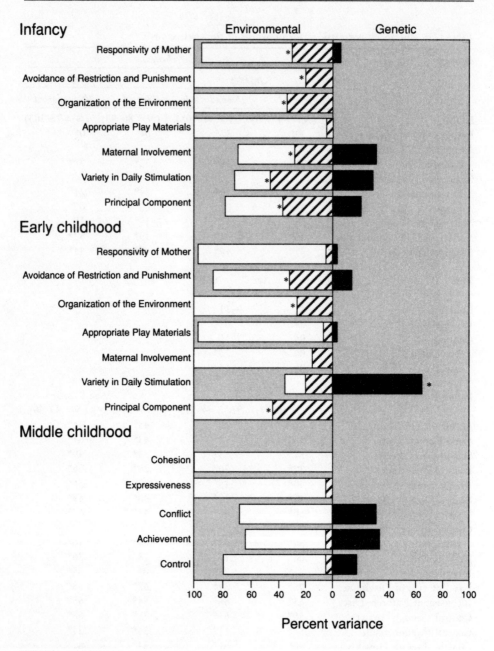

Figure 17.1　Components of variance based on maximum-likelihood model-fitting for the HOME and FES.

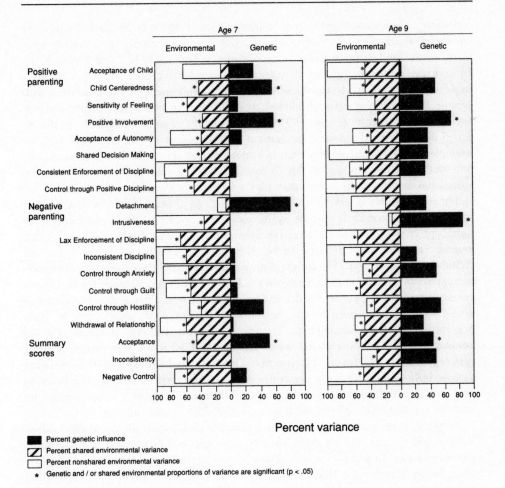

Figure 17.2 Components of variance based on maximum-likelihood model-fitting for Parent Report at 7 and 9.

did not yield significant changes in chi-squares from full models, results on the HOME during infancy are interesting for descriptive purposes. In addition, shared environmental contributions were present for all but Appropriate Play Materials. Greater similarities for nonadoptive siblings' scores on HOME scales may be related to mothers responding to genetic similarities in their children. Alternatively, because nonadoptive infants are genetically related to their parents, greater similarities between siblings on some of the HOME scales may reflect greater parent–offspring resemblance. For example, the HOME may be related to parental characteristics, which in turn are related genetically to infant attributes. This is a form of "passive genotype–environment correlation" in which genes and the environment are correlated. That is, because nonadoptive infants share genes and environments with their parents, children can passively "inherit" environments as well as genotypes (Bergeman & Plomin, 1988).

Similar to correlation patterns in infancy, Variety in Daily Stimulation in early childhood appears to be significantly genetically influenced – approximately two-thirds of the total variance is due to genetic effects (see Table 17.3 and Figure 17.1). In contrast to the pattern of nonadoptive and adoptive correlations in infancy, however, Maternal Involvement and the principal component are not associated with genetic factors during early childhood. It is difficult to conclude whether discrepancies between results on the HOME from infancy to early childhood are due to actual developmental changes, instrument differences, or chance. It is possible, for example, that maternal involvement is a more salient construct during infancy, and one which may be more susceptible to genetic influence. During childhood, however, situational or contextual factors may play a larger role in maternal behavior.

Three of the five scales for children's perceptions of their family environment at age 7 yielded correlations that were somewhat greater (not significantly) for nonadoptive siblings than for adoptive siblings (Table 17.3). Figure 17.1 shows that genetic variance was estimated to be 32% for the Conflict scale, 34% for the Achievement scale, and 18% for the Control scale. Shared environmental effects do not appear to contribute to variance on the FES. Greater resemblance for nonadoptive siblings' perceptions of conflict and achievement in the home may be related to actual shared experiences in the family environment or to similarities in their perceptions of these dimensions. A parent–offspring analysis of the 7-year-old version of the FES, correlated with the standard FES for the parent, yielded different results (Chipuer, Merriwether-DeVries, & Plomin, 1992). Significant genetic influence emerged for Control, Cohesion, and Expressiveness. The discrepancy between these results may be due to differences in genetic effects shared by siblings versus those between adult parents and their young children.

As noted, parents reported on their parenting style for each sibling when children were 7 and 9 years old on eight socially desirable and eight socially undesirable parenting styles. In addition, three factors portray overall styles of parenting: Acceptance, Inconsistency, and Negative Control. Table 17.3 and Figure 17.2 show that many facets of parenting behavior – positive and negative – are genetically mediated.

At age 7, three of the 16 subscales showed significant genetic variance: Child Centeredness (60%), Positive Involvement (61%), and Detachment (83%). The Acceptance factor also yielded significant genetic effects (53%). Patterns of correlations for other dimensions of parenting also suggested genetic influence (e.g. Acceptance of Child – 34%) but were non-significant given the present sample size. The average percentage of genetic variance across the 16 subscales was 22%; similarly, the average percentage of genetic variance for the three summary factors was 25%. All but two of the parenting scales (Acceptance of Child and Detachment) were significantly influenced by shared environmental effects at age 7.

Interestingly, even though the sample size at age 9 is lower than at age 7, several scales showed genetic effects. Genetic effects were significant for Positive Involvement (71%) and Intrusiveness (86%). Again, the Acceptance factor was significantly genetically influenced at age 9 (45%); Inconsistency also showed substantial genetic contributions (48%). The average genetic estimate across the 16 subscales was 33%, and the average for the summary factors was 31%. In addition, genetic estimates for 14 of the 19 scales at age 9 were greater than 30%, whereas only six scales at age 7 yielded genetic effects greater than 30%. Genetic influence on parenting behavior suggests that genetically influenced attributes in children may elicit different parenting styles. Increases from age 7 to 9 in the number of scales that yielded moderate to large genetic contributions may be related to developmental issues. For example, three scales that showed substantial genetic influence at age 9, but not at age 7, were Acceptance of Autonomy (e.g. "I am aware of his need for privacy"), Shared Decision Making (e.g. "I let him help me decide about things that affect him"), and Control through Anxiety (e.g. "I warn him about future punishments to prevent him from acting badly"). Such parenting behavior may be less relevant for younger, less cognitively-skilled children; thus, parents may be less inclined to behave according to genetic differences in children. It is also interesting that the summary factor for Control did not show significant genetic influence at either age. This finding has been demonstrated elsewhere (Rowe, 1981, 1983). Parents may have preconceived ideas about how much control should be exerted on children – regardless of the child's style.

Is there an overlap between children's genetic variance on temperament and genetic components of family environment?

The second goal of this study was to examine whether temperament plays a role in children's family environments, and if so, to what degree genetic influences on antecedent and contemporaneous measures of children's temperament overlap with genetic components on environmental measures. Based on model-fitting analyses of temperament data, genetic influence on the IBR traits was non-significant for Affect-Extraversion. However, model-fitting for Activity and Task Orientation during infancy and childhood yielded significant genetic effects: 58% of the variance was due to genetic contributions for Activity during infancy and 62% during childhood; for Task Orientation, 36% of the total variance was genetic during infancy and 42% of the variance during childhood was due to genetic effects. In contrast to tester ratings, genetic influence on maternal ratings of temperament during infancy or childhood was non-significant. Thus, the remainder of the results examining genetic covariance between temperament and environmental measures are limited to Activity and Task Orientation from the IBR during infancy and childhood.

The next step in examining bivariate relations involves the examination of phenotypic correlations between measures of temperament and the environment. Rather than presenting all possible combinations of correlations, the exploration of associations was limited to using the summary scores on the HOME during infancy and early childhood (principal components) and the three general factors representing parenting styles at ages 7 and 9 (Acceptance, Inconsistency, and Negative Control). In addition, because the HOME subscale, Variety in Daily Stimulation, showed consistent genetic variance over time, it too was included in the phenotypic analyses. The child's report on the FES at age 7 does not include a summary score; therefore, correlations were examined for all five subscales (Cohesion, Expressiveness, Conflict, Achievement, and Control). Table 17.4 presents the phenotypic correlations between Activity and Task Orientation with the HOME, FES, and Parent Report measures. Unless phenotypic correlations are observed in nonadoptive families, genetic mediation is unlikely to emerge from bivariate sibling analyses. Correlations that are lower in adoptive than in nonadoptive families suggest passive genotype–environment correlation. Although similar correlations in adoptive and nonadoptive families suggest that passive genotype–environment correlation is unimportant, reactive and active types of genotype–environment correlation may mediate the association. All three types of genotype–environment correlations are assessed by bivariate sibling analyses (Plomin, in press).

Table 17.4 Phenotypic correlations for adopted and nonadopted children for measures of activity (Act.) and Task Orientation (Task) during infancy and early childhood with measures of the environment

Measure	Adoptive				Nonadoptive			
	Infancy		Early Childhood		Infancy		Early Childhood	
	Act.	Task	Act.	Task	Act.	Task	Act.	Task
Infancy (HOME)	(N = 195–213)				N = (214–232)			
Principal component	.27*	.27*	.01	.06	.23*	.28*	.21*	.15*
Variety in Daily Stimulation	.16*	.14*	.12	.17*	.04	.26*	.20*	.19*
Early Childhood (HOME)	(N = 187–199)				(N = 198–224)			
Princ. comp.	.14*	.20*	.02	.17*	.12	.18*	.08	.23*
Var. Stim.	.07	.07	−.06	.05	−.12	.04	.13	.15*
Age 7 (FES)	(N = 151–156)				(N = 178–179)			
Cohesion	−.10	−.09	−.07	−.15	.10	−.07	−.05	−.08
Expressivity	−.09	.04	−.11	−.07	.03	.00	−.03	−.03
Conflict	.00	.09	−.03	.16	−.05	.11	.10	.08
Achievement	.03	−.01	.05	.11	.05	−.08	−.05	−.13
Control	−.07	−.02	.06	.15	−.07	.08	.13	.06
Age 7 Parent Report	(N = 130–138)				(N = 146–155)			
Acceptance	−.09	−.10	−.06	−.09	−.03	.12	.11	.04
Inconsistency	−.06	.08	.01	−.10	−.02	.07	−.08	−.02
Negative Control	−.06	−.14	−.07	.03	.03	−.08	.01	−.06
Age 9 Parent Report	(N = 113–117)				(N = 123–127)			
Acceptance	.02	−.05	−.01	.02	−.06	.02	.26*	.14
Inconsistency	−.01	.11	−.08	−.07	.14	−.04	−.28*	−.06
Negative Control	−.08	.04	−.05	.01	.04	−.03	−.05	−.05

*$p < .05$.

Nonadopted and adopted children's scores on Activity and Task Orientation during infancy and early childhood were significantly correlated with Variety in Daily Stimulation and the principal component during infancy. Activity in early childhood was related to HOME scores for nonadopted but not for adopted children. The correlations are positive, indicating that children who exhibit greater levels of activity and task orientation during a semi-structured situation are more likely to receive greater stimulation and experience a higher overall quality of the home environment. In addition, Task Orientation during infancy and early childhood was consistently positively correlated with the HOME principal-component score during early childhood. Finally, greater levels of Activity during childhood were associated with more acceptance and less inconsistent parenting at age 9 – for nonadopted children. It is somewhat surprising that children who are higher in activity receive greater acceptance and less inconsistent parenting, because activity is often perceived as a negative attribute in children. However, during a testing situation, moderately high activity may reflect excitement, whereas low levels of activity may be indicative of passivity or disinterest.

As mentioned above, passive genotype–environment correlation is implied when phenotypic associations between child temperament and parenting are greater in nonadoptive than in adoptive families. That is, associations between children's activity and parenting may reflect characteristics that are genetically shared by parents and their nonadopted offspring. For example, greater levels of acceptance may reflect enthusiasm for parenting; items in the Acceptance factor include "I tell him how happy he makes me," "I give him a lot of care and attention," and "I like to hug and kiss him." Such enthusiasm may be genetically correlated with children's active, .enthusiastic styles.

Phenotypic correlations between temperament and the FES and Parent Report at age 7 were non-significant. It is somewhat puzzling that associations exist for temperament and parenting of 9-year-old children, but do not exist for temperament and parenting of 7-year-old children. As previously mentioned, however, it is possible that certain aspects of parenting are less relevant to 7-year-old children and more salient for older children. Thus, parents may respond less to individual differences in children for certain parenting behaviors.

The degree to which genetic and environmental factors mediate the covariation between temperament and environmental measures can be assessed by fitting matrices of data that include phenotypic and sibling correlations for temperament and environmental measures. Such bivariate model-fitting techniques estimate several common and specific genetic and environmental components of variance. The model that was used in the present study, pictured in Figure 17.3, estimated six parameters: genetic, shared environmental, and nonshared environmental variance that is common to temperament and the environmental measure, as well as genetic, shared environmental, and nonshared environmental variance that is unique to the environmental measure.

Figure 17.4 depicts the percentages of components of variance for the bivariate associations between the temperament and environmental measures. Each block of four bars corresponds to the total genetic and environmental components of variance for the environmental measure. For example, the first set of bars shows that 22% of the variance on the HOME is genetic (segment of bar that appears on the right side of the axis), whereas the portion of the bar that lies left of the axis indicates that 78% of the variance on the HOME principal-component score during infancy is environmental. In addition, approximately half of the environmental variance is due to shared effects, and the remainder of the environmental component of variance is associated with nonshared environmental effects. Note that these percentages converge with the aforementioned univariate model-fitting results for each environmental measure.

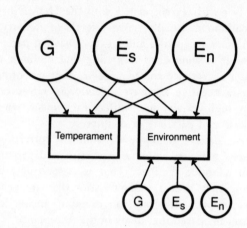

Figure 17.3 Bivariate path model for temperament–environment associations. The three large circles represent factors common to measures of temperament and the environment, and the three smaller circles signify factors unique to the environmental measure. G = genetic factor, E_s = Shared environmental factor, and E_n = Nonshared environmental factor.

The four smaller horizontal bars within each larger bar represent the breakdown of variance for a given environmental measure that is common with variance on the four temperament dimensions (Activity and Task Orientation during infancy and during early childhood). For example, bivariate model-fitting results for the HOME principal-component score during infancy and Activity during infancy indicates that the genetic variance on the HOME is primarily unique (17%), rather than in common with genetic variance on Activity (5%). In contrast, all of the genetic variance on the HOME principal-component score during infancy is shared with genetic variance for Task Orientation during infancy. For Variety in Daily Stimulation, most of the genetic variance is unique to the environmental measure when links are examined with temperament during infancy. However, bivariate model-fitting results for Variety in Daily Stimulation and Activity or Task Orientation during childhood indicate that all 30% of the genetic variance is shared with genetic variance on the temperament scales.

During early childhood, the HOME principal-component score showed no genetic influence; thus, there can be no common or unique genetic variance with temperament indices. In contrast, 66% of the total variance for Variety in Daily Stimulation was genetic. However, almost all of that variance was unique to the measure, rather than shared with genetic variance on the IBR.

During middle childhood, a pattern similar to that of early childhood emerged: 45% of the variance for the Acceptance scale for parenting was genetic, and most of that variance is unique to the Acceptance scale (44% for associations with infant Activity, 45% for Task Orientation, 41% for Activity during childhood, and 32% for associations with childhood Task Orientation). Similarly, Inconsistency yielded 48% genetic effects, with most of those effects unique to Inconsistency (42%, 46%, 48%, and 36% for bivariate associations with Activity and Task Orientation during infancy and childhood, respectively).

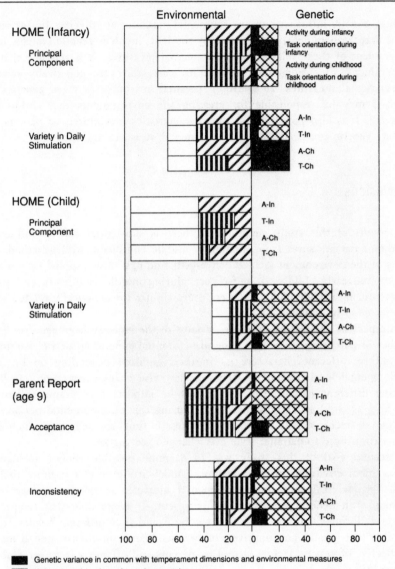

Figure 17.4 Components of variance based on bivariate maximum-likelihood model-fitting analyses for temperament dimensions and environmental measures. Blocks comprising four bars correspond to the total genetic and environmental components of variance for the environmental measure; total genetic variance on the environmental measure appears on the right side of the axis; total shared environment for the environmental measure is depicted by patterned portion of the bar on the left side of the axis; the white segment of the bar signifies nonshared environmental variance for the environmental measure. The four smaller horizontal bars within each larger block represent the covariance for Activity during infancy, Task Orientation during infancy, Activity during childhood, and Task Orientation during childhood with each environmental measure.

Thus, genetic influences on environmental measures are diverse. For example, it seems that temperament, as measured by the IBR, involves genetic systems that are independent of those related to environmental measures. It is possible that other temperament dimensions not included in this study are genetically related to environmental measures. In addition, parental attributes that are genetically influenced may be responsible for creating the environments that children then "inherit." It is also possible that the characteristics of children or parents which mediate genetic contributions to environmental measures are manifold.

Conclusions

The results of this study suggest that there is substantial genetic influence on environmental measures from infancy to middle childhood. Multi-method assessments of the environment yielded convergent findings: observational ratings during infancy and early childhood, self-reports during middle childhood, and parental reports during middle childhood were more similar for genetically related siblings than for adoptive siblings.

An especially interesting outcome pertains to the increase from age 7 to 9 in the number of parenting scales that showed substantial genetic influence. Parents may respond to different characteristics in their children depending on the child's developmental level. Children's behavior may also change from age 7 to 9, thus eliciting different parenting styles. It will be important to examine changes in parenting as children enter adolescence. During this developmental period, which has been characteristically identified as a stressful time, the degree to which genetic factors contribute to parental style may change even further.

A second goal of this study was to examine possible genetic mediation of temperament–environment relationships. Model-fitting results indicate that most of the genetic influence on environmental measures is specific, rather than in common with genetic factors on temperament. It is possible that temperament measured during situations other than testing would yield different results. It is also plausible that greater nonadoptive sibling resemblance on environmental measures is reflective of genetic factors shared by parents and offspring (e.g. parental IQ or personality).

In conclusion, the longitudinal CAP design and data set have facilitated an examination of the nature of nurture from infancy through middle childhood. The results of this study contribute to the growing body of research that suggests that genetic factors cannot be ignored when examining facets of family environments.

18 Family Environment in Early Childhood and Outcomes in Middle Childhood: Genetic Mediation

Jenae M. Neiderhiser

The previous chapter summarized evidence that measures of the family environment can be influenced by genetic factors. Hundreds of investigations have related such environmental measures to developmental outcomes (e.g., review by Wachs, 1992). Never do these studies of genetically-related family members consider the possibility that genetics might be involved in environment–outcome associations. Instead, it is assumed that environment–outcome associations are due to environmental influences brought about by modeling or social learning.

The investigation of genetic mediation of environment–outcome associations has been suggested as an important direction for future research on the topic of "environmental genetics" (Plomin & Neiderhiser, 1992), and the hypothesis of such mediation is especially fitting given the evidence for genetic influence on environmental measures reviewed in the previous chapter. Genetic mediation of the environment-to-outcome relationship has begun to be explored in the Colorado Adoption Project (CAP). In one of the first analyses of this kind, environment-to-outcome associations for infants were compared in nonadoptive and adoptive families. Correlations were generally higher in nonadoptive than adoptive families, suggesting genetic mediation of environment-to-outcome associations (Plomin, Loehlin, & DeFries, 1985). A multivariate extension of this approach included three environmental measures and three behavioral measures and found evidence for genetic mediation (Thompson, Fulker, DeFries, & Plomin, 1986). Another study found that the relationship between an objective measure of the home environment and infant mental development showed a substantial amount of genetic mediation (Braungart, Fulker, & Plomin, 1992). Finally, path analytic models of parent–offspring resemblance have been developed that incorporate environmental indices. Analyses of home environment and IQ using this model suggest some genetic mediation (Chapter 19; Coon, Fulker, DeFries, & Plomin, 1990; Rice, Fulker, DeFries, & Plomin, 1988).

Much of the research to date on the question of possible genetic mediation of environment–outcome associations has come from the CAP. Findings thus far certainly warrant further examination of this issue. More research is especially needed concerning noncognitive aspects of development. A preliminary CAP analysis suggested that some evidence emerges of genetic mediation between parental measures of family environment and children's behavior problems (Braungart & Rende, 1991).

The present chapter focuses on positive and negative outcomes in middle childhood outside the domain of cognition. (For an analysis of environment–outcomes in the cognitive domain, see Chapter 19.) These analyses capitalize on the longitudinal nature of the CAP to provide the first analyses of environment in early childhood and outcomes in middle childhood.

A simple model, first presented by Plomin et al. (1985), was used to examine the environment-to-outcome correlations for these relationships. As illustrated in Figure 18.1, in nonadoptive families the correlation between the environmental

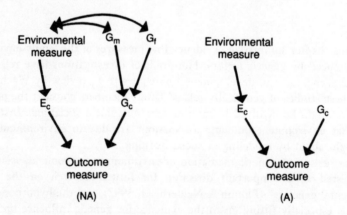

Figure 18.1 Environment–outcome model for adoptive (A) and nonadoptive (NA) families. The outcome measure is assumed to be causally determined by the genotype (G_c) and the environment (E_c). A measured feature of the environment is assumed to act on the immediate environment that affects the phenotype. In nonadoptive families the environmental measure is also allowed to relate to the child's phenotype through the genetically influenced parental characteristics.

measure as it relates to the outcome measure can be mediated either environmentally or genetically. For adoptive families, the environmental variable is only connected to the child's phenotype by the environment (see Plomin et al., 1985, for more detail). In other words, the environment-to-outcome correlation for children in adoptive homes can be due only to environmental factors, while the environment-to-outcome correlation for children in nonadoptive homes may be due to both genetic and environmental components. Thus, the difference between the environment-to-outcome correlation in nonadoptive families and adoptive families can provide an estimate of the genetic component of the correlation, while the correlation in the adoptive families is a direct estimate of the environmental component of the correlation. The present report extends this approach by employing multiple regression to summarize associations between several aspects of early childhood family environment as they relate to outcome measures in middle childhood.

Method

Sample

The present analyses included children participating in the CAP tested at ages 1, 2, 3, 4, 7, 9, and 10 years. Because the CAP is an ongoing longitudinal study, the sample sizes decrease across time, ranging from 223 nonadopted children and 271 adopted children at 1 year to 128 nonadopted and 159 adopted children at 10 years. There are approximately as many girls as there are boys in this sample (52% boys, 48% girls).

Measures

Tester ratings and parental questionnaires of the family environment in infancy and early childhood were included in order to investigate the relationship of these environmental measures with parent, teacher, and child reports on middle childhood indices of positive (competence) and negative (behavior problems) outcomes.

Family environment measures Two measures of the early childhood family environment were assessed when the children were 1, 2, 3, and 4 years of age.

Caldwell and Bradley's (1978) Home Observation for Measurement of the Environment (HOME) was administered when the children were 1, 2, 3, and 4 years of age. The HOME is a 45-item semi-structured interview rating of the home environment traditionally scored on a dichotomous rating scale. In the CAP sample, the HOME has been scored on a quantitative rating scale to represent more accurately environmental variation in middle-class homes (see Plomin & DeFries,

1985 for more detail). The six scales suggested by Caldwell and Bradley (1978) were used in these analyses: Responsivity of Mother, Avoidance of Restriction and Punishment, Organization of the Environment, Appropriate Play Materials, Maternal Involvement, and Variety in Daily Stimulation. Previous research has shown that the HOME is moderately heritable in the CAP when the children are 1 and 2 years of age (Braungart et al., 1992; see Chapter 17, this volume, for more details).

Mothers' perceptions of the family environment when the children were 1 and 3 years of age were assessed through the use of the Family Environment Scale (FES: Moos & Moos, 1981). The FES consists of three second-order dimensions: Relationship, Personal Growth, and System Maintenance. The Relationship dimension includes questions about family cohesiveness, expressiveness, and conflict. The Personal Growth dimension includes questions about the amount of emphasis that is placed on independence, achievement, recreational activities, and moral-religiousness. Finally, the System Maintenance dimension consists of questions about organization and control within the family.

Middle childhood outcome measures   Two domains of childhood outcome were explored: Competence and self-worth, and behavior problems.

Competence and self-worth were assessed through the use of two measures: the Harter (1982) Self-Perception Profile for Children, and three subscales from the CAP Social Competence Scale. Both the Harter and the CAP Social Competence Scale show evidence for genetic influence (see Chapter 10 for details).

Harter's Self-Perception Profile for Children consists of five subscales: Self-Worth, Social Acceptance, Athletic Competence, Scholastic Competence, and Physical Appearance. The version of the Harter used in the CAP is administered as part of a battery of questions about "feelings" during a telephone interview with the children when the children are 9 and 10 years of age.

The CAP Social Competence Scale is based on the Walker-McConnell Scale of Social Competence and School Adjustment (Walker & McConnell, 1988). Teacher reports at 7, 9, and 10 years on three subscales – Leadership, Confidence, and Popularity – were included as part of the competence dimension.

To assess behavior problems in middle childhood, one parent completed the parent form of the Child Behavior Checklist (CBCL: Achenbach and Edelbrock, 1983) when the children were 7, 9, and 10 years of age. Teachers completed the Teacher Report Rating Form of the CBCL (Achenbach & Edelbrock, 1986) at 7, 9, and 10 years. The broad-based factors of Internalizing and Externalizing behavior were used in these analyses. Correlations between parent and teacher reports on the CBCL were only modest (average $r = .22$ across all ages); therefore, parent and teacher ratings were analyzed separately. Previous research has demonstrated that the CBCL is influenced substantially by genetic factors (Edelbrock, Rende, Plomin, and Thompson, in press). The specific genetic and environmental characteristics of the CBCL in the CAP sample are discussed in Chapter 11.

Results

Mean differences

In order to assess mean gender differences and mean differences between adopted and nonadopted children, 2 (boys and girls) × 2 (adopted and nonadopted) analyses of variance were conducted. The means for boys and girls by adoptive status are listed in Table 18.1 for the family environment measures, and in Table 18.2 for the middle childhood outcome measures. Significant gender differences were identified for 11 of the 31 outcome measures and 2 of the environmental measures, and significant differences as a result of adoptive status were identified for only 3 of the 30 environmental measures and 9 of the outcome measures. The effects of gender were regressed out of the data for the remainder of the analyses.

Table 18.1   Means for early childhood family environment measures

Measure	Nonadopted		Adopted	
	Boys	Girls	Boys	Girls
Mother report				
FES – year 1				
Relationship	90.49	91.17	90.92	89.21
Personal Growth	92.63	91.03	91.78	92.04
System Maintenance	85.66	86.51	95.34	95.88[a]
FES – year 3				
Relationship	90.68	91.52	90.23	90.50
Personal Growth	59.47	59.72	60.94	59.92
System Maintenance	88.10	90.84	96.77	96.88[a]
Observer ratings				
HOME – year 1				
Responsivity of Mother	−1.28	−.20	.37	.16
Avoidance of Restriction and Punishment	.43	−.35	.66	−.40
Organization of the Environment	−.43	.11	−.07	.20
Appropriate Play Materials	−.91	−1.23	.08	.01
Maternal Involvement	−.30	−.24	−.20	.02
Variety in Daily Stimulation	−.58	.06	.10	−.05
HOME – year 2				
Responsivity of Mother	−1.20	−.22	.59	.16
Avoidance of Restriction and Punishment	−.11	−.18	.52	−.09
Organization of the Environment	−.31	.07	.01	−.07
Appropriate Play Materials	−1.50	−1.40	.31	−.25[a]
Maternal Involvement	−.77	−.05	.38	−.09
Variety in Daily Stimulation	−.58	−.08	.01	.13
HOME – year 3				
Responsivity of Mother	−.75	−.61	.30	.13
Avoidance of Restriction and Punishment	.27	−.29	.64	−.53[a]
Organization of the Environment	−.18	.12	−.13	.28

Table 18.1 (Cont.)

Measure	Nonadopted		Adopted	
	Boys	Girls	Boys	Girls
Appropriate Play Materials	−.81	−.92	−.48	−.11
Maternal Involvement	−.21	−.33	.09	−.23
Variety in Daily Stimulation	−.33	−.06	−.27	.39
HOME − year 4				
Responsivity of Mother	1.01	1.41	.89	.84
Avoidance of Restriction and Punishment	.82	.77	1.12	.73 [b]
Organization of the Environment	−.22	.25	.09	−.06
Appropriate Play Materials	−.52	−.02	.98	.58
Maternal Involvement	−.73	−.64	.01	.00
Variety in Daily Stimulation	1.89	1.87	2.30	2.04

[a] Indicates significant ($p < .05$) mean difference for adoptive status.
[b] Indicates significant ($p < .05$) mean difference for gender.

Table 18.2 Means for middle childhood outcome measures

Measure	Nonadopted		Adopted	
	Boys	Girls	Boys	Girls
Mother reports				
Year 7				
CBC Internalizing	5.59	5.98	5.75	4.90
CBC Externalizing	8.88	6.35	11.01	7.67
Year 9				
CBC Internalizing	5.71	6.38	5.90	5.64
CBC Externalizing	8.53	6.29	9.80	7.79
Year 10				
CBC Internalizing	8.38	6.72	5.93	6.20
CBC Externalizing	5.52	6.08	8.54	8.08
Teacher reports				
Year 7				
CBC Internalizing	3.69	3.72	4.28	4.02
CBC Externalizing	6.75	3.95	9.02	5.80 [a]
CSCS Leadership	31.08	31.85	29.94	30.12 [a]
CSCS Confidence	41.52	43.92	40.28	41.23 [a,b]
CSCS Popularity	20.08	20.00	20.17	19.75
Year 9				
CBC Internalizing	4.87	4.59	6.21	5.40 [a,b]
CBC Externalizing	6.35	4.54	9.40	5.25 [a,b]
CSCS Leadership	30.89	29.57	28.85	29.02 [a]
CSCS Confidence	40.37	42.63	39.29	40.91 [a,b]
CSCS Popularity	20.15	19.94	19.61	19.34
Year 10				
CBC Internalizing	5.52	4.96	6.42	4.54
CBC Externalizing	6.48	3.92	9.21	5.91
CSCS Leadership	31.00	32.19	29.18	30.12 [a]

Measure	Nonadopted		Adopted	
	Boys	Girls	Boys	Girls
CSCS Confidence	38.85	42.12	37.78	40.95[b]
CSCS Popularity	19.76	19.82	19.58	19.35
Self-reports:				
Year 9				
Harter Self-Worth	16.67	17.44	16.82	16.26
Harter Athletic Competence	12.70	11.76	13.15	11.98[b]
Harter Physical Appearance	15.03	14.94	15.50	14.22[b]
Harter Behavior/Conduct	14.60	15.84	13.93	14.70[a,b]
Harter Scholastic Competence	14.27	15.22	14.08	13.74
Harter Social Acceptance	15.50	15.28	15.92	15.04
Year 10				
Harter Self-Worth	17.24	17.53	17.50	17.02
Harter Athletic Competence	12.63	11.90	13.23	12.05[b]
Harter Physical Appearance	15.96	15.22	15.98	14.11[b]
Harter Behavior/Conduct	14.80	15.59	14.78	15.75[b]
Harter Scholastic Competence	15.43	15.59	15.09	14.79
Harter Social Acceptance	15.34	15.61	16.54	15.57

[a] Indicates significant ($p < .05$) mean difference for adoptive status.
[b] Indicates significant ($p < .05$) mean difference for gender.

Environment–outcome associations

Multiple regression analyses were conducted using the six HOME scales at ages 1, 2, 3, and 4 years and the three second-order dimensions of the FES at 1 and 3 years to predict positive and negative outcomes in middle childhood separately in nonadoptive and adoptive families. The results will be presented separately for the HOME and FES and separately for the positive outcomes (Harter self-reports and teacher ratings of competence) and negative outcomes (parent and teacher reports on the CBCL).

Table 18.3 lists multiple correlations between the HOME in infancy and early childhood and the positive outcome measures in middle childhood. Overall, the correlations are positive, suggesting that high HOME scores predict later positive outcomes. However, the magnitude of the correlations is low and few correlations are significant. This is an important finding because environmental research often focuses on significant relationships but neglects to consider whether more signific-ant correlations emerge than expected by chance, and the effect size for the significant relationships. Twelve of the 96 correlations are significant for the self-report Harter measure and five of the 72 correlations for the teacher ratings of competence are significant, with the highest correlations being .26 for the Harter and .23 for the competence measure. This small effect size presents a special problem for the analysis of genetic mediation of environment–outcome associations. If the magnitude of the association between environment and outcome is low, there is not much covariance to partition into genetic and environmental components.

Table 18.3 Multiple correlations, adjusted for attenuation, between the six HOME scales at years 1, 2, 3, and 4 and positive outcomes in middle childhood

Measure	HOME							
	Year 1		Year 2		Year 3		Year 4	
	Nonadopted	Adopted	Nonadopted	Adopted	Nonadopted	Adopted	Nonadopted	Adopted
Harter – year 9								
Self-Worth	.12	.07	.23*	.09	.08	.10	.04	.03
Athletic Competence	.11	.18*	.24*	.14	.04	.01	.06	.14
Physical Appearance	.01	.13	.10	.20*	.14	.15	.15	.10
Behavior/Conduct	.15	.11	.24*	.20*	.16	.11	.10	.21*
Social Acceptance	.15	.11	.01	.09	.12	.10	.10	.04
Scholastic Competence	.06	.13	.18	.26*	.13	.05	.09	.09
(N)	(168)	(215)	(159)	(205)	(143)	(203)	(151)	(192)
Harter – year 10								
Self-Worth	.13	.08	.09	.15	.13	.17	.16	.10
Athletic Competence	.03	.14	.23*	.09	.08	.12	.02	.10
Physical Appearance	.13	.14	.14	.09	.08	.22*	.18	.09
Behavior/Conduct	.17	.16	.20	.08	.13	.12	.10	.05
Social Acceptance	.10	.14	.08	.15	.11	.26*	.03	.10
Scholastic Competence	.15	.11	.06	.23*	.14	.19*	.13	.14
(N)	(158)	(191)	(147)	(182)	(132)	(180)	(142)	(173)
CAP Social Competence: Teacher report – year 7								
Leadership	.03	.06	.11	.23*	.14	.07	.10	.07
Popularity	.05	.02	.17	.20*	.12	.14	.16	.11
Confidence	.10	.11	.09	.09	.15	.13	.15	.10
(N)	(193)	(228)	(180)	(218)	(162)	(214)	(162)	(197)
CAP Social Competence: Teacher report – year 9								
Leadership	.15	.23*	.23*	.08	.11	.16	.18	.14
Popularity	.17	.18	.21	.12	.09	.11	.16	.18
Confidence	.17	.17	.12	.15	.10	.16	.15	.15
(N)	(147)	(183)	(135)	(177)	(123)	(176)	(128)	(167)
CAP Social Competence: Teacher report – year 10								
Leadership	.10	.04	.19	.15	.19	.10	.14	.07
Popularity	.09	.11	.06	.22*	.20	.16	.06	.12
Confidence	.04	.17	.15	.10	.22	.04	.12	.17
(N)	(132)	(151)	(122)	(142)	(113)	(141)	(117)	(136)

*p < .05.

Table 18.4 Multiple correlations, adjusted for attenuation, between the six HOME scales at years 1, 2, 3, and 4 and negative outcomes in middle childhood

Measure	HOME							
	Year 1		Year 2		Year 3		Year 4	
	Nonadopted	Adopted	Nonadopted	Adopted	Nonadopted	Adopted	Nonadopted	Adopted
CBCL: Parent report								
Year 7								
Internalizing	.05	.11	.10	.28*	.10	.07	.23*	.23*
Externalizing	.20*	.09	.30*	.27*	.16	.26*	.19	.24*
(N)	(182)	(219)	(174)	(210)	(151)	(208)	(155)	(193)
Year 9								
Internalizing	.12	.19*	.15	.09	.19	.08	.11	.27*
Externalizing	.19	.09	.14	.08	.14	.22*	.19	.25*
(N)	(161)	(198)	(154)	(190)	(136)	(189)	(144)	(177)
Year 10								
Internalizing	.11	.21*	.11	.16	.11	.14	.04	.03
Externalizing	.18	.24*	.25*	.30*	.22	.23*	.09	.23*
(N)	(147)	(178)	(135)	(168)	(123)	(169)	(132)	(162)
CBCL: Teacher report								
Year 7								
Internalizing	.14	.10	.11	.20*	.14	.08	.08	.15
Externalizing	.08	.16	.10	.36*	.12	.36*	.15	.04
(N)	(182)	(214)	(168)	(205)	(149)	(207)	(151)	(187)
Year 9								
Internalizing	.18	.04	.25*	.20*	.12	.27*	.15	.21*
Externalizing	.07	.04	.14	.32*	.13	.27*	.09	.33*
(N)	(142)	(170)	(130)	(165)	(119)	(164)	(124)	(156)
Year 10								
Internalizing	.18	.02	.15	.13	.21	.34*	.24	.18
Externalizing	.15	.13	.15	.09	.18	.44*	.13	.13
(N)	(123)	(141)	(113)	(131)	(104)	(129)	(111)	(127)

*$p < .05$.

Nonetheless, the results suggest that these weak phenotypic associations between the HOME and the positive outcomes are mediated environmentally in that the correlations are on average similar in nonadoptive and adoptive families. For example, the average correlations between the HOME and Harter Self-Perception Profile are .12 in nonadoptive families and .13 in adoptive families. Moreover, correlations that are greater in nonadoptive than in adoptive families do not show a consistent trend across the four years of the HOME and both years of the Harter or the three years of the competence measure.

As shown in Table 18.4, similar results emerge for correlations between the HOME and parent and teacher ratings of behavior problems on the CBCL. Although approximately half of the 64 correlations between the HOME in infancy and early childhood and ratings of behavior problems in middle childhood are significant, there is no clear pattern of genetic influence on this relationship. In fact, more often than not, the correlation between the HOME and negative outcome is significant for adopted children and not for nonadopted children.

Table 18.5 lists multiple correlations between mother reports on the FES second-order factors in infancy and early childhood and positive-outcome measures in middle childhood. Only 11 of the 84 correlations are significant. In general, the pattern of correlations between the FES and positive outcomes in middle childhood is similar to the pattern of the HOME and middle childhood outcome in that no clear pattern of genetic influence emerged.

Table 18.5 Multiple correlations, adjusted for attenuation, between the three FES scales at years 1 and 3 and positive outcome measures in middle childhood

Measure	FES			
	Year 1		Year 3	
	Nonadopted	*Adopted*	*Nonadopted*	*Adopted*
Harter – year 9				
Gl. self-worth	.14	.10	.09	.04
Athletic Competence	.10	.04	.08	.15
Physical Appearance	.09	.09	.11	.11
Behavior Conduct	.11	.06	.14	.06
Social Acceptence	.12	.09	.12	.10
Scholastic Competence	.17*	.07	.08	.04
(N)	(217)	(170)	(219)	(160)
Harter – year 10				
Gl. self-worth	.03	.09	.15	.06
Athletic Competence	.07	.02	.10	.10
Physical Appearance	.10	.10	.28*	.05
Behavior/Conduct	.10	.02	.05	.09
Social Acceptance	.09	.05	.17	.08
Scholastic Competence	.12	.04	.17	.06
(N)	(192)	(160)	(193)	(146)
CAP Social Competence:				
Teacher report – year 7				
Leadership	.05	.16*	.11	.15*

Measure	FES			
	Year 1		Year 3	
	Nonadopted	Adopted	Nonadopted	Adopted
Popularity	.04	.14	.03	.06
Confidence	.13*	.16*	.09	.17*
(N)	(231)	(194)	(235)	(185)
CAP Social Competence:				
Teacher report – year 9				
Leadership	.20*	.16	.06	.08
Popularity	.17	.07	.17	.09
Confidence	.17	.14	.15	.10
(N)	(193)	(149)	(235)	(185)
CAP Social Competence:				
Teacher report – year 10				
Leadership	.16	.20*	.04	.14
Popularity	.10	.14	.10	.19*
Confidence	.20*	.14	.04	.04
(N)	(155)	(133)	(155)	(124)

* $p < .05$.

The only evidence for genetic influence on the relationship between early childhood environment and middle childhood outcome is seen in Table 18.6, which contains the multiple correlations between the FES and the CBCL. Parental ratings of externalizing behavior at 7, 9, and 10 years are significantly related to the FES at 3 years of age for nonadoptive families and are non-significant for adoptive families, suggesting genetic influence on this relationship. The findings for parental reports of externalizing behavior are replicated for the FES at 1 year of age with the exception of externalizing at 10 years, in which there are no differences in the correlations for nonadopted and adopted children. For teacher reports of externalizing behavior the results are similar, though not as strong. For example, the FES at 1 year is significantly related to teacher reports of externalizing at 7 and 9 years, but not at 10 years.

Table 18.6 Multiple correlations, adjusted for attenuation, between the three FES scales at years 1 and 3 and negative outcome measures in middle childhood

Measure	FES			
	Year 1		Year 3	
	Nonadopted	Adopted	Nonadopted	Adopted
CBCL: Parent report				
Year 7				
Internalizing	.14	.10	.20*	.11
Externalizing	.36*	.06	.33*	.04
(N)	(222)	(181)	(231)	(174)

Table 18.6 (Cont.)

Measure	FES			
	Year 1		Year 3	
	Nonadopted	*Adopted*	*Nonadopted*	*Adopted*
Year 9				
Internalizing	.03	.22*	.15	.17*
Externalizing	.17*	.07	.33*	.09
(N)	(207)	(164)	(207)	(154)
Year 10				
Internalizing	.03	.17*	.36*	.15
Externalizing	.10	.11	.37*	.04
(N)	(181)	(150)	(184)	(138)
CBCL: Teacher report				
Year 7				
Internalizing	.10	.07	.02	.17*
Externalizing	.24*	.08	.06	.12
(N)	(218)	(181)	(224)	(173)
Year 9				
Internalizing	.09	.10	.26*	.17*
Externalizing	.25*	.11	.26*	.22*
(N)	(180)	(142)	(183)	(135)
Year 10				
Internalizing	.11	.10	.01	.08
Externalizing	.10	.22*	.24*	.20*
(N)	(145)	(124)	(147)	(115)

*$p < .05$.

Conclusions

Two broad conclusions can be drawn from the results of these analyses. First, relationships between the early environment and middle childhood outcome tend to be weak. This conclusion is supported by earlier CAP results in which longitudinal relationships between the environment at year 1 and developmental outcome during early childhood were found to be modest, especially for noncognitive measures of development (Plomin, DeFries, & Fulker, 1988). The second general conclusion that can be drawn from these results is that when environment-to-outcome correlations are significant, they do not appear to show genetic mediation. With the exception of the environment–outcome relationship between the FES and parental reports of externalizing behavior, no well-defined pattern of significant correlations emerged for nonadopted but not adopted children.

Several factors may lead to underestimates for the role of genetic mediation in environment–outcome associations in this study. In addition to the usual limitations of the measures (i.e., the HOME and FES) and the sample (e.g., a normal sample), a more interesting possibility is that the design limits identification of genetic effects on environment–outcome to genotype–environment correlations of the passive kind (Plomin, Loehlin, & DeFries, 1985). In other words, because nonadopted children share genes as well as family environment with their parents they are passively exposed to environments that are correlated with their genetic propensities. The other two types of genotype–environment correlations that could play a role in this relationship are reactive and active genotype–environment correlations. The experiences of an individual that are the result of others' reactions to that individual's genetic propensities cause reactive genotype–environment correlations. Active genotype–environment correlations occur when individuals actively select or create environments that are correlated with their genetic characteristics. If either reactive or active genotype–environment correlations were important in the environment-to-outcome relationships, they would affect both adoptive and nonadoptive families, which would result in the underestimation of genetic mediation of that relationship (Plomin et al., 1988).

In other words, the environment–outcome design used in the present analyses, although powerful for detecting passive genotype–environment correlation, is unable to detect genotype–environment correlation of the reactive or active types. In contrast, multivariate genetic analyses of the association between an environmental measure and an outcome measure, for example using the sibling adoption design, can detect genotype–environment correlation of any type (Plomin, in press). That is, the sibling adoption design can be used to decompose the covariance between an environmental measure and an outcome measure into its genetic and environmental components of covariance. Thus, if reactive or active genotype–environment correlation is important, a multivariate genetic analysis of this type should yield evidence for genetic mediation even if the environment–outcome design used in the present analyses does not because this latter design only detects passive genotype–environment correlation. For this reason, multivariate genetic analyses of this type using the sibling adoption design are underway.

19 Home Environmental Influences on General Cognitive Ability

Stacey S. Cherny

The relationship between measures of the home environment and children's cognitive development has been a major focus of research in developmental psychology (e.g., Wachs, 1992). Although all but a few of these studies involve families in which children are genetically related to their parents, rarely is the possibility considered that genetic factors mediate associations between home environment and children's development. In the case of associations between measures of the home environment and children's cognitive development, it seems plausible that genetic influences on parental intelligence might mediate such associations.

The purpose of this chapter is to incorporate measures of the home environment into model-fitting adoption analyses of parent–offspring resemblance for IQ, in order to assess the direct environmental effects of home environment independent of genetic and environmental influences mediated via parental IQ. These analyses extend research described in the previous two chapters. Chapter 17 applies the sibling adoption design to measures of the home environment treated as dependent variables, in order to ask whether genetic factors influence these measures. Chapter 18 applies a simple method described by Plomin, Loehlin, and DeFries (1985) to investigate the contributions of passive genotype–environment correlations to associations between environmental measures and measures of development. The approach merely compares environment–development correlations in nonadoptive and adoptive families. If correlations between environmental measures and developmental outcomes are greater in nonadoptive than in adoptive families, genetic mediation is implied. The analyses in Chapter 18 are limited by the problem that our environmental measures showed only weak associations with developmental measures such as self-esteem and behavioral problems, even in nonadoptive families.

It is difficult to detect genetic mediation of environment–development associations when the associations themselves are meager. In contrast, the present analyses involve cognitive development, which is associated with home environment.

The present analyses represent a multivariate extension of earlier CAP work that incorporated an environmental index in parent–offspring adoption analyses. Although DeFries, Plomin, Vandenberg, and Kuse (1981) found few associations between home environment and cognitive development at 1 year of age, Rice, Fulker, DeFries, and Plomin (1988) showed that parental IQ increasingly mediated associations between home environment and cognitive development in early childhood at 3 and 4 years. Coon, Fulker, DeFries, and Plomin (1990) reported that genetic mediation via parental IQ can account for most of the association between home environment and children's IQ at 7 years of age. The present approach simultaneously examines the influence of a set of home environmental indicators (Home Observation for Measurement of the Environment (Caldwell & Bradley, 1978) and the Family Environment Scale (Moos & Moos, 1981)) in parent–offspring adoption analyses, when the children are 7 years of age and, for the first time in the CAP, when the children are 9.

Method

Subjects and measures

The first principal component from the specific cognitive abilities test battery (GFAC) was used as the measure of general cognitive ability in the parents (see Chapter 4). General cognitive ability in the adoptive and nonadoptive probands was assessed by the Wechsler Intelligence Scale for Children–Revised (WISCR) at age 7, and from the first principal component from the telephone administered specific cognitive abilities test battery (SCATPC) at age 9.

Two measures of the home environment were used. The Home Observation for Measurement of the Environment (HOME: Caldwell & Bradley, 1978) was used as the measure of the home environment at ages 1 and 2. The Family Environment Scale (FES: Moos & Moos, 1981) was used as a contemporary measure of the home environment at ages 5, 7, and 9 years. The HOME contains items rated by both the CAP testers and the parents. Because the correlation between the HOME at 1 and 2 years of age is low, the HOME tests at each age were used in separate analyses of 7- and 9-year-old cognitive ability. The HOME consists of six scales: Responsivity of Mother, Avoidance of Restriction and Punishment, Organization of the Environment, Appropriate Play Materials, Maternal Involvement, and Variety in Daily Stimulation. A total score for the HOME was also computed by summing the six scale scores for those subjects who had scores on all six scales.

The FES is a parental questionnaire that assesses the quality of the social relationships in the family. It includes 10 scales: Cohesion, Expressiveness, Conflict, Independence, Achievement, Intellectual–Cultural, Active–Recreational, Moral–Religious, Organization, and Control. In an attempt to obtain items with more desirable statistical properties, these items were rated on a 5-point Likert scale, rather than the original true/false format used by Moos & Moos (1981). The FES manifests high longitudinal correlations in our sample. In order to use it as a contemporary measure of the environment and also to capitalize on increased reliability by summing over repeated measurements, the analyses of general cognitive ability at age 7 included the average of standardized FES items obtained at ages 5 and 7, and the year 9 analyses included the average of standardized FES items obtained at ages 7 and 9. Factor scores on two orthogonal factors were also created. At year 7, the first factor, labeled Personal Growth, had substantial positive loadings on Cohesion, Expressiveness, Independence, Intellectual–Cultural, and Active–Recreational, with a negative loading on Conflict. The second factor, labeled Traditional Organization, had substantial positive loadings on Achievement, Moral-Religious, Organization, and Control. The factor pattern was highly similar at year 9, although the loadings on Independence and Achievement were more moderate in magnitude than those at year 7. The year 7 factor pattern is identical to that reported by Coon et al. (1990), even though our sample is somewhat larger. The factor pattern is also similar to that reported by Plomin and DeFries (1985) for the FES at year 1.

Model

The model chosen for the analyses is a multivariate extension of a model first presented by Rice et al. (1988) and later by Coon et al. (1990). Their model is the univariate parent–offspring model, discussed in Chapter 3, in which the observed phenotypic value (P) for each family member is completely determined by that individual's additive genetic value (G) and environmental deviation (E). G and E are correlated s. Assortative mating, that is, the correlation between the parents' phenotypes, is p. The extension of this model by Rice et al. and Coon et al. allowed for the child's environmental deviation to be influenced directly (f) by a measure of the home environment (H). This measure of the home environment can be correlated with each rearing parent's genotype and environmental deviation (r_G and r_E respectively). The multivariate version of this model allows for multiple home environmental measures, arranged in a vector, $\mathbf{H}$, which are each correlated with the rearing parent's genotype and environmental deviation ($\mathbf{r}_G$ and $\mathbf{r}_E$, respectively). Each home environmental measure could also exert a direct effect on the child's environmental deviation ($\mathbf{f}$). This model is depicted in a multivariate path diagram in Figure 19.1 for adoptive families and in Figure 19.2 for nonadoptive families.

Since previous analyses of the CAP dataset indicated that there was no significant difference between maternal and paternal cultural transmission, a single cultural transmission parameter, z, was estimated in the present model for both mothers and fathers. In addition, since there was no significant difference between nonadoptive, adoptive, and biological spousal correlations, only a single assortment parameter, p, was estimated. Furthermore, since the maternal and paternal correlations between the home environmental measures and general intelligence were highly similar and not significantly different, single r_G and r_E vectors were estimated for mothers and fathers. The expected variances and covariances implied by the model are presented in Table 19.1, along with the constraints necessary for identification of the model.

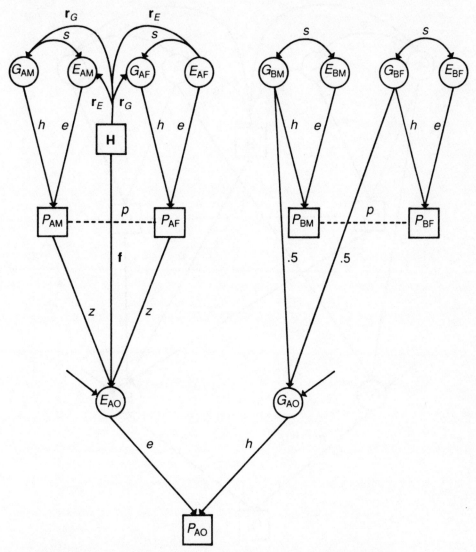

AM = adoptive mother; AF = adoptive father; BM = biological mother; BF = biological father
AO = adopted offspring.
h = square root of h^2 (heritability).
e = square root of e^2 (environmental variance).
s = G–E correlation.
p = spousal correlation.
Explanations of other symbols appear in the text.

Figure 19.1 Full model of genetic (G) and environmental (E) transmission with home environmental measures for adoptive families (dashed lines represent conditional paths).

Models were fitted using the maximum-likelihood pedigree procedure described in Chapter 3.

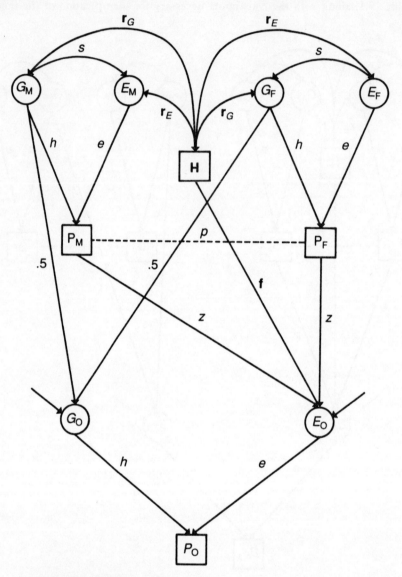

M = mother; F = father; O = offspring.
h = square root of h^2 (heritability).
e = square root of e^2 (environmental variance).
s = G–E correlation.
p = spousal correlation.
Explanations of other symbols appear in the text.

Figure 19.2 Full model of genetic (G) and environmental (E) transmission with home environmental measures for nonadoptive families (dashed line represents a conditional path).

Table 19.1 Expected adoptive and nonadoptive variance and covariances

Variable	Expectation
	Nonadoptive families
$\sigma_{M,F}$	$[p]\sigma_P^2$
$\boldsymbol{\sigma}_{H,M} = \boldsymbol{\sigma}_{H,F}$	$[\mathrm{diag}(\boldsymbol{\Sigma}_H)]^{1/2}\,[\mathbf{r}_G h + \mathbf{r}_E e]\sigma_{Pc}$
$\sigma_{M,C} = \sigma_{F,C}$	$[\frac{1}{2}h(h + se)(1 + p) + ez(1 + p) + e\mathbf{f}'(\mathbf{r}_G h + \mathbf{r}_E e)]\sigma_P\,\sigma_{Pc}$
$\boldsymbol{\sigma}_{H,C}$	$[\mathrm{diag}(\boldsymbol{\Sigma}_H)]^{1/2}\,[\mathbf{f}_e + (\mathbf{r}_G h + \mathbf{r}_E e)2ez + \mathbf{r}_G h]\sigma_{Pc}$
$\sigma_M^2 = \sigma_F^2$	$[h^2 + c^2 + 2hse]\sigma_P^2$
$\boldsymbol{\Sigma}_H$	$\boldsymbol{\Sigma}_H$
σ_C^2	$[h^2 + c^2 + 2hse]\sigma_{Pc}$
	Adoptive families
$\sigma_{AM,AF} = \sigma_{BM,BF}$	$[p]\sigma_P^2$
$\boldsymbol{\sigma}_{H,AM} = \boldsymbol{\sigma}_{H,AF}$	$[\mathrm{diag}(\boldsymbol{\Sigma}_H)]^{1/2}\,[\mathbf{r}_G h + \mathbf{r}_E e]\sigma_{Pc}^2$
$\boldsymbol{\sigma}_{H,BM} = \boldsymbol{\sigma}_{H,BF}$	0
$\sigma_{AM,C} = \sigma_{AF,C}$	$[ez(1 + p) + e\mathbf{f}'(\mathbf{r}_G h + \mathbf{r}_E e)]\sigma_P\,\sigma_{Pc}$
$\sigma_{BM,C} = \sigma_{BF,C}$	$[\frac{1}{2}h(h + se)(1 + p)]\sigma_P\,\sigma_{Pc}$
$\boldsymbol{\sigma}_{H,C}$	$[\mathrm{diag}(\boldsymbol{\Sigma}_H)]^{1/2}\,[\mathbf{f}e + (\mathbf{r}_G h + \mathbf{r}_E e)2ez]^{1/2}\sigma_{Pc}$
$\sigma_{AM}^2 = \sigma_{AF}^2 = \sigma_{BM}^2 = \sigma_{BF}^2$	$[h^2 + c^2 + 2hse]\sigma_P^2$
$\boldsymbol{\Sigma}_H$	$\boldsymbol{\Sigma}_H$
σ_C^2	$[h^2 + e^2]\sigma_{Pc}^2$
	Constraints
s	$z(h + se)(1 + p) + \mathbf{f}'\mathbf{r}_G$
se	$[zeh(1 + p) + \mathbf{f}'\mathbf{r}_G]/[1 - ze(1 + p)]$

Results

Home environment/IQ correlations

Table 19.2 contains correlations of the six HOME scales and total score at year 1 with rearing parents' general cognitive ability and adopted and nonadopted child general cognitive ability at years 7 and 9. The HOME total score is correlated significantly with both mothers' and fathers' IQ. There is also a relatively small but consistent relationship between parental IQ and the six HOME scales, with similar patterns of correlations in mothers and fathers. The correlations between child IQ and the HOME scales and total score are less consistent and do not strongly suggest either direct or indirect home environmental influences on cognitive ability at either 7 or 9 years.

The HOME total score at year 2 shows the same high maternal and paternal correlations (see Table 19.3). There is also a significant correlation with WISC-R at year 7 for nonadopted probands and an essentially zero correlation for adopted

Table 19.2 Correlations of HOME scales at year 1 with parental general intelligence and with child general intelligence at years 7 and 9

HOME scale	Parental GFAC[a]		WISC-R year 7		SCATPC[b] year 9	
	Mother	Father	Nonadoptive	Adoptive	Nonadoptive	Adoptive
Total	.17** (319)	.20** (318)	.13 (157)	-.13 (124)	.01 (119)	.04 (109)
Responsivity of Mother	-.00 (411)	.06 (407)	.10 (192)	.02 (161)	-.00 (154)	.17* (146)
Avoidance of Restriction	.15** (428)	.11* (425)	.10 (194)	-.12 (173)	.02 (151)	-.12 (159)
Organization of Environment	.05 (442)	.02 (439)	.14* (203)	-.09 (175)	.10 (161)	-.03 (162)
Appropriate Play Materials	.08 (460)	.12** (456)	-.01 (210)	-.04 (185)	-.01 (166)	.00 (172)
Maternal Involvement	.07 (451)	.06 (448)	.08 (205)	-.05 (182)	-.02 (162)	.08 (167)
Variety in Daily Stimulation	.08 (477)	.06 (473)	.01 (214)	-.08 (193)	.03 (171)	.01 (179)

Note: Sample sizes appear in parentheses.

[a] First principal component score from the specific cognitive abilities test battery.

[b] First principal component score from the telephone administered specific cognitive abilities test battery.

*$p < .05$, **$p < .01$.

Table 19.3 Correlations of HOME scales at year 2 with parental general intelligence and with child general intelligence at years 7 and 9

HOME scale	Parental GFAC[a]		WISC–R year 7		SCATPC[b] year 9	
	Mother	Father	Nonadoptive	Adoptive	Nonadoptive	Adoptive
Total	.18** (410)	.20** (409)	.22** (197)	.01 (166)	.08 (156)	.04 (153)
Responsivity of Mother	.06 (439)	.16** (437)	.21** (209)	-.06 (177)	.12 (166)	.03 (163)
Avoidance of Restriction	.17** (437)	.15** (435)	.03 (208)	-.04 (177)	-.06 (165)	-.12 (163)
Organization of Environment	-.10* (445)	-.06 (442)	.07 (208)	.10 (182)	.01 (165)	.14 (171)
Appropriate Play Materials	.09* (433)	.13** (431)	.12 (206)	-.08 (177)	.18* (164)	-.03 (163)
Maternal Involvement	.11* (441)	.06 (440)	.14* (208)	-.00 (181)	-.03 (164)	.01 (165)
Variety in Daily Stimulation	.11* (461)	.09* (458)	.09 (213)	.10 (191)	.11 (170)	.13 (177)

Note: Sample sizes appear in parentheses.

[a] First principal component score from the specific cognitive abilities test battery.

[b] First principal component score from the telephone administered specific cognitive abilities test battery.

*p < .05, **p < .01.

Table 19.4 Correlations of FES scales with parental general intelligence and with child general intelligence at years 7 and 9

FES measure	FES 5/7				FES 7/9			
	Parental GFAC[a]		WISC-R year 7		Parental GFAC[a]		SCATPC[b] year 9	
	Mother (N = 429)	Father (N = 427)	Nonadopted (N = 208)	Adopted (N = 188)	Mother (N = 411)	Father (N = 409)	Nonadopted (N = 168)	Adopted (N = 178)
Scale								
Cohesion	.00	.03	.05	.07	.02^c	.06^e	.10^g	.05
Expressivity	.04	.09	.05	−.01	.06^c	.05^e	.08^g	−.04
Conflict	.02	.05	.06	−.21**	−.01	.04	−.05	−.13
Independence	−.02	.05	.09	.06	.01^d	.09^f	.05^h	−.06^i
Achievement	−.12*	−.09	.05	.06	−.14**	−.10*	.02	−.02
Intellectual–Cultural	.17**	.27**	.21**	.03	.12*	.23**	.15*	.10
Active–Recreational	.08	.04	−.05	−.10	.07	.01	−.03	−.01
Moral–Religious	−.05	−.16**	−.11	−.01	−.05^d	−.19^f	−.00^h	−.04^i
Organization	−.05	−.09	−.07	.09	−.04	−.08	−.05	.16*
Control	−.05	−.06	−.10	−.00	−.02^c	−.09^e	−.08^g	.17*
Factor								
Personal Growth	.06	.11*	.09	.05	.07^d	.12*	.10^h	−.00^i
Traditional Organization	−.09	−.15**	−.10	.07	−.09^d	−.15**	−.02^h	.13^i

Note: Sample sizes appear in parentheses.
[a] First principal component score from the specific cognitive abilities test battery.
[b] First principal component score from the telephone administered specific cognitive abilities test battery.
^{c} N = 412, ^{d} N = 392, ^{e} N = 410, ^{f} N = 410, ^{g} N = 390, ^{g} N = 169, ^{h} N = 158, ^{i} N = 169.
*p < .05, **p < .01.

probands, suggesting genetic mediation. This pattern is also suggested for age 9 SCATPC, although neither correlation is significant. The correlations with child IQ at 7 and 9 years are slightly higher for the nonadopted children at year 2 than at year 1. Genetic mediation of the home environmental influences is also suggested for Responsivity of Mother and Appropriate Play Materials at both years 7 and 9, and Maternal Involvement at year 7, since the nonadoptive correlations are higher than the corresponding adoptive correlations.

In Table 19.4, the IQ/FES correlations are presented. Fathers' IQ, but not mothers' IQ, correlated with both FES factors. Parental IQ is related consistently only to Achievement (mothers only), Intellectual–Cultural, and Moral–Religious (fathers only), for both the year 5/7 and year 7/9 FES measures. The nonadoptive correlation is substantially higher than the adoptive correlation for WISC-R with FES Intellectual–Cultural, suggesting that this relationship may be genetically mediated. The other correlations do not show a consistent pattern.

A clearer picture emerges from the results of the model-fitting procedures, whereby we can quantify the impressions formed from simple inspection of the correlations.

Model-fitting

The full model presented in Figures 19.1 and 19.2 was fitted to the HOME scales at year 1 with IQ at years 7 and 9, HOME scales at year 2 with IQ at 7 and 9, FES scales at 5 and 7 with IQ at 7, and FES scales at 7 and 9 with IQ at 9. Univariate models were fitted to the HOME total scores for the same age combinations as for the multivariate models and bivariate models were fitted to the two FES factors similarly. Tests of individual parameters were obtained via comparison of a model omitting that parameter against the full model, using the likehood ratio χ^2. Simultaneous 2 df tests of r_G and r_E for a home environmental measure were also performed, allowing an overall test of the correlation of the home measure with parental IQ.

Tables 19.5 through 19.10 contain parameter estimates from the 12 full models, along with χ^2 tests of those parameters. Estimates of heritability from the multivariate models, under the assumption that IQ measures at 7 and 9 are isomorphic with the adult measure of IQ, appear below each table. All estimates of heritability were highly significant and involved a simultaneous test of all r_G in the models. Estimates of h^2 ranged from .33 to .34 at year 7 and .30 to .31 at year 9. Estimates of s ranged from .02 to .03 at year 7 and .00 to .01 at year 9, indicating that there was little or no genotype–environment correlation for IQ. Estimates of z ranged from .02 to .03 at year 7 and .00 to .01 at year 9, implying little or no cultural transmission for IQ. Finally, assortative mating was estimated at .28 in all models, and was highly significant. Estimates from the univariate and bivariate models were virtually identical.

Examining the parameter estimates for the HOME at year 1 with IQ at year 7 in Table 19.5, we see that there was substantial r_G for most scales, although none was statistically significant. The total score showed high r_G, although it was also not

Table 19.5 HOME at year 1 with IQ at year 7: Parameter estimates

Parameter	Scale[a]						
	Total	RSPNS	RSTRC	ORGNZ	PLAY	INVLV	VRTY
r_G	.22	.15	.17	.24	.03	.13	.10
r_E	.08	-.07	.04	-.14	.11	-.01	.03
f	.00	.06	-.02	-.00	-.02	.01	-.06
$\chi^2_1(r_G = 0)$	2.551	1.503	1.793	3.404	0.056	1.098	0.626
$\chi^2_1(r_E = 0)$	0.655	0.477	0.163	1.945	1.391	0.011	0.081
$\chi^2_2(r_E = r_G = 0)$	17.986***	1.966	11.786**	3.611	7.584*	3.653	4.254
$\chi^2_1(f = 0)$	0.000	1.057	0.145	0.000	0.078	0.023	0.789

Note: $h^2 = .33^{**}$, $s = .03$, $e^2 = .65$, $p = .28^{**}$, $z = .03$.
[a] RSPNS = Responsivity of Mother; RSTRC = Avoidance of Restriction and Punishment; ORGNZ = Organization of the Environment; PLAY = Appropriate Play Materials; INVLV = Maternal Involvement; VRTY = Variety in Daily Stimulation.
*$p < .05$, **$p < .01$, ***$p < .001$.

Table 19.6 HOME at year 2 with IQ at year 7: Parameter estimates

Parameter	Scale[a]						
	Total	RSPNS	RSTRC	ORGNZ	PLAY	INVLV	VRTY
r_G	.39	.36	.07	-.07	.30	.13	.01
r_E	-.02	-.11	.15	-.06	-.08	.03	.13
f	.00	.01	-.03	.11	-.03	.12	.13
$\chi^2_1(r_G=0)$	3.737	7.250**	0.267	0.272	4.746*	1.118	0.007
$\chi^2_1(r_E=0)$	0.059	1.110	2.385	0.359	0.550	0.068	1.748
$\chi^2_2(r_E=r_G=0)$	27.011***	15.043***	17.284***	5.131	11.392**	7.187*	8.335*
$\chi^2_1(f=0)$	0.000	0.022	0.131	2.059	0.257	2.483	2.807

Note: $h^2 = .34***$, $s = .02$, $e^2 = .64$, $p = .28***$, $z = .02$
[a] RSPNS = Responsivity of Mother; RSTRC = Avoidance of Restriction and Punishment; ORGNZ = Organization of the Environment; PLAY = Appropriate Play Materials; INVLV = Maternal Involvement; VRTY = Variety in Daily Stimulation.
*$p < .05$, **$p < .01$, ***$p < .001$.

Table 19.7 HOME at year 1 with IQ at year 9: Parameter estimates

Parameter	Total	Scale[a]					
		RSPNS	RSTRC	ORGNZ	PLAY	INVLV	VRTY
r_G	.00	−.03	.02	.20	−.02	−.05	.09
r_E	.23	.07	.14	−.10	.14	.12	.03
f	.04	.10	−.06	.03	−.01	.03	−.00
$\chi^2_1(r_G = 0)$	0.000	0.089	0.024	1.773	0.019	0.152	0.475
$\chi^2_1(r_E = 0)$	3.648	0.565	1.931	0.845	1.791	1.474	0.116
$\chi^2_2(r_E = r_G = 0)$	18.034***	0.988	10.958**	2.163	7.899*	3.646	4.511
$\chi^2_1(f = 0)$	0.246	2.243	0.708	0.340	0.019	0.193	0.002

Note: $h^2 = .30*$, $s = .01$, $e^2 = .69$, $p = .28***$, $z = .01$
[a] RSPNS = Responsivity of Mother; RSTRC = Avoidance of Restriction and Punishment; ORGNZ = Organization of the Environment; PLAY = Appropriate Play Materials; INVLV = Maternal Involvement; VRTY = Variety in Daily Stimulation.
*$p < .05$, **$p < .01$, ***$p < .001$.

significant. The 2 *df* test of r_G and r_E simultaneously was statistically significant, however. The direct effect (f) of the HOME total score on the environment of the child was estimated at zero. The 2 *df* tests for the scales indicated that Avoidance of Restriction and Punishment and Appropriate Play Materials showed significant mediation through the parental phenotype. There was no significant direct effect of the HOME measures (**f**) at year 1 on IQ at 7, which is consistent with the correlations presented in Table 19.2.

The genetic mediation was stronger for the HOME measures at year 2 with IQ at 7 (Table 19.6). For the total score, r_G was relatively high and r_E was essentially zero. However, only the 2 *df* test was significant. The direct effect (f) was again zero. Responsivity of Mother and Appropriate Play Materials both showed significant genetic mediation of their influences on IQ at 7. All but one (Organization of the Environment) of the 2 *df* tests of mediation via the parental phenotype were statistically significant. None of the r_E nor **f** parameters was statistically significant, although there was a suggestion of environmental mediation for Avoidance of Restriction and Punishment and Variety in Daily Stimulation. There was also a suggestion of a direct effect (f) of Organization of the Environment, Maternal Involvement, and Variety in Daily Stimulation on IQ at 7.

The HOME scales at year 1 with IQ at year 9 suggest more environmental mediation and less genetic mediation than with IQ at year 7 (Table 19.7). This is most notable for the total score, where r_E was substantial and r_G zero. However, only the 2 *df* test was significant. None of the scales showed statistically significant genetic or environmental mediation. The 2 *df* tests for both Avoidance of Restriction and Punishment and Appropriate Play Materials did suggest significant mediation, which appears to be largely environmental via the parental phenotype. Except possibly for Responsivity of Mother, no suggestion of a direct home environmental effect was found.

The HOME scales at year 2 with IQ at year 9 show a similar pattern to that with IQ at year 7, although the slightly reduced sample size at year 9 reduced power accordingly (Table 19.8). The r_E for the total score was statistically significant, while r_G was again estimated at zero. There appears to be substantial genetic mediation for Responsivity of Mother and Appropriate Play Materials, although these effects could only be detected via the stronger 2 *df* tests. There is significant environmental mediation for Avoidance of Restriction and Punishment and Maternal Involvement. Finally, there appears to be a suggestion of a direct effect for Organization of the Environment and Variety in Daily Stimulation, although neither was statistically significant.

Estimates for the FES at years 5 and 7 with IQ at year 7 appear in Table 19.9. Both factors show some genetic mediation and no environmental mediation, although only the 2 *df* test for Traditional Organization was statistically significant. In examining the scales, both Conflict and Intellectual–Cultural show evidence for genetic mediation. The effect of Achievement appears to be largely environmentally mediated, although only the 2 *df* test for mediation via the parental phenotype was significant. Moral–Religious showed some genetic and environmental mediation that resulted in a significant 2 *df* test. Finally, a significant direct effect (f) of Conflict was found.

Table 19.8 HOME at year 2 with IQ at year 9: Parameter estimates

Parameter	Scale[a]						
	Total	RSPNS	RSTRC	ORGNZ	PLAY	INVLV	VRTY
r_G	.00	.22	−.07	−.12	.30	−.14	.04
r_E	.25	.00	.24	−.03	−.07	.22	.10
f	.09	.05	−.09	.12	.01	.08	.14
$\chi^2_1(r_G = 0)$	0.000	1.991	0.214	0.684	3.765	0.992	0.080
$\chi^2_1(r_E = 0)$	25.788***	0.000	5.681*	0.075	0.386	3.996*	1.003
$\chi^2_2(r_E = r_G = 0)$	25.788****	11.089**	17.993***	5.574	10.776**	7.826**	7.932*
$\chi^2_1(f = 0)$	0.864	0.448	1.412	2.619	0.026	1.103	3.030

Note: $h^2 = .31**$, $s = .00$, $e^2 = .69$, $p = .28***$, $z = .00$.
[a] RSPNS = Responsivity of Mother; RSTRC = Avoidance of Restriction and Punishment; ORGNZ = Organization of the Environment; PLAY = Appropriate Play Materials; INVLV = Maternal Involvement; VRTY = Variety in Daily Stimulation.
*$p < .05$, **$p < .01$, ***$p < .001$.

Table 19.9 FES at years 5 and 7 with IQ at year 7: parameter estimates

Parameter	Scale[a]										Factor	
	COHSN	EXPR	CONF	INDP	ACH	INTCL	ACTRC	MRELG	ORG	CONT	Personal Growth	Tradition organization
r_G	.10	.09	.22	.07	.06	.28	.09	-.11	-.09	-.10	.13	-.18
r_E	-.05	.03	-.12	-.02	-.18	.09	.02	-.07	-.02	-.00	.02	-.03
f	.04	.05	-.14	.08	.05	.07	-.11	-.02	.04	-.03	.05	.03
$\chi_1^2(r_G = 0)$	0.769	0.497	3.995*	0.301	0.238	5.120*	0.468	0.835	0.529	0.638	1.057	2.071
$\chi_1^2(r_E = 0)$	0.301	0.104	2.110	0.034	3.555	1.014	0.044	0.543	0.070	0.002	0.052	0.125
$\chi_2^2(r_E = r_G = 0)$	0.866	3.659	4.203	0.584	8.310*	38.195***	2.915	9.366**	3.199	2.441	5.779	10.490**
$\chi_1^2(f = 0)$	0.425	0.580	4.542*	1.329	0.609	1.752	2.921	0.099	0.268	0.228	0.576	0.576

Note: $h^2 = .33**$, $s = .02$, $e^2 = .65$, $p = .28***$, $z = .02$

[a] COHSN = Cohesion; EXPR = Expressiveness; CONF = Conflict; INDP = Independence; ACH = Achievement; INTCL = Intellectual–Cultural; ACTRC = Active–Recreational; MRELG = Moral–Religious; ORG = Organization; CONT = Control.

*$p < .05$, **$p < 0.1$, ***$p < .001$.

Table 19.10 FES at years 7 and 9 with IQ at year 9: Parameter estimates

Parameter	Scale[a]										Factor	
	COHSN	EXPR	CONF	INDP	ACH	INTCL	ACTRC	MRELG	ORG	CONT	Personal Growth	Traditional Organization
r_G	.22	.20	.02	-.07	.12	.18	.02	.04	-.14	-.22	.19	-.07
r_E	-.11	-.06	.00	.10	-.23	.12	.04	-.18	.03	.08	-.01	-.10
f	.06	.01	-.11	-.04	-.04	.13	-.02	.02	.10	.08	.03	.08
$\chi^2_1(r_G=0)$	3.126	1.920	0.012	0.195	0.643	1.796	0.025	0.064	1.255	2.754	2.147	0.374
$\chi^2_1(r_E=0)$	1.307	0.402	0.001	0.807	4.777*	1.398	0.157	3.334	0.080	0.786	0.013	1.798
$\chi^2_2(r_E=r_G=0)$	3.500	3.247	0.075	1.324	10.753**	23.939***	1.329	10.022**	2.874	4.000	6.890*	8.539*
$\chi^2_1(f=0)$	1.018	0.012	2.531	0.258	0.316	4.115*	0.084	0.068	2.200	1.538	0.374	1.301

Note: $h^2 = .30*$, $s = .00$, $e^2 = .70$, $p = .28***$, $z = .00$

[a] COHSN = Cohesion; EXPR = Expressiveness; CONF = Conflict; INDP = Independence; ACH = Achievement; INTCL = Intellectual–Cultural; ACTRC = Active–Recreational; MRELG = Moral–Religious; ORG = Organization; CONT = Control.

*$p < .05$, **$p < .01$, ***$p < .001$.

For the FES at year 9 (Table 19.10), the Personal Growth factor appears to show some genetic mediation, while both genetic and environmental mediation appears for Traditional Organization. However, only the 2 *df* tests were statistically significant. Cohesion, Expressiveness, Intellectual–Cultural and Control effects all appear to be genetically mediated; however, only Intellectual–Cultural was statistically significant, even with the 2 *df* test. The effect of Achievement on IQ was significantly environmentally mediated. Finally, only Intellectual–Cultural showed a statistically significant direct effect on IQ.

Discussion

For the present analysis, the parent–offspring full adoption design was employed to assess the nature of home environmental influences on general cognitive ability. The model allowed us to separate the direct effect of the home environmental measures on child IQ (via child's environmental deviation) from the indirect influences mediated by the parental genotype or environmental deviation. That is, the direct effect of the home environment was estimated while controlling or partialling out genetic and other environmental influences originating from the parents. Estimates of heritability, G–E correlation, cultural transmission, and assortment were obtained and found to be relatively unaffected by the particular home environmental measures chosen for analysis.

In general, the relationships between the measures of the home environment and child IQ were small, with a trend toward higher correlations in nonadoptive than in adoptive families. This suggests some genetic mediation of home environmental influence on IQ. Model-fitting results confirmed the general impression of greater genetic than environmental mediation of home environmental influences. The HOME total scores showed substantial genetic mediation at year 7, but not at year 9. The FES Personal Growth factors had genetically mediated effects on IQ at both years 7 and 9, but the Traditional Organization factors showed genetic mediation only at year 7.

Results from the environmental measures and year 7 IQ analyses were quite consistent with those reported by Coon et al. (1990), although the magnitude of the effects was not as great. In the case of the FES, this might be a result of using the average of the FES at ages 5 and 7, while Coon et al. averaged the FES scores obtained at ages 1, 3, 5, and 7 years. Since the present sample is larger than the one on which Coon et al. reported (both as a result of employing the pedigree approach to analysis and due to continued testing of subjects), results might also be expected to differ. Finally, the multivariate approach may also contribute to any differences.

It is puzzling that some of the results for the HOME appear different at year 9 than at year 7. For the year 1 HOME total score, the analysis at year 7 suggested primarily genetic mediation, but the year 9 analysis indicated environmental mediation. Only the 2 *df* tests were significant in both cases, indicating that resolving the nature of the parental mediation was not possible in either case. This pattern was also seen for Avoidance of Restriction and Punishment.

For the HOME at year 2, again there was genetic mediation for the total score at year 7 while environmental mediation was found at year 9. In this case, the environmental mediation at year 9 was statistically significant. Maternal Involvement at year 2 showed exactly the same pattern as the total score.

Differences between the year 7 and year 9 results for the FES were found for Conflict, with genetic mediation present at year 7 but not at 9. The other differences were merely of magnitude of effects, resulting in no major changes in the conclusions. It will be interesting to replicate the results at year 9 with the full sample and to chart the trend through early adolescence.

We conclude that the relationships between measures of the home environment and child IQ, although relatively small, are generally mediated through the parental genotype for IQ. However, in some instances, mediation is also via the parental environmental deviation. Only for a few measures was there any evidence of direct transmission of home environmental influences to child IQ. Thus, these findings suggest that home environmental influences, to the extent that they exist, are mediated primarily via parental IQ. Assuming that the measures of the home environment that we employed are relevant to the development of general cognitive ability, the finding of little direct influence of home environmental measures on the children's IQ is consistent with our estimates of little environmental variance shared by siblings for IQ (see Chapter 4). A direct effect of home environment on child IQ would contribute to shared environmental influences in sibling pairs. Of course, home environmental influences are more likely to be detected when extreme negative or positive environments are present. However, such influences, at least as measured by the HOME and FES, appear to be small and often genetically mediated in the normal range of individual differences represented by the CAP families.

20 Genotype–Environment Interaction and Correlation

Scott L. Hershberger

Traditional behavioral genetic analysis has been concerned largely with the estimation of genetic and environmental influences within a single sample at one point in time. The most frequent question asked under this traditional perspective is: "How much of the variation in a trait can be accounted for by individual differences in genotypes and how much by individual differences in environments?" The most common response to this question is given in terms of heritability and environmentality coefficients.

Although this is a reasonable first step in understanding the etiology of individual differences, it is important to go beyond this basic nature–nurture question. Genotype–environment interaction (G×E) and genotype–environment correlation (CovGE) extend the information contained in simple heritability and environmentality coefficients by addressing the relationship between genes and environments. From this point of departure, the researcher may then search for the developmental processes responsible for the creation of the interaction or correlation, the probability of their discovery increasing with the longitudinal assessment of G×E and CovGE.

G×E refers to the possibility that different genotypes may develop different phenotypes in response to environmental conditions (Plomin & Hershberger, 1991). For example, a child possessing a high genotype for introversion may be moderately interactive with a small coterie of friends but quite inhibited within a large group of unfamiliar children, whereas a child with a high genotype for extraversion may be uninhibited in both situations. The meaning of "interaction" in G×E is identical to its meaning in the familiar analysis-of-variance context: a nonadditive relationship exists between the two main effect variables, here genes and environments.

CovGE occurs when different genotypes are selectively exposed to different environments (Plomin, DeFries, & Loehlin, 1977). As one example, CovGE may be of importance when adolescents differing in intelligence select different educational alternatives. If the more intelligent adolescents are exposed to more difficult educational demands, then a positive association is found between genotype and environment. But CovGE may be negative as well, a situation that may occur, for instance, when children of lower intelligence are given enhanced educational opportunities to improve their scholastic performance. A CovGE is literally a statistical association between genotypes and environments, and should be given the same interpretation as any correlation between two observed variables.

At present, no one has proposed a developmental theory as to how G×E might influence individual differences. This is not the case for CovGE. Scarr and McCartney (1983) proposed a theory of development that incorporated three types of CovGE, each distinguishable by the process inducing the correlation. *Passive CovGE* refers to the transmission of both genes and environments from parents to children. By parents transmitting both genes and environments, children are passively exposed to environments correlated with their genetic endowments. *Evocative CovGE* (referred to as "reactive" by Plomin et al. (1977)) occurs when a child evokes experiences that derive from the reactions of others to the child's genetically influenced behavior. *Active CovGE* refers to the child's selection of an environment correlated with the child's genotype. According to Scarr and McCartney, the relative importance of the three CovGEs changes across childhood. The influence of passive CovGE declines from infancy to adolescence, whereas the importance of evocative and active CovGE increase over the same period. Thus, the period of middle childhood, the focus of this book, should witness the decline of passive CovGE and the rise of evocative and active CovGE.

The Colorado Adoption Project (CAP) provides an opportunity to assess the importance of both G×E and CovGE across development, and indeed previous attempts in the CAP have done so for infancy (ages 1 and 2; Plomin & DeFries, 1985) and early childhood (ages 3 and 4; Plomin, DeFries, & Fulker, 1988). A hierarchical regression model has been used to detect the presence of significant G×E on a diverse number of child outcome measures. Under this regression model, the genotype of the adopted child is represented by the phenotypic score obtained from the biological mother. The environment of the adopted child may be represented as either the phenotypic score obtained by an adoptive parent or some direct measure of the adoptive home environment. As a first step, the main effects of genes and environments are entered into the regression equation. Then, the change in R^2 is examined upon entry of an interaction term. If the change in R^2 is significant, evidence exists for the significance of G×E.

Plomin and DeFries (1985) have reported the results of applying this regression approach at 1 and 2 years of age to the prediction of a host of child outcome measures, using a number of biological and adoptive parent/home measures. Of the

80 regression equations computed at each age, only one at each age was statistically significant at $p < .05$: at year 1, the interaction of EASI Emotionality-Anger and HOME Avoidance of Restriction and Punishment in the prediction of adopted child CCTI Emotionality, and at year 2, the interaction of EASI Activity and FES Traditional Organization in the prediction of adopted child CCTI Activity. Plomin et al. (1988) have reported G×E results from the CAP for children 3 and 4 years of age. Of the many G×Es explored in this analysis, not one, including the two significant interactions found in the earlier years, was significant at $p < .05$. Thus, to date, G×E has proven difficult to detect in the CAP data.

The CAP data have revealed more promising results for the significance of CovGE. Two analytic approaches have been used to assess the magnitude of passive CovGE. The first, and more complicated approach, uses a parent–offspring design, wherein the correlations between adoptive parents and adopted children are compared with the correlations between biological parents and their children. This "model-fitting" approach typically obtains maximum-likelihood parameter estimates of passive CovGE.

For the traits examined to date, passive CovGE assessed through model-fitting has been found to be generally minute but occasionally significant. For example, Plomin, Coon, Carey, DeFries, and Fulker (1991) found for CCTI temperament traits the following values: Emotionality (.00), Activity (.01), Sociability (.04), and Impulsivity (.00). Values for passive CovGE for general intelligence range from .01 to .05 for ages 1 through 7 with no discernible pattern (Fulker, DeFries, & Plomin, 1988). In the realm of specific cognitive abilities, passive CovGE for perceptual speed (.02) at age 3 and for verbal ability (.04) at age 4 are significant (Bergeman, Plomin, DeFries, & Fulker, 1988). Plomin et al. (1988) also report very small values of passive CovGE (less than .05) for height and shyness for ages 1 through 4.

The second approach to the assessment of passive CovGE, and the approach taken in this chapter, is to compare the variances of adopted children with those of children reared by their biological parents. When positive, passive CovGE increases the phenotypic variance of nonadopted children but not adopted children, for adoptive parents and adopted children share no genetic variance and thus passive CovGE cannot occur. If passive CovGE is significant, significant variance differences must exist between the two types of children. When the variance for nonadopted children significantly exceeds the variance for adopted children, positive passive CovGE is in evidence; when the magnitude of the difference favors adopted children, evidence for negative passive CovGE has been obtained.

Using this approach for detecting passive CovGE, Plomin and DeFries (1985) and Plomin et al. (1988) compared the variances for adopted and nonadopted children, ages 1 through 4, for a number of cognitive, personality, and physical variables measured in the CAP. Their general conclusion was that outside the cognitive realm, little evidence existed for passive CovGE. Nevertheless, evidence was found to support Scarr and McCartney's (1983) hypothesis that passive CovGE does decline from infancy. The variance-comparison approach to the detection of significant passive CovGE, in comparison with the model-fitting approach, is less powerful and subject to higher standard errors (Loehlin & DeFries, 1987). Yet, with sample sizes of sufficient magnitude, the comparison of variances can be an effective means of uncovering significant passive CovGE (Eaves, 1976). Certainly, the variance difference test has the advantage of not requiring the inclusion of both parents and children within the same analysis.

The CAP design also provides a means of identifying evocative CovGE using specific measures of the environment (Plomin et al., 1977). Evocative CovGE can be assessed by the correlation between the phenotypic score of the biological mother and an environmental measure of the adoptive home of the adopted-away child. A biological mother's score serves as an index for the genotype of the child, while the environmental measure of the adoptive home indicates the influence of the child on the environment. Most of the evocative CovGEs examined in the CAP are not significant (Plomin & DeFries, 1985; Plomin et al., 1988). The highest found was −.17 between the Activity dimension of the EASI Temperament Survey and the first principal component extracted from a factor analysis of the HOME. The approach taken to the assessment of evocative CovGE in this chapter extends the previously used method by employing a multivariate design: Canonical correlations are computed between a *set* of measures obtained from biological mothers and a *set* of adoptive-home environmental measures reactive to the behavior of the child.

Active CovGE will not be assessed in this chapter, for the CAP design provides no unambiguous method by which active CovGE may be examined. In theory, the estimation of active CovGE could proceed in a fashion similar to evocative CovGE: A set of environmental measures sensitive to the tendency of children to select certain stimuli from the environment could be correlated with a set of relevant cognitive and personality variables measured on biological mothers. However, environmental measures employed in the CAP seem more reasonably interpreted as measures of the reactive environment.

Both G×E and passive and evocative CovGE will be examined for the period of middle childhood. In the case of CovGE, special attention will be given to the developmental hypotheses proposed by Scarr and McCartney (1983).

Results

Genotype–environment interaction

Due to the large number of variables available for children and adults and for adoptive home environments in the CAP, the number of possible G×Es to explore is overwhelming. For this reason, some restrictions are required on the number of G×Es analyzed. A major restriction employed in this chapter is that child outcome measures (e.g., parental ratings of the children's activity) were only analyzed with comparable adult measures (e.g., parental self-report of activity). (The only exception is that parental Extraversion and Neuroticism were employed because of their importance for adult personality; they were studied in relation to the temperament outcome measures of the children.) The environmental measures were also limited. The major measures of the adoptive family environment were the three second-order factors of the Family Environment Scale (FES: Moos & Moos, 1981). In addition, parental measures for the adoptive parents were also used as "environmental" measures in the adoptive homes. In all, 78 G×E interactions were analyzed at age 7 (the only one of the four ages analyzed that included height, weight, and the interaction of biological mother's IQ with the IQ of the adoptive mother and the three scales of the FES); and 72 G×E interactions at ages 9, 10, and 11. The measures used to assess the parents, the environment, and the children's outcomes are summarized in Table 20.1.

Table 20.1 Summary of measures used for the assessment of genotype–environment interaction

Measure	Reference
Adult	
Extraversion	Cattell's 16 Personality Factor Test (Cattell, Eber, & Tatsuoka, 1970)
Neuroticism	
Emotionality–Fear	EASI self-report (Buss & Plomin, 1975)
Emotionality–Anger	
Activity	
Sociability	
General Intelligence	Plomin and DeFries (1985)
Depression	
Sociopathy	
Artistic Interests	
Group Sports Interests	
Individual Sports Interests	
Mechanical Interests	
Height/Weight	
Environment	
Relationship	Family Environment Scale
Personal Growth	(FES: Moos & Moos, 1981)
System Maintenance	
Child	
Emotionality	Colorado Childhood Temperament Inventory
Activity	(CCTI: Rowe & Plomin, 1977)
Sociability	
Attention Span	
Anxious/Depressed	Child Behavior Checklist (CBCL: Achenbach & Edelbrock, 1983)
Delinquent	
General Intelligence	WISC-R (Wechsler, 1974)
Artistic Interests	Plomin and DeFries (1985)
Group Sports Interests	
Individual Sports Interests	
Mechanical Interests	
Height/Weight	

Table 20.2 Use of genotype–environment interaction (G×E) to predict child behavior

Variables			R^2 change				
Genetic	Environment	Child	Age 7 (N = 171 – 191)	Age 9 (N = 161 – 185)	Age 10 (N = 158 – 177)	Age 11 (N = 127 – 145)	Other ages
Sociability	Relationship	Sociability	.01	.00	.02**	.00	3
Neuroticism	Personal Growth	Sociability	.02**	.03**	.01	.00	None
Extraversion	Relationship	Sociability	.02**	.00	.02**	.01	1, 2, 4, 5
Neuroticism	Personal Growth	Attention	.01	.02**	.01	.00	5, 6
Extraversion	Personal Growth	Attention	.01*	.02	.02*	.00	2
Extraversion	Maintenance	Attention	.01	.02**	.01	.00	1
Depression	Relationship	Anxious	.00	.02**	.00	.01	None
Depression	Personal Growth	Anxious	.05***	.02**	.00	.00	None
Sociopathy[a]	Maintenance	Delinquent	.00	.00	.04*	.04*	None
Intelligence	Relationship	Intelligence	.04***	—	—	—	None
Artistic	Artistic	Artistic	.01	.03**	.04**	.01	5, 6
Artistic	Relationship	Artistic	.01	.03**	.00	.03*	5
Artistic	Maintenance	Artistic	.00	.00	.03**	.00	4
Ind. Sports	Ind. Sports	Ind. Sports	.00	.00	.00	.05***	None
Ind. Sports	Personal Growth	Ind. Sports	.00	.00	.01	.04***	None
Ind. Sports	Maintenance	Ind. Sports	.03***	.00	.00	.01	None
Mechanical	Mechanical	Mechanical	.01	.00	.03**		None

[a] The sample sizes for the Sociopathy/Maintenance interaction were smaller than for the other regression equations: age 7 (98); age 9 (87); age 10 (80); age 11 (56).

*p < .10, **p < .05, ***p < .01

With two exceptions, the results for those G×Es that were identified as significant at a probability level of less than .05 are summarized in Table 20.2. Although the two exceptions, the interaction between Extraversion and Personal Growth and the interaction between Sociopathy and System Maintenance, obtained probability levels exceeding .05, their appearance at two ages suggests that the results are not artifactual.

Several remarks are in order concerning Table 20.2. First, the number of G×Es significant at each age is not notably different: Age 11 has the smallest number of significant interactions, 4, in contrast with age 10, which has 7. If this is a reliable difference between the two ages, it cannot be ascribed wholly to sample size differences, for the magnitude in R^2 change is also smaller at age 11. In addition, within any of the four ages, the number of significant interactions hardly differs from what would have been expected by chance. For example, at age 7, the significant interactions are approximately 6% of the total number analyzed; at age 9, approximately 9%; at age 10, approximately 10%, and at age 11, approximately 6%. Moreover, 11 of the 17 interactions occur at only one age, with the remaining 6 appearing at two ages. Perhaps the greatest confidence should be placed in those interactions that appeared at more than one age. When the 17 interactions significant in middle childhood were examined for ages 1–6, 8 were found to be significant ($p < .10$), as indicated in Table 20.2. However, no matter what the age, little additional variance in the dependent variable is accounted for by the inclusion of a significant interaction, the R^2 change ranging from .01 to .05.

As a particularly interesting example of a G×E, Figure 20.1 depicts the interaction between biological mother Depression and adoptive home FES Relationship in the prediction of the adopted child's CBCL Anxious/Depressed score at age 9.

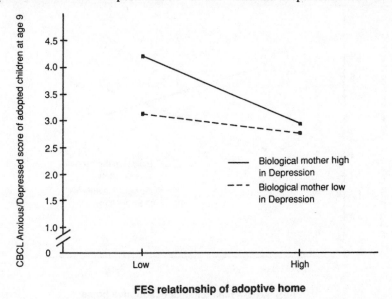

Figure 20.1 Adopted children's CBCL Anxious/Depressed scores at age 9 as a function of biological mothers' Depression and adoptive families' FES Relationship.

According to Figure 20.1, genetic differences in depression are more important in families low on Relationship. In these families, children with a higher genetic predisposition for depression experience greater anxiousness than children with a lower genetic predisposition for depression. Conversely, in families high on Relationship, genetic differences among children do not exert an influence on individual differences in anxiousness. Further, for children with a lower genetic predisposition for depression, it does not appear to matter whether the child is reared in a family high or low on Relationship, for the predicted level of anxiousness is about the same. In summary, if this interaction is not spurious, then high Relationship may act as a buffer against the development of severe anxiousness for children with a genetic propensity for depression.

As another example, Figure 20.2 presents the interaction between biological mother Sociopathy and adoptive home FES System Maintenance in the prediction of the adopted child's CBCL Delinquent score at age 10. The form of the interaction at age 11 is identical. The effect of genetic differences in sociopathy among adopted children is most apparent in families with low System Maintenance: Children with a higher genetic propensity for sociopathy exhibit a greater degree of delinquent behavior than children with a lower genetic propensity for sociopathy. But in families with high System Maintenance, delinquency is not associated with genetic differences in sociopathy. Interestingly, while the level of delinquency for children with a high genetic propensity for sociopathy declines when they are reared in a high System Maintenance home, the level increases for children with a low genetic propensity for sociopathy. The familial imposition of structure and order on a child not previously inclined to delinquent behavior may be counterproductive, even if beneficial for those children who are so inclined.

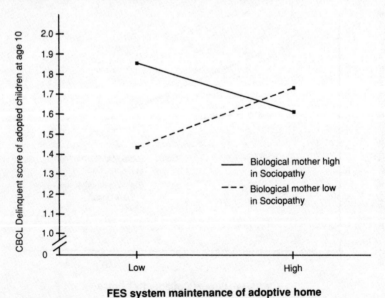

Figure 20.2 Adopted children's CBCL Delinquent scores at age 10 as a function of biological mothers' Sociopathy and adoptive families' FES System Maintenance.

While there are 17 significant interactions in middle childhood, this number does not exceed what would have been expected by chance alone. Consistent with previous CAP findings at earlier ages, the evidence of G×E for individual differences remains weak.

Passive genotype–environment correlation

The selection of measures for the assessment of passive CovGE was guided by their presence in each of the ages under study, 7 through 11. However, for one measure, the scales of the Self-Perception Profile (SPP: Harter, 1985), data were not available at 7 years. For two measures, data were available at ages younger than 7 years, thus permitting a test of Scarr and McCartney's (1983) hypothesis of a decline in passive CovGE across childhood: At ages 1 through 6, the scales of the Colorado Childhood Temperament Inventory (CCTI: Rowe & Plomin, 1977), and at age 4, the scales of the Child Behavior Checklist (CBCL: Achenbach & Edelbrock, 1983). An additional measure, the CAP Social Competence Scale (CSCS), was available only for children 7 through 11 years of age.

Table 20.3 presents the results, in terms of standard deviations, from comparing the variances of adopted (A) and nonadopted (B) children. At 7 years of age, 7 variance differences were significant; at 9 years, 6, at 10 years, 3, and at 11 years, 6. Thus support is not given to the Scarr and McCartney (1983) hypothesis of a decline in the influence of passive CovGE, for the number of significant variance differences among the age groups does not greatly differ. Indeed, for three of the scales of the CCTI, Emotionality, Activity, and Attention Span, significant variance differences appeared at ages 5 and 6, but not earlier. In one respect, though, Emotionality and Activity do somewhat support the hypothesis of a decline in passive CovGE, for significant variance differences do not occur for these scales after age 6. The results from the scales of the CBCL are the most suggestive for significant passive CovGE, for significant variance differences occur for three scales (Anxious/Depressed, Social Withdrawal, and Delinquent) at age 4, as well as for most of the other scales during middle childhood. Of greatest interest, however, is the appearance, in nearly every case where a significant difference occurs, of the adopted child variance exceeding the nonadopted child variance, a phenomenon that suggests the effects of negative passive CovGE on nonadopted children.

In summary, some evidence emerges for passive CovGE in middle childhood, especially for the CBCL, although it is negative in direction. Comparisons with earlier ages indicate no clear downward trend in the number of significant passive CovGEs.

Evocative genotype–environment correlation

For the computation of the evocative canonical CovGE, 15 adult measures were used for the biological mothers, and the three second-order FES scales were used at each age as measures of the adoptive home environment. In addition to the adult

Table 20.3 Passive genotype–environment correlation: comparisons of standard deviations for adopted (A) and nonadopted (B) children

	Ages							
	7		9		10		11	
	A	B	A	B	A	B	A	B
Measure	(N=249–258)	(N=251–269)	(N=212–248)	(N=197–230)	(N=171–222)	(N=165–205)	(N=145–188)	(N=137–175)
CBCL								
Aggressive	5.92**	4.95	5.85**	4.87	5.42	5.52	5.67	5.04
Anxious/Depressed	3.02*	2.82	3.49	3.16	3.43	3.60	3.62	3.46
Attention	3.17***	2.47	3.17***	2.46	3.02	2.70	3.10***	2.32
Delinquent Problems	1.78***	1.39	1.73***	1.39	1.57***	2.18	1.74	1.26
Social Problems	1.88	1.71	2.13***	1.73	2.15	2.26	2.13	2.03
Somatic Problems	1.44	1.39	1.43	1.51	1.65	1.80	1.45*	1.70
Thought Problems	.85***	.64	.67	.57	.60***	1.12	.61**	.49
Withdrawn	1.85	1.76	1.92	1.73	2.01	2.09	2.24	2.33
Internalizing	5.15	4.62	5.36	5.12	5.64	6.27	5.80	6.26
Externalizing	7.28**	5.96	7.06**	5.90	6.61	7.24	6.95*	5.87
CCTI								
Activity	2.98	2.74	3.92	3.58	3.70	3.76	4.03	4.18
Attention Span	3.72	3.41	4.09*	3.56	4.17*	3.64	4.29*	3.63
Emotionality	4.28	4.17	4.41	4.29	4.10	4.27	4.23	4.18
Sociability	3.99	3.96	4.25	4.02	4.51	3.99	4.50	3.91
CSCS								
Confidence	7.19	6.65	7.05	7.17	7.50	7.79	7.26	7.87
Leadership	6.63	6.77	5.22	5.09	7.19	6.92	7.16	7.04
Popularity	3.59	3.46	3.73	3.65	3.88	4.42	3.70	4.21
Problem Behavior	5.28**	4.39	3.96	3.87	4.87	4.35	4.57	4.41
Harter								
Physical Appearance			3.49	3.44	3.90	3.63	3.59**	2.90
Athletic Competence			3.36	3.49	3.39	3.55	3.64	3.68
Behavior Conduct			3.32	3.37	3.17	3.05	3.00	3.10
Self-Worth			2.95	2.75	2.66	2.64	2.43	2.22
Scholastic Competence			3.66	3.54	3.44	3.28	3.36	3.35
Social Acceptance			3.13	3.52	3.40	3.73	3.49	3.65

*p < .05, **p < .01, ***p < .001.

measures listed in Table 20.1, a few other measures were employed: 16PF Independence and Tough Poise (Cattell, Eber, & Tatsuoka, 1970), EASI Impulsivity, and measures of parental talent rather than interests.

Table 20.4 presents the unrotated first canonical correlations obtained at each age. Along with the correlations computed between biological mothers and adoptive homes, correlations for "controls," or nonadoptive parents rearing their nonadopted children in control homes are provided for comparison. Results indicated in Table 20.4 clearly confirm Scarr and McCartney's (1983) hypothesis of an increase in evocative CovGE throughout childhood. An increase in the canonical correlations for the adopted children occurs between the ages of 7 and 9, with the correlation becoming significant at age 10. Indeed, by age 11 the difference between the adopted and control childrens' canonical correlations is miniscule. Of note is the stability of the control mother – control child canonical correlations between the ages of 1 and 11. This stability is surprising, for an increase in the contribution of evocative CovGE would be expected to occur for control mother – control home as well as for biological mother – adoptive home pairs.

Table 20.4 Canonical evocative genotype–environment correlation

	Biological mother – adoptive home			Control mother – control home		
Age	*R*	*F*	*N*	*R*	*F*	*N*
1	.36	1.12	201	.58	3.81***	203
3	.32	.88	194	.56	2.92***	184
5	.34	.91	165	.58	2.58***	171
7	.33	.99	166	.55	2.32***	179
9	.40	1.15	145	.63	2.81***	150
10	.45	1.39*	143	.53	1.95***	136
11	.58	1.23	63	.59	1.30	62

Note: BM = biological mother; AH = adoptive home; CM = control mother; CH = control home.
*$p < .05$, ***$p < .001$

In order to obtain a sense of the measures contributing most to the significance of the evocative CovGEs at ages 9, 10, and 11, Table 20.5 presents the canonical structure at ages 9, 10, and 11 for the biological mother – adoptive home correlations. Examining the canonical structure at each age, one finds that the talent of Domestic Arts has the highest correlation with the biological mother variate at ages 9 (.59) and 10 (.68) but not at 11 (–.09), where it is replaced in importance by the talent of Mechanical Arts (.59). The explanation for this pattern of correlations lies with the relationship between Domestic Arts and Personal Growth. Personal Growth dominates the evocative CovGE environment variate at age 9, with a correlation of .87, and at age 10, with a correlation of .92, but does not at age 11, where the correlation is –.26. At age 11, System Maintenance has the highest correlation with the environmental variate (.83). At the bivariate correlation level, Domestic Arts and Personal Growth share one of the highest correlations at ages 9

(.24) and 10 (.32) of any of the correlations between the biological mother and adoptive home environment, thus leading these two variables to define their respective canonical variates. Analogously, the bivariate correlation at age 11 between Mechanical Arts and System Maintenance (.25) is among the highest, thus causing these two variables to define their respective canonical variates. General Intelligence is negligibly correlated with the FES factors at each of the ages, and thus contributes little to the definition of the biological mother canonical variate.

Table 20.5 Canonical structure for Biological mother – adoptive home correlations

Variable	Biological mother set			Variable	Environmental set		
	Age				Age		
	9	10	11		9	10	11
Intelligence	.05	−.04	.01	Relationship	−.05	.18	.40
Extraversion	.13	.30	.56	Personal Growth	.87	.92	−.26
Neuroticism	.61	.28	−.31	System Maintenance	.54	.63	.83
Tough Poise	−.04	−.13	.64				
Independence	−.21	−.11	.14				
Activity	.15	.22	.18				
Anger	.42	.34	−.00				
Fear	.33	.05	−.39				
Sociability	.13	.25	.56				
Impulsivity	.04	−.02	.06				
Artistic	−.11	−.03	.33				
Group Sports	−.04	−.10	.46				
Individual Sports	−.02	.13	.59				
Domestic Arts	.59	.68	−.09				
Mechanical Arts	−.00	−.09	.59				

Note: Numbers in the body of the table are correlations.

Discussion

The purpose of this chapter was to investigate the contribution of genotype–environment interaction and two types of genotype–environment correlations, passive and evocative, to the development of individual differences in middle childhood. G×E refers to the differential sensitivity of different genotypes to the same environment, whereas CovGE refers to the selective exposure of genotypes to different environments. As pointed out earlier, no developmental theory has been proposed to explain how genotypes and environments interact to produce development, or what the magnitude of the interaction might be at particular points along the developmental trajectory. On the other hand, Scarr and McCartney (1983) have proposed three mechanisms for the inducement of a correlation between genotypes and environments, and have stipulated the relative importance of each throughout development.

When G×E was examined for ages 7 through 11, 17 significant interactions were found. Within the period of 7 to 11 years, six of these 17 occurred at more than one age; if the period of time is extended to the first year, eight of these 17 appear more than once. However, the total number of significant interactions found at each age did not differ from the number expected by chance alone, thus diminishing confidence in the veracity of the findings.

A likely reason for the failure to uncover a greater number of significant G×Es results from the well-known low power of the interaction term in the general linear model (Wahlsten, 1990). Finding the smallest number of significant interactions at age 11 could have occurred in part from the smaller sample size avaliable at that age. Theoretical reasons have been advanced as well for the low probability of discovering significant G×E. McCall (1991) has pointed out that in naturalistic contexts, person and environmental harmony can often be expected to prevail, thus leading to CovGE rather than G×E. For example, the moderate heritability of temperament often leads to similarity in the dispositions of parents and children. At the species level, Bell (1990) has hypothesized that periods of intense selection are accompanied by G×E, but CovGE is the rule during periods of relative quiescence. Person–environment interactions are undoubtedly important for human development (e.g., Rutter, 1983), but the question remains as to how important are genotype–environment interactions.

Passive CovGE, referring to the correlated transmission of parental genotypes and environments to offspring, appeared to be negative, but no evidence for a clear trend in its contribution to development was uncovered. The finding of significant negative passive CovGE is unlikely to have resulted from two methodological artifacts often mentioned with regard to studying adopted children. The first artifact, selective placement, would have reduced the differences in adopted and control children's variances. In any event, little evidence for selective placement has been obtained in the CAP (Plomin & DeFries, 1985). The second artifact, the tendency for adoptive homes to be more environmentally homogeneous, would have served to increase the differences in the variances, but adoptive and control homes are similar for means and variances for most environmental indices (Plomin & DeFries, 1985).

Evocative CovGE, the tendency for genotypes to elicit different reactions from the environment, was significant during middle childhood, but not earlier, a result consistent with Scarr and McCartney's (1983) hypothesis. Indeed, by age 11, the canonical evocative CovGE was .58, hardly different from the value of .59 found for the canonical correlation computed for control families. Here, selective placement would have increased spuriously the canonical correlations computed between the set of biological mother measures and adoptive home measures, but as noted above, selective placement is not evident in the CAP. Somewhat puzzling was the finding of stability in the control mother–control child canonical correlations within the age period of 1 through 11 years. Although the control mother–control home canonical correlations do not directly assess evocative CovGE as the biological mother–adoptive home correlations do, it would have been expected that evocative CovGE would also have increased the control mother–control home canonical correlations.

One consequence of the finding of significant evocative CovGE may be a redoubled effort to construct measures of the environment sensitive to evocative effects. Fortunately, the FES scales partially fulfill the requirements of an evocative measure (i.e., reactivity to the behavior of the individual), but better measures are needed. If there is more than one child in the family, the FES responses provided by the parents represent, at best, a concatenation of the effects of all the children's behaviors. Measures for the detection of active effects are even more notable for their absence, given the large body of evidence (Lerner & Busch-Rossnagel, 1981) for the active role individuals take in constructing their development.

21 Applied Issues

Sally-Ann Rhea
Robin P. Corley

Analyses of data from the Colorado Adoption Project (CAP) that assess genetic and environmental influences on children's development are reported in the previous chapters of this book. In this chapter, we focus on applied issues relevant to adoption. Specifically, we compare various indices of adjustment of adopted and nonadopted children, and explore possible etiologies of adjustment differences. We regard adjustment, broadly defined, as the ability to function adequately in a variety of situations such as home, school, and with peers. Although the CAP data are also relevant to other applied issues, such as selective placement and timing of placement (Plomin and DeFries, 1985), adjustment of adoptees during middle childhood is the focus of this chapter.

Findings reported from previous research on adjustment of adoptees have been mixed. Results of several studies (Brodzinsky, 1990; Deutsch et al., 1982; Hersov, 1985; Kotsopoulos et al., 1988) suggest that adoptees are over-represented in clinical populations. Although results obtained from research on some non-clinical populations also indicate that adoptees may have more problems than nonadoptees (Brodzinsky, 1987; Brodzinsky, Schechter, Braff, & Singer, 1984; DiGiulio, 1988; Lindholm & Touliatos, 1980; Sharma & Benson, 1992), the various factors that may influence adjustment were not always assessed in these studies. For example, successful adjustment may vary as a function of age at placement and the extent to which the children are placed transracially or through international agencies (Hersov, 1990).

Adjustment to adoption may also change as a function of age at assessment. Brodzinsky (1990) noted that very young children do not understand the concept of adoption and, therefore, may have a positive outlook toward it. However, upon reaching an age of greater comprehension, at about the time of entry to school, the same children may experience doubts about their status. Other researchers have noted that early adolescence is a time of questioning and turmoil which may be particularly troublesome for adopted children (Grotevant & McRoy, 1990). For females, sexual maturation may cause special difficulties for adoptees (Schechter & Bertocci, 1990).

In order to assess the long-term consequences of adoption, Bohman and Sigvardsson (1978, 1985, 1990) conducted a prospective longitudinal study of three cohorts of children who had been registered for adoption in Sweden. One group consisted of 208 children whose mothers had intended to place them but did not; a second group included 203 children who were placed in long-term foster care (which eventually led to adoption by the foster parents), whereas the 168 children in the third group were adopted in infancy by well-prepared parents. Various outcomes for these children were compared to those from a control group at 11, 15, 18, and 23 years of age. At each age, girls in the three groups did not differ significantly from the controls. In contrast, at age 11, the boys in the three placement groups had a higher incidence of behavior problems than did the controls, suggesting that adopted boys are at risk, although not more so than male children of unplanned pregnancies whose mothers chose to rear them or who were placed in foster care. However, by age 15, the boys adopted by well-prepared parents were nearly as well adjusted as the controls. The boys who had not been placed for adoption and those who had been in long-term foster care continued to manifest greater maladjustment than the controls. Outcomes were assessed only from military records at age 18 and from criminal and alcohol records at age 23. Nonetheless, outcomes continued to be favorable for adoptees.

A recent meta-analysis of 62 studies of adoptees' psychological adjustment showed greater maladjustment overall in adoptees. However, larger effect sizes occurred in studies with older subjects and clinic samples (Wierzbicki, 1993).

In addition to providing longitudinal data concerning the outcomes of adoption, the CAP design facilitates an analysis of the etiology of adjustment differences. For example, differences between adopted and nonadopted children may be due to heritable influences. That is, if adopted and nonadopted children differ for some measure of adjustment, this could be due to differences between the birth parents and control parents. The CAP is one of the few adoption studies with extensive data on birth parents, and the only study with any information on birth fathers. Alternatively, differences between adopted and nonadopted children may be due to home environmental influences. Because the CAP has extensive information about the biological, adoptive, and nonadoptive parents, as well as about the adoptive and control home environments, the etiology of adjustment differences can be assessed. If differences are negligible between birth parents and adoptive parents, or between home environments in adoptive and control families, any observed differences between adopted and control children may be attributable to the adoption process itself.

The present chapter addresses several questions concerning the extent and etiology of adjustment differences in adopted children. First, are there developmental differences between adopted and nonadopted children? If so, do adoptive parents or adoptive home environments differ from those of control families? Do relinquishing birth parents differ from control parents? Are the prenatal and perinatal experiences of adoptees different from control children? And finally, is differential placement of adoptees in homes that are either well or poorly matched to their biological propensities important?

Method

Sample

Subject recruitment has been described in detail elsewhere (Plomin & DeFries, 1985; see also Chapter 2). In brief, birth and adoptive parents were recruited by social workers at Denver area adoption agencies. Upon release from the hospital (at 4.1 days on the average), the infants were taken to foster homes; they were placed in their adoptive homes at an average of 28.5 days. Consequently, these children did not undergo a traumatic separation from the birth mother or other relatives, eliminating a potential confound in assessing differences between adoptive and nonadoptive families.

Measures

For these analyses we use key CAP measures drawn from the domains of cognition, temperament, and problem behaviors, as indices of the children's adjustment in middle childhood. Data regarding their parents' temperament and cognition and their home environment are analyzed to assess the etiology of any differences between the adopted and nonadopted children.

Children's cognitive and scholastic abilities were measured during the laboratory session after the completion of the first grade of school, when the WISC-R (Wechsler, 1974), the Reading Recognition subtest of the Peabody Individual Achievement Test, and several subtests of KeyMath were administered to the children. Descriptions of these cognitive and school achievement measures and results from analyses of differences between adopted and nonadopted children are presented more fully in Chapter 7. Scholastic competence is measured at age 9 in response to questions on the Self-Perception Profile for children (Harter, 1982). The adoptive and control parents were administered the WAIS-R (Wechsler, 1981), typically when they accompanied their children to the laboratory session.

Temperament, personality, and problem behaviors were measured by questionnaires administered to the parents and teachers when the children were 7 and 9, from ratings in the laboratory at age 7, and via interviews with the children at age 9. During the laboratory session, we administered a version of Coddington's (1972) questionnaire for stressful life events affecting children, and another questionnaire specific to first-grade experiences, to both the children and their parents. An additional rating is the number of illnesses the child had in the past year. Except for the instrument specific to first grade, these questionnaires are administered again at age 9. These instruments are described in more detail in Chapter 2.

Personality for birth, adoptive, and control parents was assessed using Cattell's 16PF (Cattell, Eber, & Tatsvoka, 1970) which was administered as part of the larger paper-and-pencil test upon entry to the study.

Although we do not make direct assessments of the environment in the home during middle childhood, parents in the CAP completed modified versions of the Family Environment Scale (FES: Moos & Moos, 1981) when the children were 7 and 9 years old. At age 7, we added Dibble and Cohen's Parent Report (PR), (1974) to the CAP battery of parent-rated environmental assessments.

Prenatal and perinatal environmental circumstances were assessed from hospital records and maternal reports. Questions regarding use of coffee, alcohol, and nicotine were included in the initial paper-and-pencil test for all parents. Five-minute APGAR scores and birthweight data were abstracted from hospital records provided by the agencies for the adopted children and by the hospitals for the control children.

Results

In an earlier book, Plomin and DeFries (1985) reported comparisons among adoptive, birth, and control parents and between adoptive and control home environments and children. Overall, few differences between the groups were found. In addition, little evidence for selective placement and no important differences between adopted and nonadopted children emerged through age 2. In this chapter, we report group differences from a larger sample (more than 100 families have been added to the CAP) tested at later ages (focusing on 7- and 9-year-olds), and we consider the effects of placement in homes which are well matched or unmatched to the child's biological propensities.

Outcomes in middle childhood

As reported in Chapter 4, the CAP adopted and nonadopted children score significantly higher and exhibit less variance on IQ than the national sample (Wechsler, 1974). At 7 years, mean scores for the 108 adopted boys and 89 girls are 113.6 ($SD = 11.0$) and 109.3 ($SD = 9.8$), whereas they are 115.4 ($SD = 11.5$ and 113.7 ($SD = 10.6$) for the 115 control boys and 100 girls. For the Reading Recognition test, differences between adopted and nonadopted children are not significant. Although the difference approaches significance for the KeyMath Numeration test, adoption does not appear to increase the risk of problems on achievement tests.

The means and standard deviations for adopted and nonadopted female and male children's temperament and behavior problem scores are presented in Tables 21.1–21.4 for parent, teacher, and tester reports, and children's self-reports, respectively. Gender differences are found in multiple domains, especially activity, but in no case do they account for more than 4% of the variance. Most tests for gender-by-status interactions are not significant; the exceptions are differences in perception of scholastic skills and self-esteem, with adopted girls at age 9 rating themselves lowest on this scale.

Tables 21.1 and 21.2 report means for parent and teacher ratings of temperament and behavior problems. Parents and teachers rate adopted children higher on the CBCL Externalizing scale at 7 and 9 years of age. Other significant differences in Tables 21.1 and 21.2 involve components of Externalizing problems. For example, at both 7 and 9, teachers rate adoptees as less persistent and having a shorter attention span than nonadopted children. Adopted children are also rated as significantly less confident at 7 and 9. In contrast, at both 7 and 9, parents rate adopted children as more sociable than nonadopted children. It should be emphasized, however, that although these differences are statistically significant given the relatively large sample sizes, the differences account for only 2 to 8% of the variance. Ratings of stressful events and illness at age 7 showed no significant differences between adopted and nonadopted children.

Table 21.1 Parent reports of children's behavior and temperament at years 7 and 9

Measure	Girls				Boys			
	Adopted		Nonadopted		Adopted		Nonadopted	
	Mean	SD	Mean	SD	Mean	SD	Mean	SD
	(N = 89 – 93)		(N = 92 – 95)		(N = 103 – 106)		(N = 110 – 112)	
Age 7 CBCL[a]								
Internalizing	.05	(1.1)	.14	(1.0)	.13	(1.2)	−.05	(.9)
Externalizing[1]	.22	(1.1)	−.04	(.9)	.13	(1.0)	−.15	(.9)
Age 7 CCTI[b]								
Sociability[1]	20.0	(4.3)	18.5	(4.1)	19.8	(3.8)	18.5	(3.7)
Emotionality	14.2	(4.3)	14.3	(4.1)	13.3	(3.7)	14.0	(3.6)
Activity[2]	18.8	(4.1)	18.7	(3.4)	20.3	(3.1)	19.6	(2.8)
Attention Span	18.3	(3.3)	18.1	(3.2)	18.8	(3.2)	18.5	(3.1)
Age 7 CSCS[c]								
Leadership	34.3	(5.7)	33.1	(5.8)	33.3	(4.7)	32.6	(4.8)
Problem Behavior[2]	22.0	(5.4)	21.6	(4.7)	23.6	(5.8)	23.2	(5.3)
Popularity	23.8	(3.4)	23.4	(3.5)	23.2	(3.7)	23.3	(3.4)
Confidence	46.9	(5.3)	45.9	(5.1)	44.1	(5.6)	44.9	(5.3)
	(N = 81 – 82)		(N = 74)		(N = 96 – 97)		(N = 92 – 93)	
Age 9 CBCL[a]								
Internalizing	.02	(.9)	.12	(1.0)	.08	(1.0)	.00	(1.1)
Externalizing[1]	.15	(1.0)	−.07	(.8)	.10	(1.1)	−.09	(.9)
Age 9 CCTI[b]								
Sociability[1]	20.3	(4.1)	19.1	(4.1)	20.1	(4.1)	19.0	(4.1)
Emotionality	13.8	(4.4)	14.1	(4.4)	13.4	(3.9)	13.6	(4.1)
Activity[1,2]	18.7	(3.9)	17.9	(3.6)	19.9	(3.8)	18.8	(3.4)
Attention Span	18.5	(4.1)	18.5	(3.2)	18.0	(4.1)	18.1	(3.8)

[a] Achenbach's Child Behavior Checklist.
[b] Colorado Childhood Temperament Inventory.
[c] CAP Social Competence Scale, parent and teacher version.
[1] Significant status difference. ($p < .05$)
[2] Significant gender difference ($p < .05$).

Table 21.2   Teacher reports of children's behavior and temperament at years 7 and 9

Measure	Girls				Boys			
	Adopted		Nonadopted		Adopted		Nonadopted	
	Mean	SD	Mean	SD	Mean	SD	Mean	SD
	(N = 70 – 85)		(N = 92 – 95)		(N = 103 – 106)		(N = 110 – 112)	
Age 7 CBCL[a]								
Internalizing	.04	(.9)	−.06	(1.1)	.05	(.9)	−.04	(1.1)
Externalizing[1]	.11	(1.2)	−.19	(.9)	.21	(1.0)	−.12	(1.0)
Age 7 CCTI[b]								
Sociability	18.8	(4.4)	18.2	(4.1)	18.6	(3.4)	17.5	(4.3)
Emotionality	10.2	(3.9)	10.3	(4.2)	11.0	(4.2)	11.2	(4.6)
Activity[1,2]	18.4	(4.1)	17.6	(4.0)	19.7	(3.8)	18.7	(3.7)
Attention Span[1,2]	18.3	(3.9)	19.4	(3.8)	16.7	(4.2)	18.0	(3.6)
Age 7 CSCS[c]								
Leadership	30.6	(6.6)	31.9	(7.0)	29.3	(6.2)	31.1	(6.3)
Problem Behavior[1,2]	20.5	(6.5)	19.0	(5.5)	24.2	(7.6)	21.2	(6.3)
Popularity	21.9	(4.2)	22.4	(4.1)	21.5	(4.6)	22.0	(4.4)
Confidence[1,2]	43.5	(7.5)	45.9	(7.0)	40.7	(7.7)	43.3	(7.4)
	(N = 75 – 77)		(N = 69 – 74)		(N = 90 – 91)		(N = 76 – 79)	
Age 9 CBCL[a]								
Internalizing	.02	(1.0)	−.04	(1.1)	.09	(1.0)	−.06	(.9)
Externalizing[1]	.16	(1.1)	−.07	(1.0)	.24	(1.1)	−.22	(.8)
Age 9 CCTI[b]								
Sociability	18.8	(4.3)	18.7	(4.3)	18.6	(3.5)	17.8	(3.9)
Emotionality	10.2	(4.2)	10.7	(4.6)	11.5	(4.3)	10.7	(4.6)
Activity[2]	18.2	(3.8)	17.7	(4.0)	19.0	(4.1)	18.6	(4.1)
Attention Span[1]	17.4	(4.1)	18.4	(4.1)	16.7	(4.6)	18.0	(3.8)
Age 9 CSCS[c]								
Leadership[1]	29.4	(7.1)	30.6	(7.4)	29.6	(6.9)	32.3	(5.9)
Problem Behavior[2]	21.4	(5.7)	20.4	(6.3)	23.7	(7.5)	22.1	(7.4)
Popularity	21.4	(4.3)	22.0	(4.8)	21.2	(5.0)	21.7	(4.5)
Confidence[1]	41.8	(8.1)	44.0	(7.8)	40.0	(7.7)	42.2	(7.4)

[a] Achenbach's Child Behavior Checklist.
[b] Colorado Childhood Temperament Inventory.
[c] CAP Social Competence Scale, parent and teacher version.
[1] Significant status difference ($p < .05$).
[2] Significant gender difference ($p < .05$).

As shown in Table 21.3, tester ratings from the laboratory session at 7 years of age yield significant status differences for Modified Behavior Record (MBR) Attention Span, CCTI Emotionality, Activity, and Attention Span, and CSCS Self-Esteem, findings which generally agree with the teacher data. Again, however, these mean differences account for less than 4% of the variance of these measures.

Table 21.3 Tester ratings of behavior and temperament at year 7

Behavior and temperament scales	Girls				Boys			
	Adopted		Nonadopted		Adopted		Nonadopted	
	Mean	SD	Mean	SD	Mean	SD	Mean	SD
	(N = 66 – 91)		(N = 80 – 100)		(N = 82 – 108)		(N = 83 – 116)	
MBR[a]								
Attention span[1]	22.0	(3.4)	22.6	(3.2)	21.4	(3.7)	22.4	(3.7)
Fear	8.8	(3.1)	8.4	(2.5)	8.4	(2.6)	8.1	(2.7)
Sociability[2]	6.3	(1.6)	6.6	(1.4)	6.0	(1.4)	6.2	(1.5)
Impulsivity	13.2	(2.2)	13.0	(1.7)	13.6	(1.8)	13.2	(1.9)
CCTI[b]								
Sociability	3.7	(1.0)	3.7	(.8)	3.7	(.8)	3.7	(1.0)
Emotionality[1]	1.7	(.6)	1.4	(.6)	1.5	(.7)	1.4	(.7)
Activity[1,2]	3.6	(.7)	3.4	(.7)	3.8	(.7)	3.6	(.7)
Attention Span[1]	3.2	(.8)	3.5	(.9)	3.3	(.9)	3.5	(1.0)
CSCS[c]								
Self-Esteem[1]	20.3	(2.6)	21.1	(2.7)	20.4	(2.8)	20.5	(2.8)
Independence	16.3	(3.2)	16.7	(2.9)	16.6	(2.9)	16.8	(3.0)
Competence	17.2	(3.0)	17.8	(2.9)	17.1	(2.7)	17.1	(2.7)
Leadership	17.2	(2.9)	17.6	(2.9)	16.7	(2.6)	17.3	(2.7)
Adult Interaction	19.6	(3.6)	20.5	(2.9)	19.0	(3.5)	19.6	(3.7)

[a] Modified Behavior Record.
[b] Colorado Childhood Temperament Inventory.
[c] CAP Social Competence Scale, tester version.
[1] Significant status difference ($p < .05$).
[2] Significant gender difference ($p < .05$).

Table 21.4 lists mean comparisons for children's self-ratings from the telephone interview at 9 years of age. Few differences emerged between adopted and nonadopted children. Adopted children perceive themselves to be more active, less scholastic, more likely to have conduct problems, and more depressed. For all of these differences except activity, however, there is less than half a standard deviation's difference among the mean scores.

Table 21.4 Children's self-perceptions at year 9

Measure	Girls				Boys			
	Adopted		Nonadopted		Adopted		Nonadopted	
	Mean	SD	Mean	SD	Mean	SD	Mean	SD
	(N = 85 – 87)		(N = 80 – 81)		(N = 103)		(N = 90 – 93)	
CCTI[a]								
Sociability[1]	14.5	(3.3)	14.7	(3.3)	15.6	(3.0)	15.7	(3.1)
Emotionality[1]	10.6	(4.6)	10.4	(3.8)	9.6	(3.9)	10.0	(3.9)

Table 21.4 (Cont.)

Measure	Girls				Boys			
	Adopted		Nonadopted		Adopted		Nonadopted	
	Mean	SD	Mean	SD	Mean	SD	Mean	SD
Activity[1,2]	14.4	(2.8)	14.1	(2.6)	16.1	(3.0)	14.9	(3.3)
Attention Span	15.2	(3.1)	15.0	(3.2)	15.5	(3.2)	15.9	(2.9)
Harter[b]								
Scholastic Competence[2,3]	13.5	(3.8)	15.3	(3.6)	14.1	(3.8)	14.2	(3.4)
Social Acceptance	15.3	(3.3)	15.3	(3.4)	16.0	(2.9)	15.5	(3.4)
Athletic Competence[1]	11.8	(3.3)	11.8	(3.5)	13.2	(3.4)	12.7	(3.7)
Physical Appearance	14.3	(3.7)	14.9	(3.4)	15.5	(3.2)	15.0	(3.2)
Behavior × Conduct[1,2]	14.7	(3.3)	15.8	(3.3)	14.1	(3.3)	14.5	(3.2)
Self-Worth[3]	16.2	(3.2)	17.4	(2.6)	16.7	(2.7)	16.6	(2.8)
SIQYA[c]								
Family Relations	33.2	(5.3)	34.5	(4.2)	33.0	(5.2)	32.9	(5.2)
FES[d]								
Cohesion	16.4	(3.1)	16.7	(2.6)	16.3	(2.7)	16.3	(2.8)
Expressiveness	12.7	(3.4)	13.2	(3.1)	12.2	(3.0)	12.4	(3.0)
Conflict	9.8	(4.6)	9.4	(4.6)	10.3	(4.6)	10.5	(5.0)
Achievement[1]	15.6	(3.4)	15.2	(3.0)	16.3	(3.0)	15.8	(2.7)
Control[1]	12.9	(3.5)	12.4	(3.2)	13.6	(3.4)	13.1	(3.6)
Ascher[e]								
Loneliness	13.7	(5.4)	12.4	(4.9)	13.0	(5.5)	12.9	(5.0)
Kandel[f]								
Depression[2]	15.1	(5.1)	13.0	(4.8)	13.8	(5.0)	13.1	(4.8)

[a] Colorado Childhood Temperament Inventory.
[b] Harter's Self-Perception Profile for Children.
[c] Petersen's Self-Image Questionnaire for Young Adolescents.
[d] Family Environment Survey.
[e] Ascher's Loneliness Questionnaire.
[f] Kandel's Depressive Mood Inventory.
[1] Significant gender difference ($p < .05$).
[2] Significant status difference ($p < .05$).
[3] Significant gender-by-status interaction ($p < .05$).

Of the 72 adjustment measures that we examined, 24 manifest a significant adoption status difference. Table 21.5 summarizes the mean scores of adopted and nonadopted children for these scales. Adopted children appear to be more active and less attentive and more likely to have conduct problems, as some of the previous literature would suggest. Using clustering techniques, CAP researchers previously identified a subsample of children at risk for conduct disorder (Coon, Carey, Corley, & Fulker, 1992). There were too few girls identified to be included in the sample, but of the 15 boys identified, 11 were adopted. However, when we examine our entire sample of adopted and nonadopted children, the differences attributed to adoption status remain small, with a median effect size of .28.

Table 21.5 Scales with mean differences for adopted and nonadopted children

	F^a	Adoptive mean[a]	Control mean[b]	Overall SD^c	Effect size
Test session, age 7					
WISC-R	7.85**	111.38	114.41	11.00	.28
MBR Attention Span	5.52*	21.60	22.53	3.42	.27
CCTI Emotionality	9.66**	1.57	1.35	.63	−.36
CCTI Activity	4.29*	3.71	3.54	.72	−.24
CCTI Attention Span	6.84**	3.20	3.48	.90	.30
CSCS Self-esteem	4.65*	21.21	20.90	2.76	.25
Parent ratings					
CBCL Externalizing, 7	8.14**	.18	−.10	.99	−.29
CCTI Sociability, 7	11.82***	19.86	18.47	4.02	−.35
CBCL Externalizing, 9	4.48*	.14	−.09	.96	−.23
CCTI Sociability, 9	6.37*	20.16	19.03	4.12	−.27
CCTI Activity, 9	5.62*	19.29	18.33	3.75	−.25
Teacher ratings					
CBCL Externalizing, 7	11.22***	.25	−.15	1.05	−.38
CSCS Problem Behavior, 7	14.11***	22.91	20.03	6.88	−.42
CSCS Confidence, 7	11.84***	41.81	44.65	7.38	.39
CCTI Activity, 7	5.60*	19.10	18.05	3.87	−.27
CCTI Attention Span, 7	9.44**	17.28	18.69	4.07	.35
CBCL Externalizing, 9	9.02**	.22	−.14	1.04	−.34
CSCS Leadership, 9	5.23*	29.54	31.73	7.00	.26
CSCS Confidence, 9	5.75*	40.99	43.17	7.95	.28
CCTI Attention Span, 9	5.42*	17.07	18.20	4.22	.27
Self-ratings, age at 9					
CCTI Activity	6.51*	15.26	14.45	3.04	−.27
Harter Scholastic	7.38**	13.74	14.79	3.65	.29
Harter Behavior conduct	5.12*	14.43	15.22	3.29	.24
Kandel Depression	6.54*	14.53	13.17	4.96	−.27

[a] *df* for univariate tests range from 294–400.
[b] Pooled across gender.
[c] Pooled across gender and adoptive status.
*($p < .05$), **($p < .01$), ***($p < .001$)

In the following sections, we explore possible sources of these differences between adopted and nonadopted children by comparing adoptive and control family environments, parental characteristics, and pre- and perinatal circumstances. Finally, we assess the influence of differential placement – whether placement in a home which is well matched or unmatched to the child's biological propensities affects outcomes.

Comparisons between adoptive and control family environments

Multivariate tests of significance of the FES scales show that there are differences between the adoptive and control family environments at 7 years of age. However, as is shown in Table 21.6, these are primarily attributable to the higher scores in adoptive families on the Moral–Religious scale, with group status explaining 6% of the variance, and on the Organization scale, where group status explains less than 2% of the variance. These differences are probably due to the religious affiliation of the adoption agencies.

Table 21.6 Family environment at years 7 and 9

Parenting scale	Year 7				Year 9[a]			
	Adopted		Nonadopted		Adopted		Nonadopted	
	Mean	SD	Mean	SD	Mean	SD	Mean	SD
FES	(N = 194 – 197)		(N = 202 – 204)		(N = 178 – 179)		(N = 166 – 167)	
Cohesion	37.3	(4.3)	36.8	(4.5)	20.6	(3.1)	21.2	(3.2)
Expressiveness	32.0	(4.2)	32.5	(4.4)	19.0	(2.9)	19.1	(2.9)
Conflict	21.0	(4.9)	20.9	(5.3)	13.6	(3.2)	13.2	(3.6)
Independence	32.4	(3.7)	32.4	(3.6)	—	—	—	—
Achievement	30.0	(4.8)	29.4	(4.3)	16.3	(2.6)	16.1	(2.4)
Intellectual–Cultural	33.1	(6.0)	34.4	(5.1)	17.5	(4.0)	18.4	(3.5)
Active–Recreational	32.8	(5.5)	33.0	(5.1)	19.4	(3.6)	19.8	(3.3)
Moral–Religious	35.9	(6.1)	32.5	(7.9)	—	—	—	—
Organization	33.3	(4.8)	32.1	(5.1)	18.7	(3.3)	17.9	(3.6)
Control	29.4	(3.7)	28.4	(4.3)	17.6	(3.3)	16.8	(3.6)
PR Acceptance	73.2	(6.5)	72.5	(6.0)	73.1	(7.3)	71.6	(6.6)
PR Inconsistency	22.6	(5.4)	23.6	(5.3)	22.7	(6.1)	24.1	(5.8)
PR Negative Control	27.1	(7.2)	25.4	(6.6)	24.8	(7.8)	25.0	(8.2)

[a] FES reduced from 90 items to 40, with fewer items per scale and dropping Independence and Moral–Religious scales.

Again, at age 9 for the reduced version of the FES, multivariate tests of significance indicate that there are differences between adoptive and control families, attributable to lower scores in the adoptive families for Intellectual–Cultural orientation and to higher scores in these families on the Organization and Control scales. However, the distributions of scores for the two types of families overlap substantially and these mean differences account for only 2% of the variance in these environmental variables.

For the Parent Report at age 7, results of multivariate tests of significance indicate a difference between adoptive and control families, largely attributable to the third factor, Negative Control, a finding that replicates the FES data. However, at age 9, there are no significant differences between the two groups on these scales.

In summary, differences between the adoptive and nonadoptive home environments are not substantial; thus, any differences between the CAP adopted and nonadopted children are not likely to be due to home environmental influences assessed by these scales.

Parental characteristics

Comparisons of birth, adoptive, and control parents are reported in previous chapters in this book. In general, the differences are minimal, especially considering the relative youth of the birth parents. As shown in Tables 21.7, 21.8, and 21.9, average personality scales of the adoptive, control, and birth parents are similar to the normative samples from Cattell's 16PF manual (Cattell et al., 1970). Both types of rearing parents, all of whom were originally tested in Colorado between 1976 and 1983, exhibit few differences from samples which were racially and geographically representative of the US in 1974. Univariate tests of significance indicate that the adoptive and control parents differ on four scales – with the control parents being more assertive and liberal and the adoptive parents being more conscientious and controlled on the average. These group differences are probably due at least in part to the fact that adoptive parents were recruited by religious agencies, whereas control parents were recruited through area hospitals.

Table 21.7 CAP adoptive and control mothers and 16PF norms

16PF Scale	Adoptive mothers ($N = 233 - 236$)		Control mothers ($N = 238 - 214$)		16PF norms ($N = 729^a$)	
	Mean	SD	Mean	SD	Mean	SD
A. Outgoing	10.0	(3.3)	9.5	(3.1)	11.3	(3.2)
B. Bright	8.4	(2.0)	9.0	(1.9)	7.0	(2.2)
C. Emotionally Stable	16.3	(3.7)	15.8	(3.9)	15.6	(4.0)
E. Assertive	11.3	(4.9)	12.7	(4.5)	11.3	(4.6)
F. Happy-Go-Lucky	13.9	(4.3)	13.9	(4.4)	13.5	(4.3)
G. Conscientious	13.9	(2.9)	12.8	(3.4)	12.8	(3.3)
H. Venturesome	14.2	(6.0)	13.4	(6.1)	12.9	(5.6)
I. Tender-Minded	14.1	(2.9)	13.4	(3.3)	13.4	(3.4)
L. Suspicious	6.8	(3.2)	7.2	(3.4)	6.2	(3.4)
M. Imaginative	12.5	(3.5)	12.9	(3.3)	13.1	(3.9)
N. Astute	9.3	(3.0)	9.2	(3.2)	10.4	(2.9)
O. Apprehensive	10.7	(4.1)	10.9	(3.9)	10.7	(4.0)
Q_1. Experimenting	6.4	(2.7)	7.7	(3.3)	7.7	(3.1)
Q_2. Self-Sufficient	11.3	(3.8)	11.7	(3.6)	10.2	(3.6)
Q_3. Controlled	13.2	(2.9)	12.5	(3.0)	12.5	(3.3)
Q_4. Tense	14.1	(5.0)	14.7	(4.8)	12.9	(4.8)

[a] From R. B. Cattell, H. Eber, and M. M. Tatsuoka, 1970; Table 13 for 729 females based on the general population aged 30 years.

Inspection of the norms specific to the age group of the biological parents reveals that differences between these parents and the rearing parents are due largely to the difference in their ages. Replicating the findings from the previous report (Plomin & DeFries, 1985), the birth parents differ little from the national sample except for Scales A, C, and Q_2, in which they are less outgoing, more emotionally stable, and more assertive – a pattern which is somewhat similar to that of the older rearing parents.

Table 21.8 CAP adoptive and control fathers and 16PF norms

16PF Scale	Adoptive fathers (N = 228 – 230)		Control fathers (N = 238 – 241)		16PF norms (N = 2,255[a])	
	Mean	SD	Mean	SD	Mean	SD
A. Outgoing	8.0	(3.5)	7.7	(3.4)	10.2	(3.2)
B. Bright	8.8	(1.9)	9.1	(2.0)	7.0	(2.2)
C. Emotionally Stable	17.4	(3.8)	16.7	(3.7)	16.6	(4.1)
E. Assertive	14.1	(4.3)	15.7	(4.1)	12.9	(3.9)
F. Happy-Go-Lucky	13.0	(4.1)	13.6	(4.5)	14.2	(4.1)
G. Conscientious	14.2	(2.9)	13.2	(3.4)	13.4	(3.4)
H. Venturesome	13.7	(5.8)	14.6	(6.1)	14.8	(5.2)
I. Tender-Minded	8.1	(3.5)	8.6	(3.7)	9.0	(3.4)
L. Suspicious	7.3	(3.4)	7.6	(3.5)	7.4	(3.4)
M. Imaginative	13.5	(3.4)	13.8	(3.3)	13.0	(3.7)
N. Astute	8.5	(2.9)	8.1	(2.7)	9.2	(2.9)
O. Apprehensive	8.5	(3.7)	8.6	(3.8)	9.4	(4.2)
Q_1. Experimenting	8.9	(3.1)	9.9	(3.3)	9.5	(3.0)
Q_2. Self-Sufficient	12.0	(3.3)	12.3	(3.7)	10.3	(3.5)
Q_3. Controlled	14.1	(2.7)	13.0	(3.1)	13.3	(3.4)
Q_4. Tense	11.8	(4.9)	12.3	(5.0)	10.7	(4.7)

[a] From R. B. Cattell, H. Eber, and M. M. Tatsuoka, 1970; Table 16 for 2,255 males based on the general population aged 30 years.

Table 21.9 CAP birth parents and 16PF norms

16PF Scale	Biological mothers (N = 234 – 236)		16PF norms (N = 1,149)		Biological fathers (N = 48 – 50)		16PF norms (N = 1,312)	
	Mean	SD	Mean	SD	Mean	SD	Mean	SD
A. Outgoing	9.8	(3.0)	11.2	(3.0)	7.4	(2.9)	9.0	(3.0)
B. Bright	8.0	(2.1)	7.0	(2.2)	8.0	(2.4)	7.0	(2.2)
C. Emotionally Stable	15.9	(3.8)	13.7	(3.8)	15.2	(3.7)	14.0	(3.7)
E. Assertive	11.5	(4.0)	11.0	(3.8)	14.1	(3.3)	13.1	(3.7)
F. Happy-Go-Lucky	16.5	(4.4)	16.0	(4.4)	16.5	(4.1)	15.3	(4.3)
G. Conscientious	11.7	(3.4)	12.2	(3.5)	11.2	(3.8)	11.0	(3.4)
H. Venturesome	12.9	(5.8)	12.6	(5.1)	13.8	(5.3)	12.6	(5.0)
I. Tender-Minded	13.2	(2.8)	13.5	(2.8)	9.4	(3.3)	8.9	(3.5)
L. Suspicious	8.1	(3.1)	9.2	(3.1)	9.6	(3.2)	10.0	(3.1)
M. Imaginative	11.6	(3.5)	10.6	(3.8)	11.2	(4.1)	11.0	(3.5)
N. Astute	9.4	(2.6)	10.3	(2.7)	8.6	(2.3)	9.3	(2.7)
O. Apprehensive	12.1	(3.6)	13.0	(3.6)	10.9	(4.1)	11.9	(3.8)
Q_1. Experimenting	8.3	(2.9)	8.3	(3.1)	10.5	(2.5)	9.7	(3.1)
Q_2. Self-Sufficient	10.7	(3.7)	9.0	(3.3)	12.3	(3.2)	10.1	(3.5)
Q_3. Controlled	11.7	(3.0)	11.4	(3.1)	11.5	(2.4)	11.1	(3.1)
Q_4. Tense	14.1	(4.6)	14.3	(4.3)	13.9	(4.7)	13.3	(4.1)

[a] From R. B. Cattell, H. Eber, and M. M. Tatsuoka, 1970; Tables 1 and 4 for 1,149 females and 1,312 males based on high school juniors and seniors ages 17 years.

Data collected from the rearing parents during the laboratory visit show that adoptive and control parents also do not differ significantly in IQ, although both groups have higher IQs and less variance than the nationally normed sample (Wechsler, 1974). The mean scores are 107.8 (*SD* = 11.0) and 109.3 (*SD* = 10.7) for 193 adoptive and 218 control mothers, respectively. For 185 adoptive and 206 control fathers, these means are 113.6 (*SD* = 12.0) and 114.1 (*SD* = 11.5). Again, these differences may be indicative of differences in the Colorado population; or, as is more likely, it may be that parents who have the resources to pursue adoption have higher socioeconomic status and IQ as compared to the general population. The control fathers were matched to the adoptive fathers for education and occupation, which has apparently resulted in samples that are also comparable for IQ.

In general, mean differences among the parental groups in the CAP are few and not very large. The differences for the 16PF are apparently due to the age difference between the birth parents and rearing parents, and to the higher likelihood of religious affiliation for the adoptive families.

The few differences in personality among the parental groups are consistent with the findings regarding the adoptive and nonadoptive home environments. The tendency for the adoptive families to score higher on factors such as Conscientious and Controlled is again probably best explained by the greater likelihood of religious affiliation for these families. Finding no substantial group differences among the parents or their home environments suggests that observed differences between adopted and nonadopted CAP children are not due to these familial influences.

Prenatal and perinatal effects

Other possible "environmental" differences between adopted and nonadopted children may be their varying prenatal and birth experiences. Three possible prenatal factors which were systematically assessed from both relinquishing birth parents and control parents were coffee consumption, nicotine use, and alcohol use. For nicotine use, there are no significant differences between the two groups. There is a mean difference in coffee consumption, with the control parents averaging an additional cup per day – a finding that probably reflects the age difference between the birth and control parents. With regard to alcohol consumption, there is a distributional difference, but not a significant mean difference. Birth parents are over-represented in the heavy drinking group (3 or more drinks per day), and under-represented in the moderate group (between 1 and 3 drinks per day); there are no differences between the parents in the minimal drinking group.

Measures of perinatal factors include APGAR scores and birthweight. The 5-minute APGAR scores for the two groups show no significant differences – we find little variation on this measure with 98% of the sample rated at normal APGARs of 7–10. There is a significant mean difference in birthweight, however, with the adoptees ($N = 235$, $M = 7.04$ pounds, $SD = .99$) weighing less on average than nonadoptees ($N = 217$, $M = 7.32$, $SD = 1.10$). Birthweight of less than 2,500 grams is regarded as a risk factor for later developmental problems and is more likely to occur in births to either young (under 20) or older (over 35 years) mothers

(Holmes, Reich, & Pasternak, 1984). Lower birthweights, therefore, could contribute to differential outcomes for adoptees and nonadoptees. Although birthweights this low are rare in our sample, accounting for only 5% of the births, adoptees are twice as likely to be in the lower birthweight group.

However, results of analyses of covariance suggest that differences in coffee consumption, alcohol use, and birthweight do not account for group differences in outcome. These influences may not affect later outcomes in the CAP because the differences are relatively small. CAP children are not at the extreme of low birthweight (below 1,500 grams), and the levels of coffee and alcohol consumption are not extreme.

Influences of placement

Differential placement effects, if present, represent a specific type of genotype–environment interaction (G×E; see Chapter 20) seen only in adoptees. If the development or adjustment of mismatched adoptees differs from that of matched adoptees, adopted children as a group would tend to differ on average from nonadopted children.

The adoption agencies participating in the CAP did not actively practice selective placement other than for height and against locale (for example, placing children whose birth parents resided in southern Colorado in adoptive homes in northern Colorado). Previous analyses (Hardy-Brown, Plomin, Greenhalgh, & Jax, 1980; Ho, Plomin, & DeFries, 1979; Plomin & DeFries, 1985) indicate that the degree of realized selective placement for most measures in the CAP adult battery is negligible. This near random placement of adoptees results in some children being placed with adoptive parents who are similar to their birth parents and others placed with adoptive parents who are quite dissimilar. This permits investigation of the effects of differential placement for adoptees; that is, whether placement in a matched or unmatched home influences adoption outcomes. The effects of differential placement were examined for those outcome measures which showed adoptive status differences (Table 21.5).

In general the number of significant interactions did not exceed that expected on the basis of chance alone. Only for the domain of activity/attention was the multivariate test for interaction significant, with the parental classification based on 16PF Impulsivity. Moreover, only two of eight univariate tests (age 9 parent-reported CCTI Activity and teacher-reported CCTI Attention Span) were significant. Unexpectedly, adopted children reared in well-matched homes tend to be somewhat more active and less attentive than other adoptees. If important, this type of interaction could possibly yield mean differences between adoptees and children reared by their biological parents; however, it clearly did not do so in the present study.

Conclusions

In contrast to adopted children in studies of clinical populations, adoptees in the CAP fare quite well. Few significant differences are found in the outcomes of adoptees as compared to nonadoptees and those differences that are found account for only a small amount of variance. In agreement with results obtained from longitudinal studies, such as that of Bohman and Sigvardsson (1978, 1985, 1990), or cross-sectional studies using the full range of the population (Sharma, 1992), our results suggest that adopted children may have an increased risk for problems related to poor conduct, depression, and activity/attentiveness; however, the magnitude of any such difference in risk of problem behavior is relatively small.

Why do these results differ from those of many studies which find adopted children much more likely to experience psychological problems? As discussed in Chapter 7, studies of the prevalence of adoptees in clinical populations may be biased by the willingness of adoptive parents to seek special help for their children, possibly because of their previous experience with social service agencies.

It is also possible that our findings and those of other researchers who have studied children placed within the last 15 years reflect changes in placement practices. In his classic book, *Shared Fate*, Kirk (1964) recommended that social workers and adoptive parents recognize intrinsic differences of adoptive families. The children in our sample are on the cusp of changes which have taken place since then in the policies and practices of adoption agencies. Most of these adoptive families had access to information concerning the birth mother, and in some cases concerning the birth father as well. Some of the adoptive families participating in the CAP have had direct contact with birth parents – about 20 percent of the parents who adopted younger siblings met with the birth parents prior to these later adoptions. To varying degrees the rearing parents attribute some of their children's characteristics to these known aspects of the birth parents.

It is also the case that the CAP sample of adoptees experienced what may be close to the ideal adoption situation, in which placement in an adoptive home occurred very early in the child's life, with no traumatic separation from the birth parents. Thus, the CAP sample does not include adoptees removed from problem homes, or those who have been in a series of foster placements. Later placements, for whatever reason, may result in an increased risk for adjustment problems.

The results of our analyses clearly indicate that adoption does not inevitably lead to adjustment problems. The small mean differences between CAP adoptees and nonadoptees that we have noted in this chapter account for only a small fraction of the wide range of variation in various measures of adjustment and social functioning during middle childhood. Thus, any differences that may exist between adopted and nonadopted children, on average, are dwarfed by the individual differences that are manifested by both adopted children and children who are reared by their biological parents.

22 Conclusions

In this book, data from the Colorado Adoption Project (CAP) were used to begin to chart the emergence of individual differences during middle childhood – a crucially important but little-investigated period. We examined domains of interest such as cognitive abilities, personality, and psychopathology, as well as those more specific to middle childhood such as stress and school achievement in the early school years, motor development, and perceived self-competence. In addition, making use of the environmental assessments that have been featured in CAP, we explored the interaction between nature and nurture.

The purpose of this concluding chapter is to highlight some of the major CAP findings, to lay out future CAP analyses, and to point to possible new directions for behavioral genetic research in middle childhood – not to provide an encyclopedic summary.

Genetic Analyses of Behavioral Development

This section provides an overview of the major domains for which results of CAP genetic analyses are reported: general cognitive ability, specific cognitive abilities, school achievement, speech and language disorders, temperament and personality, competence, behavior problems, obesity, and motoric development. The issue of sex differences and adoption outcomes are also addressed.

General cognitive ability

Heritability of general cognitive ability appears to remain moderate from infancy to middle childhood in analyses based on the CAP sibling adoption design

(Chapter 4). In contrast, previous analyses based on the CAP parent–offspring adoption design suggested increasing genetic influence (Fulker, DeFries, & Plomin, 1988). It is possible that both sets of results are correct. As discussed in Chapters 2 and 3, biological parent–adopted child resemblance is a function of the heritability of a given character during childhood, the heritability of the same character in adulthood, and the correlation between genetic influences at the two different ages (DeFries, Plomin, & LaBuda, 1987). Thus, if the manifestations of genetic influence in childhood differ from those in adulthood, biological parents and their adopted-away offspring cannot be expected to show hereditary resemblance even if a trait is heritable both in childhood and in adulthood. For this reason, our expectation was that genetic factors would increasingly account for parent–offspring resemblance as the offspring developed from infants to adolescents. In contrast, because siblings in the CAP are tested at the same age, the sibling adoption design is not limited by changes in genetic influence from infancy to adulthood.

This interpretation is complicated by the results of the Louisville Twin Study which shows increases in heritability from early to middle childhood (Wilson, 1983). However, the results of the Louisville Twin Study may be affected by the inordinately central role of perinatal complications, so much more common in twins, which may have lowered heritability in infancy. This interpretation fits with recent results obtained from the MacArthur Longitudinal Twin Study at 14 and 20 months which, unlike the Louisville Twin Study, selected twins who were nearly full term (Plomin et al., 1993). In this study, Bayley (1969) scores showed heritabilities only slightly less than those found in the CAP. A recent analysis of CAP sibling and twin data replicates this finding and indicates that the two designs provide a highly consistent pattern of results (Fulker, Cherny, & Cardon, 1993).

State-of-the-art developmental genetic analyses of CAP cognitive data also uncovered evidence for genetic change as well as continuity. Although there is substantial continuity from infancy through middle childhood, new genetic influences on general cognitive ability emerge in middle childhood.

The sibling adoption design provides a direct test of the importance of shared environmental influence based on the resemblance of adoptive siblings. The results described in Chapter 4 suggest somewhat less influence of shared environment on general cognitive ability in middle childhood than might be expected from the literature, accounting for less than 20% of the observed variance on average. However, this literature depends exclusively on the twin method (see Chapter 2). It seems quite likely that twins share environmental influences to a greater extent than do non-twin siblings (Chipuer, Rovine, & Plomin, 1990). Moreover, estimates of shared environmental variance obtained from twin studies may be inflated by assortative mating. Longitudinal analyses in the CAP suggest that shared environmental influence is largely continuous from infancy through middle childhood, implying that some monolithic factor such as socioeconomic status may be at work. In contrast, nonshared environmental factors (including error of measurement) appear to be transitory. That is, they are unique to each age and do not persist from one year to the next.

Specific cognitive abilities

The CAP is the first genetic study of specific cognitive abilities from early childhood through middle childhood. The hierarchical, multivariate genetic analysis described in Chapter 5 investigates both general ("g") and specific cognitive abilities simultaneously. The results of this novel analysis support a strong genetic "g" at all ages. In early childhood, there is little evidence for genetic variance of specific cognitive abilities independent of "g." However, at 7 years of age, verbal, spatial, and memory abilities show ability-specific genetic influence independent of "g." This finding is exciting in suggesting that genetic differentiation of specific cognitive abilities emerges in middle childhood. Results of the longitudinal multivariate genetic analysis also described in Chapter 5 are consistent with this view. Confidence in this finding, however, is attenuated by results at 9 years, when only verbal ability shows genetic influence independent of "g." However, because the sample size at 7 years is nearly 50% larger than at 9, we predict that the results at 9 years will more closely approximate those at 7 when the sample size at 9 is larger.

School achievement

As described in Chapter 7, reading recognition showed significant genetic influence in middle childhood, whereas mathematics achievement did not. A finding with far-reaching significance for educators comes from multivariate analyses between school achievement measures and IQ: Most of the genetic influence on school achievement overlaps with genetic influence on IQ.

Estimates of the impact of shared environment on school achievement were surprisingly low in contrast to those obtained from twin studies, which generally show substantial shared environmental influence. Multivariate analyses indicated that the same shared environmental influences affect both verbal and performance IQ and reading and mathematics achievement.

Speech and language disorders

Chapter 8 reviews family studies of developmental articulation, language, and fluency disorders that indicate familial influence on these disorders. Two twin studies suggest that this observed familiality may be due at least in part to heritable influences. A preliminary genetic "at risk" analysis using CAP data on communication development at 3 years of age and IQ at 7 shows little evidence of genetic influence, but a more thorough study is underway.

Personality and temperament

Given the consistent finding in twin studies using parental ratings of children's temperament of substantial genetic influence we were surprised to find no evidence for genetic influence for parental ratings of temperament (see Chapter 9). This finding was consistent from infancy through middle childhood for both parent–offspring and sibling adoption analyses. These dramatically discrepant results could be explained by substantial nonadditive genetic variance that increases the similarity of MZ twins, but not first-degree relatives. Because MZ twins are genetically identical, they share both additive and nonadditive (dominant and epistatic) gene effects. In contrast, nonadditive genetic variance does not contribute importantly to either parent–offspring or sibling resemblance. Another possibility is that the parental ratings are problematic. This alternative is suggested by the finding that tester and teacher ratings show some evidence for genetic influence for sociability and activity.

The CAP represents the first behavioral genetic analysis using teacher ratings of temperament. Teacher ratings seem especially valuable because teachers have experience with many children and are thus better equipped than parents to compare children. Also, each sibling is rated by a different teacher, which avoids possible rater biases that arise when the same parent or tester rates both siblings. For these reasons, finding genetic influence for teacher ratings of temperament is especially noteworthy.

Another interesting finding is that shared environmental influence appears to be of negligible importance for temperament in middle childhood using the CAP sibling adoption design. This finding confirms a similar conclusion reached in twin analyses, which are much less powerful than the sibling adoption design for detecting the influence of shared environment.

Competence

Some developmentalists will be surprised by the results for perceived self-competence as assessed by the widely used Harter (1982) measure (see Chapter 10). First, although individual differences in perceived self-competence are usually assumed to be a function of environmental influences, results of analyses of CAP sibling data suggest that heritable variation is present at 9 years of age for three scales: Athletic Competence, Scholastic Competence, and Self-Worth. Second, although environmental influences on perceived self-competence are usually assumed to be familial in origin, little or no evidence was found for shared family environmental influences.

Moreover, similar results were obtained for mother ratings and teacher ratings of social competence. Evidence for genetic influence was particularly strong for teacher ratings of popularity, confidence, and leadership at 7 and 9 years, despite the fact that members of each sibling pair were rated by different teachers at different measurement occasions separated by two years on average.

Behavior problems

Chapter 11 presents the first behavioral genetic analysis of teacher reports on the widely-used Child Behavior Checklist (Achenbach & Edelbrock, 1993). The results suggest genetic influence for externalizing, but not internalizing, problems at 9 years of age. In contrast, little or no evidence of heritable variation was obtained for measures of stress during the first year of elementary school.

Obesity

Although the major dependent variables in the CAP are behavioral, data pertaining to physical measures have also been collected. For example, Chapter 12 summarizes results obtained from recent analyses of longitudinal data for the Body Mass Index (BMI), a measure of visceral body fat. Results of analyses of CAP adoptive and nonadoptive sibling data suggest that genetic influences on the BMI are substantial at birth; however, these genetic influences on BMI at birth are not correlated with those at later ages. In contrast, genetic influences at 1 year of age and later persist strongly into middle childhood. Nonshared environmental influences are important at each age, but do not contribute to longitudinal stability. Analyses of CAP parent–offspring data indicate that the genetic influences on BMI during early and middle childhood persist into adulthood.

Results of analyses of adiposity rebound data are also summarized in Chapter 12. Adiposity rebound refers to the rapid increase in body fat that occurs at about 6 years of age. When adiposity rebound occurs at an earlier age (e.g. 5.5 years or earlier), the risk of subsequent adult obesity is increased. Somewhat unexpectedly, results obtained from preliminary analyses of CAP sibling data suggest that individual differences in this risk factor are only moderately heritable.

Motor development

Chapter 13 provides the first genetic analysis of individual differences in motor development from infancy to middle childhood. The results suggest that there may be some surprising discoveries when this uncharted territory is explored systematically. The genetic results are intriguing because they are counterintuitive. Motor development scores show significant genetic influence at 1 year of age, declining influence at 2, and negligible genetic influence for gross motor development at 3 and at 7 years. Finding genetic influence at 1 year, but not the other ages, may be explained by the results of a multivariate genetic analysis. This multivariate analysis indicates that genetic influence on motor development at 1 is largely due to its overlap with mental development, a relationship which diminishes during early and middle childhood.

That individual differences in motor development are not highly heritable during early and middle childhood may seem counterintuitive. However, as discussed in Chapter 13, this is only surprising because it is easy to make the assumption that individual differences in motor development are genetic in origin since motor development seems more biological than other psychological traits. If these results are replicated, they could serve as an example that received wisdom provides a poor guide to the realities of genetic influence.

In contrast to the decreasing genetic influence on motor development, there is increasing environmental influence from infancy to middle childhood. The story of shared environmental influence is even more interesting because the longitudinal analysis indicates that environmental influences shared by siblings during infancy are also manifested in middle childhood.

Sex differences

Chapter 14 yields a clear conclusion concerning sex differences: There is little support for the hypothesis that boys and girls are differentially affected by heredity and environment.

Outcomes of adoption

An issue of contemporary concern is adoption as a risk factor for behavior problems. As discussed in Chapters 7 and 21, research indicates that adoptees are over-represented in clinical populations. Although non-clinical samples also show some adjustment problems for adopted as compared to nonadopted children, the differences are small. CAP results yield few significant differences between adopted children and nonadopted children in middle childhood. The few differences that emerge are in the direction of greater behavior problems for adopted children, especially boys, but these differences explain only a very small proportion of variance. In addition, as indicated in Chapter 7, adopted children are not at substantially increased risk for learning disabilities or school achievement problems.

Thus, CAP results show fewer differences than might be expected from the literature. However, it should be noted that the CAP design is prospective, unlike that of most other studies, and its sample consists of "easily-placed" Caucasian babies placed in their adoptive families in the first few weeks of life.

Genetic Analyses of Environmental Measures

As discussed in Chapter 1, behavioral genetic analyses can take us far beyond the rudimentary nature–nurture question of the relative magnitude of genetic and environmental components of variance. Three new directions for research were discussed: developmental, multivariate, and environmental analyses. The chapters of this book abound with developmental and multivariate analyses. Several chapters involve nongenetic analyses of predictions of outcomes in middle childhood from infancy and early childhood. For example, Chapter 6 presents results of longitudinal analyses between novelty preference assessments at 5 and 7 months of age and cognitive and school achievement measures at age 7; Chapter 7 describes developmental precursors of school achievement. These chapters instantiate the important point that CAP is not only a behavioral genetic study – it is also the largest long-term, longitudinal study of behavioral development of its kind.

In addition, all of the chapters offer environmental analyses in the sense that behavioral genetic analyses assess both genetic and environmental influences. However, these analyses often involve only anonymous components of shared and nonshared environmental variance. In contrast, the work described in Chapters 11 and 15–20 incorporates specific measures of the environment in order to address the two emerging themes of behavioral genetic analysis that were discussed in Chapter 1: Identifying specific sources of nonshared environment and investigating genetic influence on measures of the environment.

Nonshared environment

Chapter 15 attempts to identify specific sources of nonshared environment through the analysis of data for CAP siblings who were tested at the same age on different assessment occasions. In typical sibling studies, the siblings are tested on the same occasion when they are different ages, and it was hoped that the new approach might sharpen the search for nonshared environment.

The chapter considered two questions concerning nonshared environment. First, to what extent do siblings in the same family experience different environments when each child is 7 years old? Second, to what extent do these differences in experience at 7 relate to outcome measures at 9 years?

Concerning the first question, mothers report that they treat their children quite similarly when both children are 7 years. Because the maternal treatment of siblings at the same age was highly similar, it is not surprising that sibling differences in family environment showed few relationships with later perceived competence and behavioral adjustment. Although life events and illnesses differed for siblings in the same family, these differences also showed few predictions of later outcomes.

Of course, it is possible that this approach using same-age sibling environments is not appropriate or that environmental measures employed in the CAP are not sufficiently sensitive to detect differential treatment of siblings in a family. Another possibility, however, is that nonshared environmental factors are largely idiosyncratic, as appears to be the case for IQ (Chapter 4). Although it may be too early to reach that negative conclusion, the results of this study and others like it suggest that identifying specific sources of nonshared environment is not likely to be easy.

Genetic influence on environmental measures

Chapter 17 introduces the investigation of genetic influence on measures of the environment by systematically analyzing CAP environmental measures from infancy through middle childhood using the sibling adoption design. Several examples of genetic influence on environmental measures are reported, especially in middle childhood. Children's self-perceptions of their family environment at 7 years of age on a version of the FES and mothers' reports of parenting style at 7 and 9 years yield evidence for genetic influence. At both 7 and 9 years, maternal reports of parenting suggested greater genetic influence for a second-order factor of warmth than for control, which confirms similar findings in the literature (Plomin, in press).

Chapter 11 presents a genetic analysis of a measure of first-grade stress that finds no evidence for genetic influence. This result differs from other research that generally shows genetic influence for measures of the environment (Plomin & Bergeman, 1991). However, it should be noted that this is the first study of genetic influence on stress in childhood, when etiological factors might well be different from factors later in life. An environmental finding from these analyses of first-grade stress is intriguing: Children growing up in the same family have very different experiences in the first grade.

Many environmental measures developed by psychologists involve relationships, or at least interactions, with other people. Chapter 16 focuses on family relationships; in addition to addressing issues of stability and change in sibling relationships and links between sibling relationships and relationships with parents, it investigates genetic influence on children's behavior to their siblings and on mothers' behavior to siblings. Evidence for genetic influence emerged for several dimensions of sibling and mother–child interactions. In contrast, no mean group differences emerged between the relationships of adoptive siblings and nonadoptive siblings. Thus, these findings do not support the sociobiological hypothesis that genetically related siblings behave more altruistically toward one another than genetically unrelated siblings, even though the results show genetic influence on individual differences in behavior toward siblings.

One of the next steps in research on the "nature of nurture" is to investigate the mechanisms by which heredity affects measures of the environment. Chapter 17 broaches this topic by examining possible personality precursors to genetic influence on environmental measures in CAP. Results indicate that tester ratings of children's temperament are significantly associated with several scales of the HOME during infancy and early childhood. However, most of the genetic variance on environmental measures is independent of temperament.

Another strategy for investigating mechanisms by which genetic factors affect measures of environment is described in Chapter 20 as evocative genotype–environment correlation. Characteristics of biological mothers were employed as indices of their adopted-away children's genotypes and used in canonical correlation analyses with measures of the adoptive home. These analyses suggest that some characteristics of the biological mothers are related genetically to measures of the home environment. This same chapter also examined another issue at the interface between nature and nurture, genotype–environment interaction. As in previous CAP analyses, few significant genotype–environment interactions were found.

Another step in research on the nature of nurture is to explore the extent to which associations between environmental measures and developmental outcomes are mediated genetically. Chapter 18 attempts to address this issue longitudinally, by correlating early CAP environmental measures with outcome measures in middle childhood. A fundamental but generally unrecognized problem precluded success in these analyses. Despite the still-common assumption among developmentalists of the power of environmental influence (e.g., Hoffman, 1991), early environmental measures are generally unable to predict outcomes in middle childhood. Although some predictions are significant in the CAP, they are not consistent across ages and their effect sizes are generally small. Multivariate genetic analyses of such associations between environmental measures and outcomes are thereby precluded.

However, results of CAP analyses revealed two cases in which environmental measures predict outcomes in middle childhood. One involves the association between first-grade stress and later behavior problems. As discussed in Chapter 11, an interesting pattern of results emerged from regressions predicting behavior problems at 9 years of age from first-grade stress at 7 years, independent of teacher-rated behavior problems at 7 years. Predictions were significant for non-adopted children but not for adopted children. One possible interpretation of these results involves genetic mediation in which genetically influenced characteristics of parents are related to both family environment and child measures. The hallmark of such relationships are correlations between environment and outcome that are greater in nonadoptive families than in adoptive families (Plomin, Loehlin, & DeFries, 1985). Usually genetic mediation is considered only in relation to family environment. However, it is possible that such effects may spread to experiences beyond the family. For example, some genetically-influenced characteristics of parents might mediate children's responses to the stresses of beginning school. However, this interpretation is weakened by the finding that the measure of first-grade stress is not heritable. Nonetheless, the differences in correlations between first-grade stress and later behavior problems for nonadopted and adopted children are sufficiently large to warrant further investigation.

The second case involves the association between home environment in infancy and general cognitive ability at 7 years. As described in Chapter 19, although the associations were weak, correlations tend to be higher in nonadoptive than in adoptive families. Model-fitting analyses confirmed these impressions of genetic mediation of the HOME at 1 and 2 years with 7- and 9-year-old IQ.

These various results illustrate the diversity of the research questions that can be addressed with a longitudinal, multivariate adoption study of the magnitude and scope of the CAP. Although we have employed sophisticated analytical methods to analyze numerous measures and a range of developmental issues, we have only just begun to explore the full complexity of the CAP data set. To date, our multivariate genetic analyses have been restricted to a relatively few domains, and many cross-domain issues have yet to be addressed. As the CAP sample size increases and data are obtained at later ages, even more sophisticated analyses will be possible.

Future CAP Analyses

As indicated in Chapter 2, adoptive and nonadoptive probands and their younger siblings were tested in their homes at 1, 2, 3, and 4 years of age, and they are currently being tested in the laboratory at 7, 12, and 16 years. Telephone testing and interviews are conducted at 9, 10, 11, 13, 14, and 15 years, and some questionnaire information is being collected at 5, 6, and 8 years. At 16 years, the children are being administered the same tests that their parents completed more than a decade and a half earlier.

Sample

Within the next five years, all of the adoptive and nonadoptive probands and most of their younger siblings will have been tested through 12 years of age, and approximately 75% of the probands and 50% of their younger siblings will have been tested through 16 years of age. Moreover, 180 older siblings of the CAP probands will have been administered the adult test battery.

Beginning at 4 years of age, over 300 pairs of twins are also currently being administered the CAP battery. Within the next five years, approximately half of this twin sample will have been tested through 9 years of age.

The larger sample of adoptive and nonadoptive probands and siblings will obviously facilitate more powerful genetic analyses during middle childhood. Moreover, within the next five years sufficient data will be available to begin to explore the genetic and environmental etiologies of developmental change and continuity from middle childhood through the tumultuous adolescent years. The sample of older siblings tested on the adult battery will also greatly facilitate analyses of young adulthood.

Within the next five years, sufficient data will be available from the twins to undertake combined adoption–twin analyses. Such analyses of combined data sets will enhance the CAP parent–offspring and sibling designs in three ways. First, statistical power to detect genetic and environmental influences will be greatly increased. Second, combining these various research strategies will extend the range of testable hypotheses regarding the etiology of developmental continuity and change. Third, analyses of such combined data sets will facilitate more adequate tests of underlying assumptions through model fitting.

Models

CAP behavioral genetic analyses reported in this book range from the simple to the complex. The simplest analyses employed the multiple regression approach of DeFries and Fulker (1985). Although this highly versatile methodology was originally developed for application to twin data, it is also directly applicable to the analysis of data from adopted and nonadopted sibling pairs. As outlined in Chapter 3, an "augmented" multiple regression model (Eq. 14) can be used to predict one sib's score (C) from that of the other (P), together with their coefficient of relationship ($R = 0.5$ for nonadoptive siblings and 0.0 for adoptive sibs), and the product of the second sib's score and relationship (i.e., PR). The partial regression of C on P estimates the proportion of variance due to shared environmental influences (c^2), whereas the coefficient of PR estimates heritability (h^2).

When probands are selected for extreme scores such as a dimension of inhibition or school achievement, a "basic" regression model (Chapter 3, Eq. 13) can be used to predict the cosib's score (C) from the proband's score (P) and R only. In this application, the partial regression of cosib's score on R (i.e., B_2) provides a direct and statistically powerful test for genetic etiology (DeFries & Fulker, 1988). When the data are suitably transformed, this regression coefficient estimates h_g^2, a measure of the extent to which the deviant scores of probands are due to heritable influences. Moreover, when the augmented model is fitted to the same transformed data set, the regression of C on R estimates $h_g^2 - h^2$, thus, the significance of this partial regression coefficient provides a test of the hypothesis that the etiology of extreme scores differs from that of individual differences within the normal range. Future analyses of CAP sibling and twin data will employ this multiple regression methodology. Moreover, to facilitate even more powerful tests, these multiple regression equations will also be fitted to the combined sibling and twin data sets.

The multiple regression analysis of sibling and twin data has a number of advantages over alternative methods. First, it is very simple to apply, enabling the rapid screening of a large number of variables as exemplified by the analyses of CAP personality and temperament reported in Chapter 9. Second, the method has great flexibility in the specification of univariate models. This flexibility facilitates the investigation of many forms of G×E interactions when environmental variables are included in the prediction equations. Other covariates (e.g., gender or age) may also be included to detect other forms of interaction or covariance. So flexible is this method that it has recently been applied to detect quantitative trait loci through linkage to chromosomal markers (Fulker, Cardon, DeFries, Kimberling, Pennington, & Smith, 1991).

Useful as the multiple regression methodology is, the CAP longitudinal data sets warrant more searching developmental analyses. During the past several years, we have devoted considerable time and effort to the development of longitudinal models that incorporate both time constant and autoregressive processes. Results of analyses that fit such models to CAP cognitive data are reported in Chapters 4 and 5. In addition to the analysis of CAP sibling data, we have also recently applied these methods to analyze combined sibling and parent–offspring data sets (Phillips & Fulker, 1989) and data from both twins and siblings (Cardon, Fulker, DeFries, & Plomin, 1992). Although our applications of these developmental models to the analysis of CAP data have largely involved cognitive measures, the clarity with which the modeling illuminates the nature of the developmental processes is remarkable. As reported in Chapter 4, shared environmental influences on general cognitive ability appear to be global and not time specific. In contrast, there is substantial genetic continuity from infancy to middle childhood, but significant new genetic variation is manifested at the later ages. Finally, nonshared environmental influences do not contribute to observed continuity. Thus, these results suggest that individual differences in mental development are substantially influenced by the genotype, with new variation appearing at later ages, and prior variation providing the substrate for future development through age-to-age transmission. Moreover, the results of these analyses illustrate how genetic change and continuity are manifested in the context of a complex environmental background. The exciting challenge for the future will be to apply these models to analyze the CAP longitudinal data from infancy through young adulthood.

A third analytic methodology that we have developed is that of multivariate genetic analysis, with the goal of combining this approach with the longitudinal analyses outlined above. We regard this complex area as particularly important since it is only through multiple indicators that the structure of behavioral constructs can be truly understood. To date, as reported in Chapter 5, we have applied this approach to the problem of specific versus general cognitive abilities. Hierarchical genetic and environmental models that specify factors unconfounded by test-specific variance were developed. Furthermore, as noted in Chapter 5, this approach has been successfully applied to the study of developmental processes. Future CAP analyses will fit such multivariate models to sibling and twin data sets simultaneously, thereby facilitating developmental behavioral genetic analyses of unparalleled power and scope.

Data archive

In order to foster additional analyses of CAP data, the data set is being placed in the NIMH-supported archive of the Henry A. Murray Research Center at Radcliffe College. Data from adopted and nonadopted children tested from 1 to 7 years of age, their biological, adoptive, and nonadoptive parents, and home environments are now available for analysis from this source. Within the next several years, data from adopted and nonadopted probands tested throughout middle childhood will also be archived. Developmentalists and behavioral geneticists who employ other models and methodologies are strongly encouraged to exploit this unique data set.

Future Research on Middle Childhood

Questions about the etiology of individual differences during middle childhood far outnumber answers. Rather than expound on what we do not know, which could easily fill a much larger book, in this closing section we mention three strategies that developmental psychologists could employ to answer at least some of these questions. Then, out of all the possible directions for future research in developmental behavioral genetics, we focus on just one – molecular genetics. Developmentalists are probably least familiar with this direction, but we are confident that it will have a major impact on the field by the end of the century.

Developmental psychology

The obvious shift towards greater acceptance of behavioral genetics among developmentalists leads us to predict that they will increasingly be conducting this research. Developmentalists with expertise in a specific substantive domain have the tremendous advantage of asking the most interesting, theory-driven research questions and of interpreting their research findings in a way that will have the greatest impact on colleagues in their field. There are clear signs that this is beginning to happen (Plomin, 1993).

For those developmentalists who may wish to begin to use genetic strategies in their research, one possibility would be to include siblings in their samples. More than 80% of families have more than one child, and it is relatively easy to recruit the sibling of a subject. After analyzing the topic of interest, the data could then be examined from a new perspective that considers sibling similarities and differences. How similar are siblings in the same family? For most traits, siblings are not very similar, which leads to the nonshared environmental question, Why are siblings in the same family so different? Multivariate questions could also be asked: Do the familial or nonfamilial influences on one aspect of the phenomenon overlap with influences on another aspect? Developmental questions can be asked about age differences and age changes in sibling resemblance. Although sibling analyses are familial, rather than genetic, such analyses represent an important first step in understanding the etiology of individual differences.

Another possibility would be to undertake twin studies. Twin studies are relatively easy to initiate, and small twin studies could be readily accomplished even by research groups with limited resources. About 1% of all births are twins, and parents of twins are usually very willing to participate in research because their children are so obviously special. Although results of individual twin studies could be informative, collaborative analyses of combined data sets could facilitate more powerful tests of hypotheses relevant to a number of important developmental issues.

A third possibility would be to collect data from half-siblings and step-siblings. Full adoption designs like CAP would be extremely difficult to institute in middle childhood because the numbers of such adoptions declined dramatically during the 1970s. However, little use has been made in behavioral genetics of the large numbers of half-siblings and step-siblings that can be found in families in which parents divorced and remarried. Half-sibling and step-sibling resemblances could also be compared to that of full siblings in the same families and in nondivorced families (Chipuer, Plomin, Reiss, & Hetherington, in press; Reiss, 1993).

Developmental research on nurture as well as nature will benefit enormously as developmentalists increasingly use behavioral genetic designs in their research.

Molecular genetics

A reason for considerable optimism about the future of genetic research in behavioral development is that behavioral genetics will inherit the incredible advances currently being made in molecular genetics. In 1980, Botstein, White, Skolnick, and Davis proposed that the human genome could be mapped using DNA markers. Within three years, such markers were used to localize to chromosome 4 the gene that causes Huntington's disease (Gusella et al., 1983). A major goal of the Human Genome Project (Watson, 1990) is to foster the localization of other disease-causing genes by constructing a human genetic map with very closely spaced molecular markers.

It seems clear that we are at the dawn of a new era in which molecular genetic techniques will revolutionize behavioral genetics by identifying specific genes that contribute to genetic variance in behavior (e.g., Aldhous, 1992; McGuffin & Murray, 1992; Plomin, 1990, 1993). For behavior, unlike for the single-gene disorders that are quickly located on the genome map, the quest is to find not the gene, but the many genes, that affect a trait. New strategies are needed to identify genes that affect behavioral traits, even when the genes account for only a small amount of variance, when nongenetic factors are important (as they are for behavioral traits), and when traits are distributed as quantitative dimensions rather than qualitative disorders. In other words, we need to use molecular genetic techniques in a quantitative genetic framework. One promising approach is sib-pair linkage analysis in which chromosomal markers are used to assess the effects of genes that influence quantitative characters (Fulker, Cardon, DeFries, Kimberling, Pennington, & Smith, 1991). Other approaches include allelic association (Edwards, 1991) using candidate genes (Boerwinkle, Chakraborty, & Sing, 1986) and, especially, DNA polymorphisms in coding regions that result in functional differences among individuals (Sobell, Heston, & Sommer, 1992). Such approaches may eventually reveal the genetic architecture of complex behavioral characters at the actual DNA level. However, given the breathtaking pace of technological advances in molecular genetics, by the turn of the century we may be investigating multiple-gene influences for complex dimensions and disorders using completely different techniques from those in use today.

Although new developments in molecular genetics will almost certainly contribute substantially to advances in the genetic analysis of complex behavioral characters during the 21st century, quantitative genetic analyses of such characters will continue to be crucially important. When analyses employing both genetic approaches are conducted, the relative importance of individual quantitative trait loci can be measured by assessing their contributions to the genetic variance and covariance due to all genetic influences. Thus, as was noted a quarter of a century ago, single-gene and quantitative genetic analyses "are complementary, not mutually exclusive; both should be exploited in order to attain a more complete understanding of the genetic causes of individual differences in behavior" (DeFries & Hegmann, 1970, p. 53).

Epilogue

This first behavioral genetic study of middle childhood has yielded a rich harvest of findings about both nature and nurture. Future CAP analyses promise even more bountiful harvests as the adopted and nonadopted children develop into adolescents and young adults and as twins add to the power of the adoption analyses. More generally, we are convinced that the future of behavioral genetic research on development is bright. Not only is there the promise of molecular genetics shining on the horizon, but also the momentum from findings, methods, and projects such as those presented in this book will propel the field far into the next century, especially as genetic research continues to flow into the mainstream of developmental research. As Michael Rutter concluded at a watershed nature/nurture symposium at the 1993 meeting of the Society for Research in Child Development, the time has come to put the nature versus nurture controversy behind us and "to bring nature and nurture together in the study of development."

References

Preface

Plomin, R. & DeFries, J. C. (1985). *Origins of individual differences in infancy: The Colorado Adoption Project.* Orlando, FL: Academic Press.

Plomin, R., DeFries, J. C., & Fulker, D. W. (1988). *Nature and nurture during infancy and early childhood.* Cambridge, England: Cambridge University Press.

Chapter 1   Nature and Nurture in Middle Childhood

Burks, B. (1928). The relative influence of nature and nurture upon mental development: A comparative study of foster parent–foster child resemblance and true parent–true child resemblance. *Twenty-Seventh Yearbook of the National Society for the Study of Education, 27*, 219–316.

Cantwell, D. P. (1975). Genetic studies of hyperactive children: Psychiatric illness in biologic and adopting parents. In R. R. Fieve, D. Rosenthal, & H. Brill (Eds.), *Genetic research in psychiatry* (pp. 273–280). Baltimore: Johns Hopkins University Press.

DeFries, J. C., Plomin, R., & LaBuda, M. C. (1987). Genetic stability of cognitive development from childhood to adulthood. *Developmental Psychology, 23*, 4–12.

DeFries, J. C., Vandenberg, S. G., & McClearn, G. E. (1976). The genetics of specific cognitive abilities. *Annual Review of Genetics, 10*, 179–207.

Deutsch, C. K. & Kinsbourne, M. (1990). Genetics and biochemistry in attention deficit disorder. In M. Lewis & S. M. Miller (Eds.), *Handbook of developmental psychopathology* (pp. 93–107). New York: Plenum Press.

Dunn, J. & Plomin, R. (1990). *Separate lives: Why siblings are so different.* New York: Basic Books.

Edelbrock, C., Rende, R., Plomin, R., & Thompson, L. A. (in press). Genetic and environmental effects on competence and problem behavior in childhood and early adolescence. *Journal of Child Psychology and Psychiatry.*

Fisch, R. O., Bilek, M. K., Deinard, A. S., & Chang, P. N. (1976). Growth, behavioral, and psychologic measurements of adopted children: The influences of genetic and socio-economic factors in a prospective study. *Behavioral Pediatrics, 89,* 494–500.

Foch, T. T. & Plomin, R. (1980). Specific cognitive abilities in 5- to 12-year-old twins. *Behavior Genetics, 10,* 507–520.

Fulker, D. W., DeFries, J. C., & Plomin, R. (1988). Genetic influence on general mental ability increases between infancy and middle childhood. *Nature, 336,* 767–769.

Galton, F. (1876). The history of twins as a criterion of the relative powers of nature and nurture. *Royal Anthropological Institute of Great Britain and Ireland Journal, 6,* 391–406.

Garfinkle, A. S. & Vandenberg, S. G. (1981). Development of Piagetian logico-mathematical concepts and other specific cognitive abilities. In L. Gedda, P. Parisi, & W. E. Nance (Eds.), *Twin research 3: Intelligence, personality, and development* (pp. 51–60). New York: Liss.

Goldsmith, H. H. (1993). Nature–nurture and the development of personality: Introduction. In R. Plomin & G. E. McClearn (Eds.), *Nature, nurture, and psychology* (pp. 155–160). Washington, DC: APA Books.

Goldsmith, H. H. & Gottesman, I. I. (1981). Origins of variation in behavioral style: A longitudinal study of temperament in young twins. *Child Development, 52,* 91–103.

Goodman, R. & Stevenson, J. (1989a). A twin study of hyperactivity: I. An examination of hyperactivity scores and categories derived from Rutter teacher and parent question-naires. *Journal of Child Psychology and Psychiatry, 30,* 671–689.

Goodman, R. & Stevenson, J. (1989b). A twin study of hyperactivity: II. The aetiological roles of genes, family relationships, and perinatal adversity. *Journal of Child Psychology and Psychiatry, 30,* 671–689.

Horn, J. M., Loehlin, J. C., & Willerman, L. (1979). Intellectual resemblance among adoptive and biological relatives: The Texas Adoption Project. *Behavior Genetics, 9,* 177–207.

Knaack, R. (1978). A note on the usefulness of the Coloured Progressive Matrices (CPM) with preschool children. *Psichologie in Erziehung und Unterricht, 25,* 159–167.

Koch, H. L. (1966). *Twins and twin relations.* Chicago: University of Chicago Press.

Kraut, A. G. (1992). Report from the Washington Office. *SRCD Newsletter,* Fall Issue, 7.

Leahy, A. M. (1935). Nature–nurture and intelligence. *Genetic Psychology Monographs, 17,* 236–308.

Loehlin, J. C. (1992). *Genes and environment in personality development.* Newbury Park, CA: Sage Publications.

Loehlin, J. C., Horn, J. M., & Willerman, L. (1981). Personality resemblance in adoptive families. *Behavior Genetics, 11,* 309–330.

Loehlin, J. C., Willerman, L., & Horn, J. M. (1982). Personality resemblances between unwed mothers and their adopted-away offspring. *Journal of Personality and Social Psychology, 42,* 1089–1099.

Loehlin, J. C., Willerman, L., & Horn, J. M. (1985). Personality resemblance in adoptive families when the children are late adolescents and adults. *Journal of Personality and Social Psychology, 48,* 376–392.

Mangan, G. (1982). *The biology of human conduct: East–West models of temperament and personality.* Elmsford, NY: Pergamon Press.

Matheny, A. P., Jr. & Dolan, A. B. (1980). A twin study of personality and temperament during middle childhood. *Journal of Research in Personality, 14,* 224–234.

McGue, M., Bouchard, T. J., Iacono, W. G., & Lykken, D. T. (1993). Behavioral genetics of cognitive ability: A life-span perspective. In R. Plomin & G. E. McClearn (Eds.), *Nature, nurture, and psychology* (pp. 59–76). Washington, DC: APA Books.

Morrison, J. R. & Stewart, M. A. (1973). The psychiatric status of the legal families of adopted hyperactive children. *Archives of General Psychiatry, 28*, 888–891.

O'Connor, M., Foch, T. T., Sherry, T., & Plomin, R. (1980). A twin study of specific behavioral problems of socialization as viewed by parents. *Journal of Abnormal Child Psychology, 8*, 189–199.

Plomin, R. (1986). *Development, genetics, and psychology.* Hillsdale, NJ: Lawrence Erlbaum Associates.

Plomin, R. (1988). The nature and nurture of cognitive abilities. In *Advances in the psychology of human intelligence* (pp. 1–33). Hillsdale, NJ: Lawrence Erlbaum Associates.

Plomin, R. & Bergeman, C. S. (1991). The nature of nurture: Genetic influence on "environmental" measures. *Behavioral and Brain Sciences, 14*, 373–427.

Plomin, R. & Daniels, D. (1987). Why are children in the same family so different from each other? *The Behavioral and Brain Sciences, 10*, 1–16.

Plomin, R., Emde, R. N., Braungart, J. M., Campos, J., Corley, R., Fulker, D. W., Kagan, J., Reznick, J. S., Robinson, J., Zahn-Waxler, C., & DeFries, J. C. (1993). Genetic change and continuity from 14 to 20 months: The MacArthur Longitudinal Twin Study. *Child Development, 64*, 1354–1376.

Plomin, R. & Foch, T. T. (1980). A twin study of objectively assessed personality in childhood. *Journal of Personality and Social Psychology, 39*, 680–688.

Plomin, R. & Loehlin, J. C. (1989). Direct and indirect IQ heritability estimates: A puzzle. *Behavior Genetics, 19*, 331–342.

Plomin, R. & Neiderhiser, J. M. (1992). Quantitative genetics, molecular genetics, and intelligence. *Intelligence, 15*, 369–387.

Plomin, R. & Nesselroade, J. R. (1990). Behavioral genetics and personality change. *Journal of Personality, 58*, 191–220.

Plomin, R., Reiss, D., Hetherington, E. M., & Howe, G. (in press). Nature and nurture: Genetic influence on measures of the family environment. *Developmental Psychology*.

Plomin, R. & Rende, R. (1991). Human behavioral genetics. *Annual Review of Psychology, 42*, 161–190.

Plomin, R. & Vandenberg, S. G. (1980). An analysis of Koch's (1966) Primary Mental Abilities test data for 5- to 7-year-old twins. *Behavior Genetics, 10*, 409–412.

Riese, M. L. (1990). Neonatal temperament in monozygotic and dizygotic twin pairs. *Child Development, 61*, 1230–1237.

Rutter, M., Macdonald, H., Le Couteur, A., Harrington, R., Bolton, P., & Bailey, A. (1990). Genetic factors in child psychiatric disorders: II. Empirical findings. *Journal of Child Psychology and Psychiatry, 31*, 39–83.

Safer, D. J. (1973). A familial factor in minimal brain dysfunction. *Behavior Genetics, 3*, 175–186.

Scarr, S. (1966). Genetic factors in activity and motivation. *Child Development, 38*, 663–673.

Scarr, S. (1969). Social introversion-extraversion as a heritable response. *Child Development, 40*, 823–832.

Scarr, S., Webber, P. I., Weinberg, R. A., & Wittig, M. A. (1981). Personality resemblance among adolescents and their parents in biologically related and adoptive families. *Journal of Personality and Social Psychology, 40*, 885–898.

Scarr, S. & Weinberg, R. A. (1977). Intellectual similarities within families of both adopted and biological children. *Intelligence, 1*, 170–191.

Segal, N. (1986). Monozygotic and dizygotic twins: A comparative analysis of mental ability profiles. *Child Development, 56*, 1051–1058.

Skodak, M. & Skeels, H. M. (1949). A final follow-up of one hundred adopted children. *Journal of Genetic Psychology, 75*, 85–125.

Thompson, L. A., Detterman, D. K., & Plomin, R. (1991). Associations between cognitive abilities and scholastic achievement: Genetic overlap but environmental differences. *Psychological Science, 2*, 158–165.

Wachs, T. D. (1993). The nature–nurture gap: What we have here is a failure to collaborate. In R. Plomin & G. E. McClearn (Eds.), *Nature, nurture, and psychology* (pp. 375–391). Washington, DC: APA Books.

Willerman, L. (1973). Activity level and hyperactivity in twins. *Child Development, 44*, 288–293.

Wilson, R. S. (1975). Twins: Patterns of cognitive development as measured on the WPPSI. *Developmental Psychology, 11*, 126–139.

Wilson, R. S. (1983). The Louisville Twin Study: Developmental synchronies in behavior. *Child Development, 54*, 298–316.

Wilson, R. S. (1986). Continuity and change in cognitive ability profile. *Behavior Genetics, 16*, 45–60.

Chapter 2   The Colorado Adoption Project

Achenbach, T. M. & Edelbrock, C. (1983). *Manual for the Child Behavior Checklist and Revised Child Behavior Profile*. Burlington: Department of Psychiatry, University of Vermont.

Ascher, S. (1985). *Loneliness Questionnaire*. Personal communication.

Bayley, N. (1969). *Manual for the Bayley Scales of Infant Development*. New York: Psychological Corporation.

Brooks-Gunn, J. & Petersen, A. L. (1984). Problems in studying and defining pubertal events, *Journal of Youth and Adolescence, 13*, 181–196.

Bruininks, R. H. (1978). *Bruininks-Oseretsky Test of Motor Proficiency: Examiner's Manual*. Circle Pines, MN: American Guidance Service.

Buss, A. H. & Plomin, R. (1975). *A temperament theory of personality development*. New York: Wiley-Interscience.

Buss, A. H. & Plomin, R. (1984). *Temperament: Early developing personality traits*. Hillsdale, NJ: Lawrence Erlbaum Associates.

Caldwell, B. M. & Bradley, R. H. (1978). *Home Observation for Measurement of the Environment*. Little Rock: University of Arkansas.

Cardon, L. R., Corley, R. P., DeFries, J. C., Plomin, R., & Fulker, D. W. (1992). Factorial validation of a telephone test battery of specific cognitive abilities. *Personality and Individual Differences, 13*, 1047–1050.

Cattell, R. B., Eber, H., & Tatsuoka, M. M. (1970). *Handbook for the Sixteen Personality Factor Questionnaire (16PF)*. Champaign, IL: Institute for Personality and Ability Testing.

Chess, S. & Thomas, A. (1984). *Origins and evolution of behavior disorders: Infancy to early adult life*. New York: Brunner/Mazel.

Coddington, R. D. (1972). The significance of life events as etiologic factors in the diseases of children: II. A study of a normal population. *Journal of Psychosomatic Research, 16*, 205–213.

Compas, B. (1987). Stress and life events during childhood and adolescence. *Clinical Psychology Review, 7*, 275–302.

Connolly, A. J., Nachtman, W., & Pritchett, E. M. (1976). *KeyMath Diagnostic Arithmetic Test*. Circle Pines, MN: American Guidance Service.

Daniels, D., Plomin, R., & Greenhalgh, J. (1984). Correlates of difficult temperament in infancy. *Child Development, 55*, 1184–1194.

DeFries, J. C. (1975). Commentary on "Quantitative genetic perspectives: Implications for human development," by L. L. Cavalli-Sforza. In K. W. Schaie, V. E. Anderson, G. E. McClearn, & J. Money (Eds.), *Developmental human behavior genetics* (pp. 145–149). Lexington, MA: Lexington Books.

DeFries, J. C., Vandenberg, S. G., McClearn, G. E., Kuse, A. R., Wilson, J. R., Ashton, G. C., & Johnson, R. C. (1974). Near identity of cognitive structure in two ethnic groups. *Science, 183*, 338–339.

Dibble, E. & Cohen, D. (1974). Companion instruments for measuring children's competence and parental style. *Archives of General Psychiatry, 30*, 805–815.

Fletcher, R. (1990). *The Cyril Burt scandal: Case for the defence.* New York: Macmillan.

Gorsuch, R. L. (1983). *Factor analysis* (2nd edition). Hillsdale, NJ: Lawrence Erlbaum Associates.

Gottfried, A. W. (1984). *Home environment and early cognitive development: Longitudinal research.* New York: Academic Press.

Groves, R. M. & Kahn, R. L. (1979). *Surveys by telephone: A national comparison with personal interviews.* New York: Academic Press.

Harter, S. (1982). The perceived competence scale for children. *Child Development, 53*, 87–97.

Hauser, R. M. & Featherman, O. L. (1977). *The process of stratification: Trends and analysis.* New York: Academic Press.

Hearnshaw, L. S. (1979). *Cyril Burt, Psychologist.* Ithaca, NY: Cornell University Press.

Hedrick, D. L., Prather, E. M., & Tobin, A. R. (1975). *Sequenced Inventory of Communication Development.* Seattle: University of Washington Press.

Jensen, A. R. (1969). How much can we boost IQ and scholastic achievement? *Harvard Educational Reveiw, 39*, 1–123.

Joynson, R. B. (1989). *The Burt affair.* London: Routledge.

Kamin, L. J. (1974). *The science and politics of I. Q.* Potomac, MD: Lawrence Erlbaum Associates.

Kamin, L. J. (1981). Studies of adopted children. In H. J. Eysenck & L. J. Kamin (Eds.), *The intelligence controversy* (pp. 114–125). New York: Wiley-Interscience.

Kandel, D. & Davies, J. (1982). Epidemiology of depressive mood in adolescents. *Archives of General Psychiatry, 39*, 1205–1212.

Kent, J. & Plomin, R. (1987). Testing specific cognitive abilities by telephone and mail. *Intelligence, 11*, 391–400.

Lewontin, R. C. (1975). Genetic aspects of intelligence. *Annual Review of Genetics, 9*, 387–405.

Matheny, A. P., Jr. (1980). Bayley's Infant Behavior Record: Behavioral components and twin analyses. *Child Development, 51*, 1157–1167.

McDevitt, S. C. & Carey, W. B. (1978). The measurement of temperament in 3–7-year-old children. *Journal of Child Psychology and Psychiatry, 19*, 245–253.

Moos, R. H. & Moos, B. S. (1981). *Family Environment Scale manual.* Palo Alto, CA: Consulting Psychologists Press.

Petersen, A. C., Schulenberg, J. E., Abramowitz, R., Offer, D., & Jarcho, H. (1984). A Self-Image Questionnaire for Young Adolescents (SIQYA): Reliability and validity studies. *Journal of Youth and Adolescence, 13*, 93–11.

Plomin, R. (1986). *Development, genetics, and psychology.* Hillsdale, NJ: Lawrence Erlbaum Associates.

Plomin, R. & DeFries, J. C. (1981). Multivariate behavioral genetics and development: Twin studies. In L. Gedda, P. Parisi, & W. E. Nance (Eds.), *Progress in clinical and biological research, twin research 3: Part B. Intelligence, personality, and development* (Vol. 69B, pp. 25–33). New York: Alan R. Liss.

Plomin, R. & DeFries, J. C. (1985). *Origins of individual differences in infancy: The Colorado Adoption Project.* Orlando, FL: Academic Press.

Plomin, R., DeFries, J. C., & Fulker, D. W. (1988). *Nature and nurture during infancy and early childhood.* Cambridge, England: Cambridge University Press.

Reiss, A. J., Duncan, O. D., Hatt, P. K., & North, C. C. (1961). *Occupations and social status.* Glencoe, IL: Free Press.

Rice, T., Corley, R., Fulker, D. W., & Plomin, R. (1986). The development and validation of a test battery measuring specific cognitive abilities in four-year-old children. *Educational and Psychological Measurement, 46,* 699–708.

Rice, T., Plomin, R., & DeFries, J. C. (1984). Development of hand preference in the Colorado Adoption Project. *Perceptual and Motor Skills, 58,* 683–689.

Rowe, D. C. & Plomin, R. (1977). Temperament in early childhood. *Journal of Personality Assessment, 41,* 150–156.

Schaefer, E. S. & Bell, R. Q. (1958). Development of a parental research instrument. *Child Development, 29,* 339–361.

Singer, S., Corley, R., Guiffrida, C., & Plomin, R. (1984). The development and validation of a test battery to measure differentiated cognitive abilities in three-year-old children. *Educational and Psychological Measurement, 49,* 703–713.

Sklar, J. & Berkov, B. (1974). Abortion, illegitimacy, and the American birth rate. *Science, 185,* 909–915.

Terman, L. M. & Merrill, M. A. (1973). *Stanford-Binet Intelligence Scale: 1972 norms edition.* Boston: Houghton Mifflin.

Walker, H. M. & McConnell, S. W. (1988). *Walker-McConnell Scale of Social Competence and School Adjustment.* Austin, TX: Pro-Ed.

Wechsler, D. (1974). *Manual for the Wechsler Intelligence Scale for Children–Revised.* New York: Psychological Corporation.

Wechsler, D. (1981). *Manual for the Wechsler Adult Intelligence Scale–Revised.* New York: Psychological Corporation.

Chapter 3   Adoption Design Methodology

Burden, R. L. & Faires, J. D. (1989). *Numerical analysis* (4th edition). Boston: PWS-Kent.

Cardon, L. R., DiLalla, L., Plomin, R., DeFries, J. C., & Fulker, D. W. (1990). Genetic correlations between reading performance and IQ in the Colorado Adoption Project. *Intelligence, 14,* 245–257.

Cardon, L. R., Fulker, D. W., DeFries, J. C., & Plomin, R. (1992). Continuity and change in general cognitive ability from 1 to 7 years of age. *Developmental Psychology, 28,* 64–73.

Carey, G. (1986). A general multivariate approach to linear modeling in human genetics. *American Journal of Human Genetics, 39,* 775–786.

Cavalli-Sforza, L. L. (1975). Quantitative genetic perspectives: Implications for human development. In K. W. Schaie, V. E. Anderson, G. E. McClearn, & J. Money (Eds.), *Developmental human behavior genetics.* Lexington, MA: Lexington Books.

CERN (1977). *Minuit: A system for function minimization and analysis of parameter errors and correlations.* Geneva, Switzerland: CERN.

Cherny, S. S., DeFries, J. C., & Fulker, D. W. (1992). Multiple regression analysis of twin data: A model-fitting approach. *Behavior Genetics, 22,* 489–497.

Coon, H., Fulker, D. W., DeFries, J. C., & Plomin, R. (1990). Home environment and cognitive ability of 7-year-old children in the Colorado Adoption Project: Genetic and environmental etiologies. *Developmental Psychology, 26,* 459–468.

Crow, J. F. (1986). *Basic concepts in population, quantitative, and evolutionary genetics.* New York: Freeman.

DeFries, J. C. & Fulker, D. W. (1985). Multiple regression analysis of twin data. *Behavior Genetics, 15,* 467–473.

DeFries, J. C. & Fulker, D. W. (1988). Multiple regression analysis of twin data: Etiology of deviant scores versus individual differences. *Acta Geneticae Medicae et Gemellologiae, 37,* 205–216.

DeFries, J. C., Plomin, R., & LaBuda, M. C. (1987). Genetic stability of cognitive development from childhood to adulthood. *Developmental Psychology, 23,* 4–12.

Eaves, L. J., Heath, A. C., & Martin, N. G. (1984). A note on the generalized effect of assortative mating. *Behavior Genetics, 14,* 371–376.

Eaves, L. J., Long, J., & Heath, A. C. (1986). A theory of developmental change in quantitative phenotypes applied to cognitive development. *Behavior Genetics, 16,* 143–162.

Falconer, D. S. (1989). *Introduction to quantitative genetics* (3rd edition). New York: Longman Group Ltd.

Fisher, R. A. (1918). The correlation between relatives on the supposition of Mendelian inheritance. *Transactions of the Royal Society, Edinburgh, 52,* 399–433.

Fulker, D. W., Cherny, S. S., & Cardon, L. R. (1993). Continuity and change in cognitive development. In R. Plomin & G. E. McClearn (Eds.), *Nature, nurture, and psychology* (pp. 77–97). Washington, DC: APA Books.

Fulker, D. W. & DeFries, J. C. (1983). Genetic and environmental transmission in the Colorado Adoption Project: Path analysis. *British Journal of Mathematical and Statistical Psychology, 36,* 175–188.

Fulker, D. W., DeFries, J. C., & Plomin, R. (1988). Genetic influence on general mental ability increases between infancy and middle childhood. *Nature, 336,* 767–769.

Gorsuch, R. L. (1983). *Factor analysis* (2nd edition). Hillsdale, NJ: Lawrence Erlbaum Associates.

Guttman, L. (1954). A new approach to factor analysis: The radex. In P. F. Lazarsfeld (Ed.), *Mathematical thinking in the social sciences* (pp. 258–349). Glencoe, IL: Free Press.

Hegmann, J. P. & DeFries, J. C. (1970). Maximum variance linear combinations from phenotypic, genetic, and environmental covariance matrices. *Multivariate Behavioral Research, 5,* 9–18.

Hewitt, J. K., Eaves, L. J., Neale, M. C., & Meyer, J. M. (1988). Resolving causes of developmental continuity or "tracking": I. Longitudinal twin studies during growth. *Behavior Genetics, 18,* 133–151.

Jinks, J. L. & Fulker, D. W. (1970). Comparison of the biometrical genetical, MAVA, and classical approaches to the analysis of human behavior. *Psychological Bulletin, 73,* 311–349.

Jöreskog, K. G. (1973). A general method for estimating a linear structural equation system. In A. Goldberger & O. Duncan (Eds.), *Structural equation models in the social sciences* (pp. 85–112). New York: Seminar Press.

Jöreskog, K. G. & Sörbom, D. (1989). *LISREL 7: A Guide to the Program and Applications* (2nd edition). Chicago: SPSS, Inc.

Lange, K., Westlake, J., and Spence, M. A. (1976). Extensions to pedigree analysis III. Variance components by the scoring method. *Annals of Human Genetics, 39,* 485–491.

Li, C. C. (1975). *Path analysis: A primer.* Pacific Grove, CA: Boxwood Press.

Little, R. J. A. & Rubin, D. B. (1987). *Statistical analysis with missing data.* New York: Wiley and Son.

Loehlin, J. C. (1987). *Latent variable models.* Baltimore: Lawrence Erlbaum Associates.

Mather, K. & Jinks, J. L. (1982). *Biometrical genetics: The study of continuous variation* (3rd edition). London: Chapman and Hall.

Neale, M. C., Heath, A. C., Hewitt, J. K., Eaves, L. J., & Fulker, D. W. (1989). Fitting genetic models with LISREL: Hypothesis testing. *Behavior Genetics*, *19*, 37–69.

Neale, M. C. & Cardon, L. R. (1992). *Methodology for genetic studies of twins and families*. Boston: Kluwer Academic Publishing.

Numerical Algorithms Group (1990). *The NAG Fortran library manual, mark 14*. Oxford: Numerical Algorithms Group.

Phillips, K. & Fulker, D. W. (1989). Quantitative genetic analysis of longitudinal trends in adoption designs with application to IQ in the Colorado Adoption Project. *Behavior Genetics*, *19*, 621–658.

Plomin, R., DeFries, J. C., & Fulker, D. W. (1988). *Nature and nurture during infancy and early childhood*. New York: Cambridge University Press.

Rice, T., Carey, G., Fulker, D. W., & DeFries, J. C. (1989). Multivariate path analysis of specific cognitive abilities in the Colorado Adoption Project: Conditional path model of assortative mating. *Behavior Genetics*, *19*, 195–207.

Spearman, C. (1904). General intelligence objectively determined and measured. *American Journal of Psychology*, *15*, 201–293.

Vogler, G. P. (1985). Multivariate path analysis of familial resemblance. *Genetic Epidemiology*, *2*, 35–53.

Wright, S. (1921). Correlation and causation. *Journal of Agricultural Research*, *20*, 557–585.

Chapter 4   General Cognitive Ability

Bayley, N. (1969). *Manual for the Bayley Scales of Infant Development*. New York: Psychological Corporation.

Binet, A. & Simon, T. (1905). On the necessity of establishing a scientific diagnosis of the inferior states of intelligence. *L'Année Psychologique*, *11*, 163–190.

Binet, A. & Simon, T. (1908). The development of intelligence in children. *L'Année Psychologique*, *14*, 1–94.

Bouchard, Jr., T. J. & McGue, M. (1981). Familial studies of intelligence: A review. *Science*, 212, 1055–1059.

Burks, B. S. (1928). The relative influence of nature and nurture upon mental development. A comparative study of foster parent–foster child resemblance and true parent–true child resemblance. *Twenty-Seventh Yearbook of the National Society for the Study of Education*, *27*, 219–316.

Cardon, L. R., Fulker, D. W., DeFries, J. C., & Plomin, R. (1992). Continuity and change in general cognitive ability from 1 to 7 years of age. *Developmental Psychology*, *28*, 64–73.

Cattell, R. B. (1960). The multiple abstract variance analysis equations and solutions for nature–nurture research on continuous variables. *Psychological Review*, *67*, 353–372.

Cattell, R. B. (1965). Methodological and conceptual advances in evaluating heredity and environmental influences and their interaction. In S. J. Vandenberg (Ed.), *Methods and goals in human behavior genetics* (p. 95–130). New York: Academic Press.

Eaves, L. J., Long, J., & Heath, A. C. (1986). A theory of developmental change in quantitative phenotypes applied to cognitive development. *Behavior Genetics*, *16*, 143–162.

Fulker, D. W., Cherny, S. S., & Cardon, L. R. (1993). Continuity and change in cognitive development. In R. Plomin and G. E., McClearn (Eds.), *Nature, nurture, and psychology*, (pp. 77–79). Washington, DC: APA Books.

Galton, F. (1869). *Hereditary genius*. London: Macmillan.

Hewitt, J. K., Eaves, L. J., Neale, M. C., & Meyer, J. M. (1988). Resolving causes of developmental continuity or "tracking": I. Longitudinal twin studies during growth. *Behavior Genetics, 18*, 133–151.

Humphreys, L. G. & Davey, T. C. (1988). Continuity in intellectual growth from 12 months to 9 years. *Intelligence, 12*, 183–197.

Kent, J. & Plomin, R. (1987). Testing specific cognitive abilities by telephone and mail. *Intelligence, 11*, 391–400.

Newman, J., Freeman, F., & Holzinger, K. (1937). *Twins: A study of heredity and environment.* Chicago: University of Chicago Press.

Phillips, K. & Fulker, D. W. (1989). Quantitative genetic analysis of longitudinal trends in adoption designs with application to IQ in the Colorado Adoption Project. *Behavior Genetics, 19*, 621–658.

Plomin, R., DeFries, J. C., & Fulker, D. W. (1988). *Nature and nurture during infancy and early childhood.* Cambridge, England: Cambridge University Press.

Skodak, M. & Skeels, H. M. (1949). A final follow-up of one hundred adopted children. *Journal of Genetic Psychology, 75*, 85–125.

Terman, L. H. (1916). *The measurement of intelligence.* Boston: Houghton-Mifflin.

Terman, L. M. & Merrill, M. A. (1973). *Stanford-Binet Intelligence Scale: 1972 norms edition.* Boston: Houghton Mifflin.

Wechsler, D. (1974). *Manual for the Wechsler Intelligence Scale for Children–Revised.* New York: Psychological Corporation.

Wilson, R. S. (1972a). Similarity in developmental profile among related pairs of human infants. *Science, 178*, 1005–1007.

Wilson, R. S. (1972b). Synchronies in mental development: An epigenetic perspective. *Science, 202*, 939–948.

Wilson, R. S. (1972c). Twins: Early mental development. *Science, 175*, 914–917.

Wilson, R. S. (1983). The Louisville Twin Study: Developmental synchronies in behavior. *Child Development, 54*, 298–316.

Chapter 5 Specific Cognitive Abilities

Bergeman, C. S., Plomin, R., DeFries, J. C., & Fulker, D. W. (1988). Path analysis of general and specific cognitive abilities in the Colorado Adoption Project: Early childhood. *Personality and Individual Differences, 9*, 391–395.

Burt, C. (1939). The relations of educational abilities. *British Journal of Educational Psychology, 9*, 45–71.

Burt, C. (1949). Alternative methods of factor analysis and their relations to Pearson's method of Principal Axes. *British Journal of Psychological Statistics, 2*, 98–121.

Cardon, L. R. & Fulker, D. W. (in press). A model of developmental change in hierarchical phenotypes. *Behavior Genetics.*

Cardon, L. R., Corley, R. P., DeFries, J. C., Plomin, R., & Fulker, D. W. (1992). Factorial validation of a telephone test battery of specific cognitive abilities. *Personality and Individual Differences*, (pp. 1047–1050).

Cardon, L. R., Fulker, D. W., DeFries, J. C., & Plomin, R. (1992). Multivariate genetic analysis of specific cognitive abilities in the Colorado Adoption Project at age 7. *Intelligence, 16*, 383–399.

Cyphers, L., Fulker, D. W., Plomin, R., & DeFries, J. C. (1989). Cognitive abilities in the early school years: No effects of shared environment between parents and offspring. *Intelligence, 13*, 369–384.

DeFries, J. C., Plomin, R., Vandenberg, S. G., & Kuse, A. R. (1981). Parent–offspring resemblance for cognitive abilities in the Colorado Adoption Project: Biological, adoptive, and control parents and one-year-old children. *Intelligence, 5,* 245–277.

Eaves, L. J. & Gale, J. S. (1974). A method for analyzing the genetic basis of covariation. *Behavior Genetics, 4,* 253–267.

Guilford, J. P. (1967). *The nature of human intelligence.* New York: McGraw-Hill.

Humphreys, L. G. (1985). General intelligence: An integration of factor, test, and simplex theory. In B. B. Wolman (Ed.), *Handbook of intelligence: Theories, measurement, and applications* (pp. 201–244). New York: Wiley.

Humphreys, L. G. (1989). Intelligence: Three kinds of instability and their consequences for policy. In R. L. Linn (Ed.), *Intelligence: Measurement, theory, and public policy* (pp. 193–216). Chicago: University of Illinois Press.

Humphreys, L. G. & Davey, T. C. (1988). Continuity in intellectual growth from 1 to 9 years. *Intelligence, 12,* 183–197.

Kent, J. & Plomin, R. (1987). Testing specific cognitive abilities by telephone and mail. *Intelligence, 11,* 391–400.

Piaget, J. (1962). *Play, dreams, and imitation in childhood.* New York: Norton.

Plomin, R. & DeFries, J. C. (1985). A parent–offspring adoption study of cognitive abilities in early childhood. *Intelligence, 9,* 341–356.

Rice, T., Carey, G., Fulker, D. W., & DeFries, J. C. (1989). Multivariate path analysis of specific cognitive abilities in the Colorado Adoption Project: Conditional path model of assortative mating. *Behavior Genetics, 19,* 195–207.

Rice, T., Corley, R., Fulker, D. W., & Plomin, R. (1986). The development and validation of a test battery measuring specific cognitive abilities in four-year-old children. *Educational and Psychological Measurement, 46,* 699–708.

Scarr, S. (1989). Protecting general intelligence. In R. L. Linn (Ed.), *Intelligence: measurement, theory, and public policy* (pp. 74–118). Chicago: University of Illinois Press.

Schmid, J. & Leiman, J. (1957). The development of hierarchical factor solutions. *Psychometrika, 22,* 53–61.

Singer, S., Corley, R., Guiffrida, C., & Plomin, R. (1984). The development and validation of a test battery to measure differentiated cognitive abilities in three-year-old children. *Educational and Psychological Measurement, 49,* 703–713.

Spearman, C. (1927). *The abilities of man.* London: Macmillan.

Thurstone, L. L. (1938). *Primary mental abilities.* Chicago: University of Chicago Press.

Vandenberg, S. G. (1968). The nature and nurture of intelligence. In D. C. Glass (Ed.), *Genetics* (pp. 3–58). New York: Rockefeller University Press.

Vernon, P. E. (1979). *Intelligence: Heredity and environment.* San Francisco: W. H. Freeman.

Wechsler, D. (1974). *Manual for the Wechsler Intelligence Scale for Children–Revised.* New York: Psychological Corporation.

Chapter 6   Longitudinal Prediction of School-Age Cognitive Abilities from Infant Novelty Preference

Bayley, N. (1969). *Manual for the Bayley Scales of Infant Development.* New York: Psychological Corporation.

Bornstein, M. H. & Sigman, M. D. (1986). Continuity in mental development from infancy. *Child Development, 57,* 251–274.

Cohen, J. (1977). *Statistical power analysis for the behavioral sciences.* New York: Academic Press.

Colombo, J. & Mitchell, D. W. (1988). Infant visual habituation: In defense of an information-processing approach. *European Bulletin of Cognitive Psychology*, *8*(5), 455–461.

Colombo, J. & Mitchell, D. W. (1990). Individual differences in early visual attention: Fixation time and information processing. In J. Colombo and J. Fagen (Eds.), *Individual differences in infancy: Reliability, stability, prediction* (pp. 193–227). Hillsdale, NJ: Lawrence Erlbaum Associates.

Colombo, J., Mitchell, D. W., Dodd, J., Coldren, J. T., & Horowitz, F. D. (1989). Longitudinal correlates of infant attention in the paired-comparison paradigm. *Intelligence*, *13*, 33–42.

DiLalla, L., Thompson, L. A., Plomin, R., Phillips, K., Fagan, J. F., Haith, M. M., Cyphers, L. H., & Fulker, D. W. (1990). Infant predictors of preschool and adult IQ: A study of infant twins and their parents. *Developmental Psychology*, *26*, 759–769.

European Bulletin of Cognitive Psychology, *8*(5), 1988.

Fagan, J. F. (1984). The intelligent infant: Theoretical implications. *Intelligence*, *8*, 1–9.

Fagan, J. F. & Detterman, D. K. (1992). The Fagan Test of Infant Intelligence: A technical summary. *Journal of Applied Developmental Psychology*, *13*, 173–193.

Fagan, J. F. & Knevel, C. (1989). *The prediction of above-average intelligence from infancy*. Paper presented at the Society for Research on Child Development, April 1989.

Fagan, J. F. & McGrath, S. K. (1981). Infant recognition memory and later intelligence. *Intelligence*, *5*, 121–130.

Fagan, J. F. & Shepherd, P. A. (1986). *The Fagan Test of Infant Intelligence: Training manual*. Cleveland, OH: Infantest Corporation.

Fagan, J. F. & Singer, L. T. (1983). Infant recognition memory as a measure of intelligence. In L. P. Lipsitt (Ed.), *Advances in infancy research*, Vol. 2 (pp. 31–72). Norwood, NJ: Ablex.

Fagan, J. F., Singer, L. T., Montie, J. E., & Shepherd, P. A. (1986). Selective screening device for the early detection of normal or delayed cognitive development in infants at risk for later mental retardation. *Pediatrics*, *78*, 1021–1026.

Fagen, J. W. & Ohr, P. S. (1990). Individual differences in infant conditioning and memory. In J. Colombo and J. Fagen (Eds.), *Individual differences in infancy: Reliability, stability, prediction* (pp. 155–192). Hillsdale, NJ: Lawrence Erlbaum Associates.

Haith, M. M., Hazen, C., & Goodman, G. S. (1988). Expectation and anticipation of dynamic visual events by 3.5-month-old babies. *Child Development*, *59*, 505–529.

Hedrick, D. L., Prather, E. M., & Tobin, A. R. (1975). *Sequenced Inventory of Communication Development*. Seattle: University of Washington Press.

Jacobson, S. W., Jacobson, J. L., O'Neill, J. M., Padgett, R. J., Frankowski, J. J., & Bihun, J. T. (1992). Visual expectation and dimensions of infant information processing. *Child Development*, *63*, 711–724.

Lewis, M. & Brooks-Gunn, J. (1981). Visual attention at three months as a predictor of cognitive functioning at two years. *Intelligence*, *5*, 131.

O'Connor, M. J., Cohen, S., & Parmalee, A. H. (1984). Infant auditory discrimination in preterm and full-term infants as a predictor of 5-year intelligence. *Developmental Psychology*, *20*, 159–165.

Plomin, R. & DeFries, J. C. (1985). *Origins of individual differences in infancy: The Colorado Adoption Project*. Orlando, FL: Academic Press.

Plomin, R., DeFries, J. C., & Fulker, D. W. (1988). *Nature and nurture during infancy and early childhood*. New York: Cambridge University Press.

Rice, T., Corley, R., Fulker, D. W., & Plomin, R. (1986). The development and validation of a test battery measuring specific cognitive abilities in four-year-old children. *Educational and Psychological Measurement*, *46*, 699–708.

Rose, D. H., Slater, A., & Perry, H. (1986). Prediction of childhood intelligence from habituation in early infancy. *Intelligence, 10,* 251–263.

Rose, S. A. & Feldman, J. F. (1990). Infant cognition: Individual differences and developmental continuities. In J. Colombo and J. Fagen (Eds.), *Individual differences in infancy: Reliability, stability, prediction* (pp. 229–245). Hillsdale, NJ: Lawrence Erlbaum Associates.

Rose, S. A., Feldman, J. F., McCarton, C. M., & Wofson, J. (1988). Individual differences in infants' information processing: Reliability, stability, and prediction. *Child Development, 59,* 1177–1197.

Rose, S. A. & Wallace, I. F. (1985) Cross-modal and intramodal transfer as predictors of mental development in full-term and preterm infants. *Developmental Psychology, 21,* 949–962.

Rose, S. A. & Wallace, I. F. (1988). Visual recognition memory: A predictor of later cognitive functioning in preterms. *Child Development, 56,* 843–852.

Terman, L. M. & Merrill, M. A. (1973). *Stanford-Binet Intelligence Scale: 1972 norms edition.* Boston: Houghton Mifflin.

Thompson, L. A., Fagan, J. F., & Fulker, D. W. (1991). Longitudinal prediction of specific cognitive abilities from infant novelty preference. *Child Development, 62,* 530–538.

Thompson, L. A., Plomin, R., & DeFries, J. C. (1985). Parent–infant resemblance for general and specific cognitive abilities in the Colorado Adoption Project. *Intelligence, 9,* 1–13.

Chapter 7   School Achievement

Achenbach, T. M. & Edelbrock, C. (1983). *Manual for the Child Behavior Checklist and Revised Child Behavior Profile.* Burlington: Department of Psychiatry, University of Vermont.

Bayley, N. (1969). *Manual for the Bayley Scales of Infant Development.* New York: Psychological Corporation.

Brodzinsky, D. M., Schecter, D. E., Braff, A. M., & Singer, L. M. (1984). Psychological and academic adjustment in adopted children. *Journal of Consulting and Clinical Psychology, 52,* 582–590.

Brodzinsky, D. M. & Steiger, C. (1991). Prevalence of adoptees among special education populations. *Journal of Learning Disabilities, 24,* 484–489.

Brooks, A., Fulker, D. W., & DeFries, J. C. (1990). Reading performance and general cognitive ability: A multivariate genetic analysis of twin data. *Personality and Individual Differences, 11,* 141–146.

Cardon, L. R., DiLalla, L. F., Plomin, R., DeFries, J. C., & Fulker, D. W. (1990). Genetic correlations between reading performance and IQ in the Colorado Adoption Project. *Intelligence, 14,* 245–257.

Cohen, J. (1977). *Statistical power analysis for the behavioral sciences.* New York: Academic Press.

Connolly, A. J., Nachtman, W., & Pritchett, E. M. (1976). *KeyMath Diagnostic Arithmetic Test.* Circle Pines, MN: American Guidance Service.

Deutsch, C. K., Swanson, J. M., Bruell, J. H., Cantwell, D. P., Weinberg, F., & Baren, M. (1982). Overrepresentation of adoptees in children with the attention deficit disorder. *Behavior Genetics, 12,* 231–238.

Dunn, L. M. (1981). *Peabody Picture Vocabulary Test: Examiner's manual.* Circle Pines, MN: American Guidance Service.

Dunn, L. M. & Markwardt, F. C. (1970). *Examiner's manual: Peabody Individual Achievement Test.* Circle Pines, MN: American Guidance Service.

Gesell, A. (1926). *The mental growth of the preschool child*. New York: Macmillan.

Gillis, J. J. & DeFries, J. C. (1991). Confirmatory factor analysis of reading and mathematics performance measures in the Colorado Reading Project (Abstract). *Behavior Genetics, 21*, 572.

Jastak, J. F. & Jastak, S. (1978). *The Wide Range Achievement Test: Manual of instructions*. Wilmington, DE: Jastak Associates, Inc.

Jensen, A. R. (1969). How much can we boost IQ and scholastic achievement? *Harvard Educational Review, 39*, 1–123.

Jöreskog, K. G. & Sörbom, D. (1989). *LISREL 7: A guide to the program and applications* (2nd edition). Chicago: SPSS, Inc.

Judd, C. M. & McClelland, G. H. (1989). *Data analysis: A model-comparison approach*. San Diego: Harcourt Brace Jovanovich.

Kagan, J., Lapidus, D. R., & Moore, M. (1978). Infant antecedents of cognitive functioning: A longitudinal study. *Child Development, 49*, 1005–1023.

Pennington, B. F., Van Orden, G., Kirson, D., & Haith, M. (1991). What is the causal relation between verbal STM problems and dyslexia? In S. Brady & D. Shankweiler (Eds.), *Phonological processes in literacy* (pp. 173–186). Hillsdale, NJ: Lawrence Erlbaum Associates.

Plomin, R. & DeFries, J. C. (1985). *Origins of individual differences in infancy: The Colorado Adoption Project*. Orlando, FL: Academic Press.

Plomin, R., DeFries, J. C., & Fulker, D. W. (1988). *Nature and nurture during infancy and early childhood*. Cambridge, England: Cambridge University Press.

Prescott, G. A., Barlow, I. H., Hogan, T. P., & Farr, R. C. (1986). *Metropolitan Achievement Tests: MAT6*. New York: Psychological Corporation.

Roe, K., McClure, A., & Roe, A. (1983). Infant Gesell scores vs. cognitive skills at age 12 years. *Journal of Genetic Psychology, 142*, 143–147.

Silver, L. B. (1970). Frequency of adoption in children with the neurological learning disability syndrome. *Journal of Learning Disabilities, 3*, 10–14.

Silver, L. B. (1989). Frequency of adoption of children and adolescents with learning disabilities. *Journal of Learning Disabilities, 22*, 325–327.

SPSS-X (1988). *Statistical Package for the Social Sciences: User's guide*. Chicago: SPSS, Inc.

Stevenson, J., Graham, P., Fredman, G., & McLoughlin, L. (1987). A twin study of genetic influences on reading and spelling ability and disability. *Journal of Child Psychology and Psychiatry, 28*, 229–247.

Terman, L. M. & Merrill, M. A. (1973). *Stanford-Binet Intelligence Scale: 1972 norms edition*. Boston: Houghton Mifflin.

Thompson, L. A., Detterman, D. K., & Plomin, R. (1991). Associations between cognitive abilities and scholastic achievement: Genetic overlap, but environmental differences. *Psychological Science, 2*, 158–165.

Wadsworth, S. J., DeFries, J. C., & Fulker, D. W. (in press). Cognitive abilities of children at 7 and 12 years of age in the Colorado Adoption Project. *Journal of Learning Disabilities*.

Wechsler, D. (1974). *Manual for the Wechsler Intelligence Scale for Children–Revised*. New York: Psychological Corporation.

Chapter 8   Developmental Speech and Language Disorders

Andrews, G. & Harris, M. M. (1964). *The syndrome of stuttering*. (Clinics in Developmental Medicine, No. 17). London: Spastics Society Medical Education and Information Unit in association with William Heinemann Medical Books.

Andrews, G., Morris-Yates, A., Howie, P., & Martin, N. (1991). Genetic factors in stuttering confirmed (letter). *Archives of General Psychiatry, 48*, 1034–1035.

Aram, D., Ekelman, B., & Nation, J. (1984). Preschoolers with language disorders: 10 years later. *Journal of Speech and Hearing Research, 27*, 232–244.

Barnes MacFarlane, W., Hanson, M., Walton, W., & Mellon, C. D. (1991). Stuttering in five generations of a single family. *Journal of Fluency Disorders, 16*, 117–123.

Bishop, D. V. M. & Adams, C. (1990). A prospective study of the relationship between specific language impairment, phonological disorders, and reading retardation. *Journal of Child Psychology and Psychiatry, 31*, 1027–1050.

Blood, G. W. & Seider, R. (1981). The concomitant problems of young stutterers. *Journal of Speech and Hearing Disorders, 46*, 31–34.

Bloodstein, O. (1981). *A handbook on stuttering* (3rd edition). Chicago: National Easter Seal Society for Crippled Children and Adults.

Byrne, B., Willerman, L., & Ashmore, L. (1974). Severe and moderate language impairment: Evidence for distinctive etiologies. *Behavior Genetics, 4*, 331–345.

Conti-Ramsden, G. (1985). Mothers in dialogue with language-impaired children. *Topics in Language Disorders, 5*, 58–68.

Cooper, E. B. (1972). Recovery from stuttering in a junior and senior high school population. *Journal of Speech and Hearing Research, 15*, 632–638.

Cox, N., Seider, R., & Kidd, K. (1984). Some environmental factors and hypotheses for stuttering in families with several stutterers. *Journal of Speech and Hearing Research, 27*, 543–548.

Felsenfeld, S., Broen, P., & McGue, M. (1992). A 28-year follow-up of adults with a history of moderate phonological disorder: Linguistic and personality results. *Journal of Speech and Hearing Research, 35*, 1114–1125.

Felsenfeld, S., Broen, P., & McGue, M. (1993). *A 28-year follow-up of adults with a history of moderate phonological disorder: Educational and occupational results.* Manuscript submitted for publication.

Felsenfeld, S., McGue, M., & Broen, P. (1993). Familial aggregation of moderate articulation disorder: Results from a 28-year follow-up. Manuscript submitted for publication.

Gray, M. (1940). The X family: A clinical and laboratory study of a "stuttering" family. *Journal of Speech and Hearing Disorders, 5*, 343–348.

Homzie, M., Lindsay, J., Simpson, J., & Hasenstab, S. (1988). Concomitant speech, language, and learning problems in adult stutterers and in members of their families. *Journal of Fluency Disorders, 13*, 261–277.

Howie, P. (1981). Concordance for stuttering in monozygotic and dizygotic twin pairs. *Journal of Speech and Hearing Research, 24*, 317–321.

Janssen, P., Kraaimaat, F., & Brutten, G. (1990). Relationship between stutterers' genetic history and speech-associated variables. *Journal of Fluency Disorders, 15*, 39–48.

Johnson, W. (1959). *The onset of stuttering.* Minneapolis: University of Minnesota Press.

Kasprisin-Burrelli, A., Egolf, D., & Shames, G. (1972). A comparison of parental verbal behavior with stuttering and nonstuttering children. *Journal of Communication Disorders, 5*, 335–346.

Kidd, K. (1983). Recent progress on the genetics of stuttering. In C. Ludlow & J. Cooper (Eds.), *Genetic aspects of speech and language disorders* (pp. 197–213). New York: Academic Press.

Kidd, K., Heimbuch, R., & Records, M. A. (1981). Vertical transmission of susceptibility to stuttering with sex-modified expression. *Proceedings of the National Academy of Sciences, 78*, 606–610.

Kidd, K., Heimbuch, R., Records, M. A., Oehlert, G., & Webster, R. (1980). Familial stuttering patterns are not related to one measure of severity. *Journal of Speech and Hearing Research, 23,* 539–545.

Kidd, K., Kidd, J., & Records, M. A. (1978). The possible causes of the sex ratio in stuttering and its implications. *Journal of Fluency Disorders, 3,* 13–23.

Leonard, L. (1987). Is specific language impairment a useful construct? In S. Rosenberg (Ed.), *Advances in applied psycholinguistics, Volume 1: Disorders of first-language development* (pp. 1–38). New York: Cambridge University Press.

Lewis, B. (1990). Familial phonological disorders: Four pedigrees. *Journal of Speech and Hearing Disorders, 55,* 160–170.

Lewis, B. (1992). Pedigree analysis of children with phonology disorders. *Journal of Learning Disabilities, 25,* 586–597.

Lewis, B., Ekelman, B., & Aram, D. (1989). A familial study of severe phonological disorders. *Journal of Speech and Hearing Research, 32,* 713–724.

Lewis, B. & Freebairn-Tarr, L. (1992). Residual effects of preschool phonology disorders in grade school, adolescence, and adulthood. *Journal of Speech and Hearing Research, 35,* 819–831.

Lewis, B. & Thompson, L. A. (1992). A study of developmental speech and langauge disorders in twins. *Journal of Speech and Hearing Research, 35,* 1086–1094.

Ludlow, C. & Cooper, J. (1983). Genetic aspects of speech and language disorders: Current status and future directions. In C. Ludlow & J. Cooper (Eds.), *Genetic aspects of speech and language disorders* (pp. 3–20). New York: Academic Press.

Neils, J. & Aram, D. (1986). Family history of children with developmental language disorders. *Perceptual and Motor Skills, 63,* 655–658.

Parlour, S. Felsenfeld, & Broen, P. (1991). *Environmental factors in familial phonological disorders: Preliminary HOME Scale results.* Paper presented at the annual convention of the American Speech-Language-Hearing Association, Atlanta, Georgia, November 1991.

Parlour, S. Felsenfeld (1991). *Using self-report to identify adults with a history of developmental articulation disorder: A validation.* Paper presented at the 21st annual meeting of the Behavior Genetics Association, June 6–9, 1991, St. Louis, MO.

Perkins, W. (1990). What is stuttering? *Journal of Speech and Hearing Disorders, 55,* 370–382.

Plomin, R., DeFries, J. C., & Fulker, D. W. (1988). *Nature and nurture during infancy and early childhood.* Cambridge, England: Cambridge University Press.

Poulos, M. & Webster, W. (1991). Family history as a basis for subgrouping people who stutter. *Journal of Speech and Hearing Research, 34,* 5–10.

Schery, T. (1985). Correlates of language development in language-disordered children. *Journal of Speech and Hearing Disorders, 50,* 73–83.

Seider, R., Gladstein, K., & Kidd, K. (1982). Language onset and concomitant speech and language problems in subgroups of stutterers and their siblings. *Journal of Speech and Hearing Research, 25,* 482–486.

Seider, R., Gladstein, K., & Kidd, K. (1983). Recovery and persistence of stuttering among relatives of stutterers. *Journal of Speech and Hearing Disorders, 48,* 394–402.

Sheehan, J. & Costley, M. S. (1977). A reexamination of the role of heredity in stuttering. *Journal of Speech and Hearing Disorders, 42,* 47–59.

Shriberg, L. & Kwiatkowski, J. (1982). Phonological disorders I: A diagnostic classification system. *Journal of Speech and Hearing Disorders, 47,* 226–241.

Shriberg, L., Kwiatkowski, J., Best, S., Hengst, J., & Terselic-Weber, B. (1986). Characteristics of children with phonological disorders of unknown origin. *Journal of Speech and Hearing Disorders, 51,* 140–161.

Siegel-Sadewitz, V. & Shprintzen, R. (1982). The relationship of communication disorders to syndrome identification. *Journal of Speech and Hearing Disorders, 47,* 338–345.

Tallal, P., Ross, R., & Curtiss, S. (1989). Familial aggregation in specific language impairment. *Journal of Speech and Hearing Disorders, 54,* 167–173.

Templin, M. (1957). *Certain language skills in children: Their development and interrelationships.* Institute of Child Welfare Monograph No. 26. Minneapolis: University of Minnesota Press.

Tomblin, J. B. (1989). Familial concentration of developmental language impairment. *Journal of Speech and Hearing Disorders, 54,* 287–295.

Van Riper, C. (1971). *The nature of stuttering.* Englewood Cliffs, NJ: Prentice-Hall.

Vetter, D. (1980). Psychosocial factors. In P. LaBenz & E. LaBenz (Eds.), *Early correlates of speech, language, and hearing* (pp. 266–329). Littleton, MA: PSG Publishing Co.

Wepman, J. (1939). Familial incidence of stammering. *Journal of Heredity, 30,* 199–204.

West, R., Nelson, S., & Berry, M. (1939). The heredity of stuttering. *Quarterly Journal of Speech, 25,* 23–30.

Whitehurst, G., Arnold, D., Smith, M., Fischel, J., Lonigan, C., & Valdez-Menchaca, M. (1991). Family history in developmental expressive language delay. *Journal of Speech and Hearing Research, 34,* 1150–1157.

Winitz, H. (1969). *Articulatory acquisition and behavior.* New York: Appleton-Century-Crofts.

Chapter 9   Personality and Temperament

Achenbach, T. M., McConaughty, S. H., & Howell, C. T. (1987). Child/adolescent behavioral and emotional problems: Implications of cross-informant correlations for situational specificity. *Psychological Bulletin, 101,* 213–232.

Bayley, N. (1969). *Manual for the Bayley Scales of Infant Development.* New York: Psychological Corporation.

Braungart, J. M., Plomin, R., DeFries, J. C., & Fulker, D. W. (1992). Genetic influence on tester-rated infant temperament as assessed by Bayley's Infant Behavior Record: Nonadoptive and adoptive siblings and twins. *Developmental Psychology, 28,* 40–47.

Buss, A. H. & Plomin, R. (1975). *A temperament theory of personality development.* New York: Wiley-Interscience.

Buss, A. H. & Plomin, R. (1984). *Temperament: Early developing personality traits.* Hillsdale, NJ: Lawrence Erlbaum Associates.

Cattell, R. B., Eber, H. W., & Tatsuoka, M. M. (1970). *Handbook for the Sixteen Personality Factor Questionnaire (16PF).* Champaign, IL: Institute for Personality and Ability Testing.

Cherny, S. S., DeFries, J. C., & Fulker, D. W. (1992). Multiple regression analysis of twin data: A model-fitting approach. *Behavior Genetics, 22,* 489–497.

Chess, S. & Thomas, A. (1984). *Origins and evolution of behavior disorders: Infancy to early adult life.* New York: Brunner/Mazel.

Cohen, J. (1977). *Statistical power analysis for the behavioral sciences,* rev. ed. New York: Academic Press

Coon, H. & Carey, G. (1988). Bias in ratings of temperament on parents and children in the Colorado Adoption Project (Abstract). *Behavior Genetics, 18,* 711.

Coon, H., Carey, G., & Fulker, D. W. (1990). A simple method of model fitting for adoption data. *Behavior Genetics, 20,* 385–404.

Cyphers, L. H., Phillips, K., Fulker, D. W., & Mrazek, D. A. (1990). Twin temperament during the transition from infancy to early childhood. *Journal of the American Academy of Child and Adolescent Psychiatry, 29,* 392–397.

DeFries, J. C. & Fulker, D. W. (1985). Multiple regression analysis of twin data. *Behavior Genetics, 15,* 467–473.

Eaves, L. J., Eysenck, H. J., & Martin, N. G. (1989). *Genes, culture and personality: An empirical approach.* London: Academic Press.

Emde, R. N., Plomin, R., Robinson, J., Reznick, J. S., Campos, J., Corley, R., DeFries, J. C., Fulker, D. W., Kagan, J., & Zahn-Waxler, C. (1992). Temperament, emotion, and cognition at 14 months: The MacArthur Longitudinal Twin Study. *Child Development, 63,* 1437–1455.

Goldsmith, H. H., Buss, A. H., Plomin, R., Rothbart, M. K., Thomas, A., Chess, S., Hinde, R. A., & McCall, R. B. (1987). Roundtable: What is temperament? Four approaches. *Child Development, 58,* 505–529.

Goldsmith, H. H., Rieser-Danner, L. A., & Briggs, S. (1991). Evaluating convergent and discriminant validity of temperament questionnaires for preschoolers, toddlers, and infants. *Developmental Psychology, 27,* 566–579.

Jöreskog, K. G. & Sörbom, D. (1989). *LISREL 7: A guide to the program and applications* (2nd edition). Chicago: SPSS, Inc.

Judd, C. M. & McClelland, G. H. (1989). *Data analysis: A model-comparison approach.* San Diego: Harcourt Brace Jovanovich.

Loehlin, J. C. (1979). *Adoption studies of personality.* Paper presented at the BGA Meeting, Middletown, CT.

Loehlin, J. C. (1989). Partitioning environmental and genetic contributions to behavioral development. *American Psychologist, 44,* 1285–1292.

Loehlin, J. C. (1992). *Genes and environment in personality development.* Newbury Park, CA: Sage Publications.

Loehlin, J. C., Horn, J. M., & Willerman, L. (1981). Personality resemblance in adoptive families. *Behavior Genetics, 11,* 309–330.

Loehlin, J. C. & Nichols, R. C. (1976). *Heredity, environment, and personality.* Austin: University of Texas Press.

Loehlin, J. C., Willerman, L., & Horn, J. M. (1982). Personality resemblances between unwed mothers and their adopted-away offspring. *Journal of Personality and Social Psychology, 42,* 1089–1099.

Matheny, A. P. R. (1980). Bayley's Infant Behavior Record: Behavioral components and twin analyses. *Child Development, 51,* 1157–1167.

Plomin, R. (1986). *Development, genetics, and psychology.* Hillsdale, NJ: Lawrence Erlbaum Associates.

Plomin, R., Chipuer, H. M., & Loehlin, J. C. (1990). Behavioral genetics and personality. In L. A. Pervin (Ed.), *Handbook of personality–Theory and research.* New York/London: The Guilford Press.

Plomin, R., Chipuer, H. M., & Neiderhiser, J. M. (1993). Behavioral genetic evidence for the importance of nonshared environment. In E. M. Hetherington, D. Reiss, & R. Plomin (Eds.), *Separate social worlds of siblings: Importance of nonshared environment on development.* Hillsdale, NJ: Lawrence Erlbaum Associates.

Plomin, R., Coon, H., Carey, G., DeFries, J. C., & Fulker, D. W. (1991). Parent–offspring and sibling adoption analyses of parental ratings of temperament in infancy and childhood. *Journal of Personality, 59,* 705–732.

Plomin, R. & Daniels, D. (1987). Why are children in the same family so different from each other? *The Behavioral and Brain Sciences, 10,* 1–16.

Plomin, R. & DeFries, J. C. (1985). *Origins of individual differences in infancy: The Colorado Adoption Project*. Orlando, FL: Academic Press.

Plomin, R., DeFries, J. C., & Fulker, D. W. (1988). *Nature and nurture during infancy and early childhood*. Cambridge, England: Cambridge University Press.

Rowe, D. C. & Plomin, R. (1977). Temperament in early childhood. *Journal of Personality Assessment, 41*, 150–156.

Scarr, S., Webber, P. L., Weinberg, R. A., & Wittig, M. A. (1981). Personality resemblance among adolescents and their parents in biologically related and adoptive families. *Journal of Personality and Social Psychology, 40*, 885–898.

Stevenson, J. & Fielding, J. (1985). Ratings of temperament in families of young twins. *British Journal of Developmental Psychology, 3*, 143–152.

Verhulst, F. C. & Koot, H. M. (1992). *Child psychiatric epidemiology*. Newbury Park, CA: Sage Publications.

Chapter 10   Competence During Middle Childhood

Bandura, A. (1977). Self-efficacy: Toward a unifying theory of behavioral change. *Psychological Review, 84*, 191–215.

Baumrind, D. (1971). Current patterns of parental authority. *Developmental Psychology Monographs, 4*.

Cooley, C. H. (1902). *Human nature and the social order*. New York: Scribner's.

Damon, W. & Hart, D. (1982). The development of self-understanding from infancy through adolescence. *Child Development, 53*, 841–864.

Daniels, D., Dunn, J., Furstenberg, F. F., & Plomin, R. (1985). Environmental differences within the family and adjustment differences within pairs of adolescent siblings. *Child Development, 56*, 764–774.

Eisenberg, N. & Harris, J. B. (1984). Social competence: A developmental perspective. *School Psychology Review, 13*, 267–277.

Erickson, E. H. (1950). *Childhood and society*. New York: Norton.

Harter, S. (1982). The perceived competence scale for children. *Child Development, 53*, 87–97.

Harter, S. (1983). Developmental perspectives on the self-system. In E. M. Hetherington (Ed.), *Handbook of child psychology: Socialization, personality, and social development* (Vol. 4, pp. 275–385). New York: Wiley.

Harter, S. (1988). *The self-perception profile for adolescents*. Unpublished manuscript. University of Denver.

Harter, S. (1990). Causes, correlates and the functional role of global self-worth: A life-span perspective. In R. J. Sternberg and J. Kolligian (Eds.), *Competence considered*. New Haven, CT: Yale University Press.

Hoffman, L. W. (1991). The influence of the family environment on personality: Accounting for sibling differences. *Psychological Bulletin, 110*, 187–203.

Jöreskog, K. G. & Sörbom, D. (1989). *LISREL 7: A Guide to the Program and Applications* (2nd edition). Chicago: SPSS, Inc.

Kurdek, L. A. & Krile, D. (1983). The relation between third-through eighth-grade children's social cognition and parents' ratings of social skills and general adjustment. *The Journal of Genetic Psychology, 143*, 201–206.

McGuire, S., Neiderhiser, J. M., Reiss, D., Hetherington, E. M., & Plomin, R. (in press). Genetic and environmental influences on perceptions of self-worth and competence in adolescence: A study of twins, full siblings, and step siblings. *Child Development*.

McHale, S. W. & Pawletko, T. M. (1992). Differential treatment in two family contexts. *Child Development*, *63*, 68–81.

Mead, G. H. (1934). *Mind, self, and society*. Chicago: University of Chicago Press.

Nottelmann, E. D. (1987). Competence and self-esteem during the transition from childhood to adolescence. *Developmental Psychology*, *23*, 441–450.

Rende, R. & Plomin, R. (1990). Quantitative genetics and developmental psychopathology: Contributions to understanding normal development. *Development and Psychopathology*, *2*, 393–407.

Rowe, D. C. (1986). Genetic and environmental components of antisocial behavior: A study of 265 twin pairs. *Criminology*, *24*, 513–533.

Rowe, D. C. & Rogers, J. L. (1985). Behavioral genetics and adolescent deviance, and "d": Contributions and issues. In A. S. Rossi (Ed.), *Gender and the life course*. New York: Aldine Publishing Co.

Steinberg, L., Mounts, N., Lamborn, S., & Dornbusch, S. (1991). Authoritative parenting and adolescent adjustment across various ecological niches. *Journal of Research on Adolescence*, *1*, 19–36.

Walker, H. M. & McConnell, S. W. (1988). *Walker-McConnell Scale of Social Competence and School Adjustment*. Austin, TX: Pro-Ed.

Wylie, R. (1979). *The self-concept: Theory and research on selected topics* (Vol. 2, rev. edition). Lincoln: University of Nebraska Press.

Chapter 11   The Stress of First Grade and its Relation to Behavior Problems in School

Achenbach, T. M. & Edelbrock, C. (1986). *Manual for the Teacher's Report Form and Teacher Version of the Child Behavior Profile*. Burlington: Department of Psychiatry, University of Vermont.

Brown, L. P. & Cowen, E. L. (1988). Children's judgments of event upsettingness and personal experiencing of stressful events. *American Journal of Community Psychology*, *16*, 123–135.

Colton, J. A. (1985). Childhood stress: perceptions of children and professionals. *Journal of Psychopathology and Behavioral Assessment*, *7*, 155–173.

Compas, B. (1987). Stress and life events during childhood and adolescence. *Clinical Psychology Review*, *7*, 275–302.

Compas, B., Howell, D. C., Phares, V., Williams, R. A., & Ledoux, N. (1989). Parent and child stress and symptoms: an integrative analysis. *Developmental Psychology*, *25*, 550–559.

DeFries, J. C. & Fulker, D. W. (1985). Multiple regression analysis of twin data. *Behavior Genetics*, *15*, 467–473.

Dunn, J. (1988). Normative life events as risk factors in childhood. In M. Rutter (Ed.), *Studies of psychosocial risk: the power of longitudinal data* (pp. 227–244). New York: Cambridge University Press.

Dunn, J. & Plomin, R. (1990). *Separate lives: Why siblings are so different*. New York: Basic Boots.

Humphrey, J. H. (1984). Some general causes of stress in children. In J. H. Humphrey (Ed.), *Stress in childhood* (pp. 3–18). New York: AMS.

Johnson, J. H. (1986). *Life events as stressors in childhood and adolescence*. Beverly Hills, CA: Sage.

Phillips, B. N. (1978). *School stress and anxiety*. New York: Human Sciences Press.

Plomin, R., Lichtenstein, P., Pedersen, N. L., McClearn, G. E., & Nesselroade, J. R. (1990). Genetic influence on life events. *Psychology and Aging, 5,* 25–30.

Pryor-Brown, L. & Cowen, E. L. (1990). Stressful life events, support, and children's school adjustment. *Journal of Clinical Child Psychology, 18,* 214–220.

Rende, R. D. & Plomin, R. (1991). Child and parent perceptions of the upsettingness of major life events. *Journal of Child Psychology and Psychiatry, 32,* 627–633.

Rende, R. & Plomin, R. (1992a). Diathesis-stress models of psychopathology: A quantitative genetic perspective. *Applied and Preventive Psychology, 1,* 177–182.

Rende, R. D. & Plomin, R. (1992b). Relations between first-grade stress, temperament, and behavior problems. *Journal of Applied Developmental Psychology, 13,* 435–446.

Sandler, I. N., Wolchik, S. A., Braver, S. L., & Fogas, B. S. (1986). Toward the assessment of risky situations. In S. M. Auerbach & A. L. Stolberg (Eds.), *Crisis intervention with children and families* (pp. 65–83). Washington, DC: Hemisphere Publishing Company.

Schultz, E. W. & Heuchert, C. M. (1983). *Child stress and the school experience.* New York: Human Sciences Press.

Chapter 12   Height, Weight and Obesity

Brook, C., Huntley, R., & Slack, J. (1975). Influence of heredity and environment in determination of skinfold thickness in children. *British Medical Journal, 2,* 719–721.

Cardon, L. R., DeFries, J. C., & Fulker, D. W. (1993). *An objective measure of adiposity rebound for prediction of adult obesity.* Manuscript submitted for publication.

Cardon, L. R. & Fulker, D. W. (in press). Genetic influences on body fat from birth to age 9. *Genetic Epidemiology.*

Fabsitz, R. R., Carmelli, D., & Hewitt, J. K. (1992). Evidence for independent genetic influences on obesity in middle age. *International Journal of Obesity, 16,* 657–666.

Fulker, D. W., DeFries, J. C., & Plomin, R. (1988). Genetic influence on general mental ability increases between infancy and middle childhood. *Nature, 336,* 767–769.

Knittle, J. L., Timmers, K., Ginsberg-Fellner, F., Brown, R. E., & Katz, D. P. (1979). The growth of adipose tissue in children and adolescents: Cross-sectional and longitudinal studies of adipose cell number and size. *Journal of Clinical Investigation, 63,* 239–246.

Plomin, R. & DeFries, J. C. (1985). *Origins of individual differences in infancy: The Colorado Adoption Project.* Orlando, FL: Academic Press.

Plomin, R., DeFries, J. C., & Fulker, D. W. (1988). *Nature and nurture during infancy and early childhood.* Cambridge, England: Cambridge University Press.

Price, R. A., Cadoret, R. J., & Stunkard, A. J. (1987). Genetic contributions to human fatness: An adoption study. *American Journal of Psychiatry, 144,* 1003–1008.

Price, R. A., Stunkard, A. J., Ness, R., Wadden, T., Heshka, S., Kanders, B., & Cormillot, A. (1990). Childhood onset obesity has high familial risk. *International Journal of Obesity, 14,* 185–195.

Rolland-Cachera, M. F., Cole, T. J., Sempé, M., Tichet, J., Rossignol, C., & Charraud, A. (1991). Body Mass Index variations: centiles from birth to 87 years. *European Journal of Clinical Nutrition, 45,* 13–21.

Rolland-Cachera, M. F., Deheeger, M., Bellisle, F., Sempé, M., Guilloud-Bataille, M., & Patois, E. (1984). Adiposity rebound in children: A simple indicator for predicting obesity. *American Journal of Clinical Nutrition, 39,* 129–135.

Rolland-Cachera, M. F., Deheeger, M., Guilloud-Bataille, M., Avons, P., Patois, E., & Sempé, M. (1987). Tracking the development of adiposity from one month of age to adulthood. *Annals of Human Biology, 14,* 219–229.

Selby, J. V., Newman, B., Quesenberry, C. P., Fabsitz, R. R., Carmelli, D., Meaney, F. J., & Slemenda, C. (1990). Genetic and behavioral influences on body fat distribution. *International Journal of Obesity*, *14*, 593–602.

Stunkard, A. J., Foch, T. T., & Hrubec, Z. (1986). A twin study of human obesity. *Journal of the American Medical Association*, *256*, 51–54.

Stunkard, A. J., Harris, J. R., Pedersen, N. L., & McClearn, G. E. (1990). The body-mass index of twins who have been reared apart. *New England Journal of Medicine*, *322*, 1483–1487.

Tanner, J. M., Hughes, P. C. R., & Whitehouse, R. H. (1981). Radiographically determined widths of bone muscle and fat in the upper arm and calf from age 3–18 years. *Annals of Human Biology*, *8*, 495–517.

Chapter 13 Motor Development

Bayley, N. (1969). *Manual for the Bayley Scales of Infant Development*. New York: Psychological Corporation.

Bruininks, R. H. (1978). *Bruininks-Oseretsky Test of Motor Proficiency: Examiner's manual*. Circle Pines, MN: American Guidance Service.

Dewey, R. (1935). *Behavior development in infants: A survey of the literature on prenatal and postnatal activity, 1920–1934*. New York: Columbia University Press.

Gesell, A. (1954). The ontogenesis of infant behavior. In L. Carmichael (Ed.), *Manual of child psychology* (2nd edition). New York: John Wiley & Sons.

Keogh, J. K. (1977). The study of movement skill development. *Quest*, *28*, 76–88.

Malina, R. M. & Mueller, W. H. (1981). Genetic and environmental influences on the strength and motor performance of Philadelphia school children. *Human Biology*, *53*, 163–179.

Payne, V. G. & Isaacs, L. D. (1991). *Human motor development: A life span approach* (2nd edition). Mountainview, CA: Mayfield.

Plomin, R. & DeFries, J. C. (1985). *Origins of individual differences in infancy: The Colorado Adoption Project*. Orlando, FL: Academic Press.

Plomin, R., DeFries, J. C., & Fulker, D. W. (1988). *Nature and nurture during infancy and early childhood*. Cambridge, England: Cambridge University Press.

Terman, L. M. & Merrill, M. A. (1973) *Stanford-Binet Intelligence Scale: 1972 norms edition*. Boston: Houghton Mifflin.

Wechsler, D. (1974). *Manual for the Wechsler Intelligence Scale for Children–Revised*. New York: Psychological Corporation.

Wilson, R. S. and Harpring, E. B. (1972). Mental and motor development in infant twins. *Developmental Psychology*, *7*, 277–287.

Chapter 14 Sex Differences in Genetic and Environmental Influences for Cognitive Abilities

Baker, L. A. (1986). Genotype–environment covariance for multiple phenotypes: A multivariate test using adopted and nonadopted children. *Multivariate Behavioral Research*, *24*, 415–430.

Baker, L. A. Moffitt, T. E., Mack, W., & Mednick, S. A. (1989). Sex differences in property criminality in a Danish adoption cohort. *Behavior Genetics*, *19*, 355–370.

Bayley, N. & Schaefer, E. S. (1964). Correlations of maternal and child behaviors with the development of mental abilities: Data from the Berkeley Growth Study. *Monographs of the Society for Research in Child Development*, *29* (6, Serial No. 97).

Bee, J. L., Mitchell, S. K., Barnard, K. E., Eyres, S. J., & Hammond, M. A. (1984). Predicting intellectual outcomes: Sex differences in response to early environmental stimulation. *Sex Roles, 10*, 783–801.

Bradley, R. H. & Caldwell, B. M. (1980). The relation of home environment, cognitive competence, and IQ among males and females. *Child Development, 51*, 1140–1148.

Caldwell, B. M. & Bradley, R. H. (1978). *Home Observation for Measurement of the Environment*. Little Rock: University of Arkansas.

Carey, G. (1986). A general multivariate approach to linear modeling in human genetics. *American Journal of Human Genetics, 39*, 775–786.

Cliff, N. (1987). *Analyzing multivariate data*. New York: Harcourt, Brace, Jovanovich, Inc.

DeFries, J. C., Plomin, R., Vandenberg, S. G., & Kuse, A. R. (1981). Parent–offspring resemblance for cognitive abilities in the Colorado Adoption Project: Biological, adoptive, and control parents and one-year-old children. *Intelligence, 5*, 245–277.

Eaves, L. J., Eysenck, H. J., & Martin, N. G. (1989). *Genes, culture and personality: An empirical approach*. London: Academic Press.

Eaves, L. J., Last, K., Young, P. A., Martin, N. G. (1978). Model-fitting approaches to the analysis of human behaviour. *Heredity, 41*, 249–320.

Elardo, R., Bradley, R., & Caldwell, B. M. (1977). A longitudinal study of the relation of infants' home environments to language development at age three. *Child Development, 48*, 595–603.

Fulker, D. W. (1988). Genetic and cultural transmission in human behavior. In B. S. Weir, E. J. Eisen, M. Goodman, & G. Namkoong (Eds.), *Proceedings of the Second International Conference on Quantitative Genetics* (pp. 318–340). Massachusetts: Sinauer Associates, Inc.

Fulker, D. W. & DeFries, J. C. (1983). Genetic and environmental transmission in the Colorado Adoption Project: Path analysis. *British Journal of Mathematical and Statistical Psychology, 36*, 175–188.

Heath, A. C., Berg, K., Eaves, L. J., Solaas, M. H., Corey, L. A., Sundet, J., Magnus, P., & Nance, W. E. (1985). Educational policy and the heritability of educational attainment. *Nature, 314*, 734–736.

Ho, H-Z. (1987). Interaction of early caregiving environment and infant developmental status in predicting subsequent cognitive performance. *British Journal of Developmental Psychology, 5*, 183–191.

Jacklin, C. N. (1992). *The psychology of gender, volume I*. New York: New York University Press.

Lange, K., Westlake, & Spence, M. A. (1976). Extensions to pedigree analysis: III. Variance components by the scoring method. *Annals of Human Genetics, London, 39*, 485–491.

Moore, T. (1968). Language and intelligence: A longitudinal study of the first eight years: II. Environmental correlates of mental growth. *Human Development, 11*, 1–24.

Neale, M. C. (1991). *Mx: Statistical modeling*. Unpublished operator's manual.

Plomin, R. & DeFries, J. C. (1985). *Origins of individual differences in infancy: The Colorado Adoption Project*, Orlando, FL: Academic Press.

Plomin, R., DeFries, J. C., & Fulker, D. W. (1988). *Nature and nurture during infancy and early childhood*. Cambridge, England: Cambridge University Press.

Rice, T., Fulker, D. W., & DeFries, J. C. (1986). Multivariate path analysis of specific cognitive abilities in the Colorado Adoption Project. *Behavior Genetics, 16*, 107–126.

Singer, S., Corley, R., Guiffrida, C., & Plomin, R. (1984). The development and validation of a test battery to measure differentiated cognitive abilities in three-year-old children. *Educational and Psychological Measurement, 49*, 703–13.

Wachs, T. D. (1979). Proximal experience and early cognitive-intellectual development: The physical environment. *Merrill-Palmer Quarterly, 25*, 3–41.

Chapter 15 Nonshared Environment in Middle Childhood

Achenbach, T. M. (1991a). *Manual for the Child Behavior Checklist/4–18 and the 1991 Profile*. Burlington: Department of Psychiatry, University of Vermont.

Achenbach, T. M. (1991b). *Manual for the Teacher's Report Form and 1991 Profile*. Burlington: Department of Psychiatry, University of Vermont.

Anderson, K. E., Lytton, H., & Romney, D. M. (1986). Mothers' interactions with normal and conduct-disordered boys: Who affects whom. *Developmental Psychology, 22,* 604–609.

Baker, L. A. & Daniels, D. (1990). Nonshared environmental influences and personality differences in adult twins. *Journal of Personality and Social Psychology, 58,* 103–110.

Beardsall, L. & Dunn, J. (1992). Adversities in childhood: Siblings' experiences, and their relations to self esteem. *Journal of Child Psychology and Psychiatry, 33,* 349–359.

Boer, F. (1990). *Sibling relationships in middle childhood*. Leiden, Netherlands: DSWO Press, University of Leiden.

Brody, G. H., Stoneman, Z., & Burke, M. (1987). Child temperaments, maternal differential behavior, and sibling relationships. *Developmental Psychology, 23,* 354–362.

Brody, G. H., Stonemen, Z., & McCoy, J. K. (1992). Associations of maternal and paternal direct and differential behavior with sibling relationships: Contemporaneous and longitudinal analyses. *Child Development, 63,* 82–92.

Bryant, B. K. & Crockenberg, S. B. (1980). Correlates and dimensions of prosocial behavior: A study of female siblings with their mothers. *Child Development, 51,* 529–544.

Cardon, L. R., Fulker, D. W., DeFries, J. C., & Plomin, R. (1992). Continuity and change in general cognitive ability from 1 to 7 years of age. *Developmental Psychology, 28,* 64–73.

Daniels, D. (1986). Differential experiences of siblings in the same family as predictors of adolescent sibling personality differences. *Journal of Personality and Social Psychology, 51,* 339–346.

Daniels, D., Dunn, J., Furstenberg, F., & Plomin, R. (1985). Environmental differences within the family and adjustment differences within pairs of adolescent siblings. *Child Development, 56,* 764–774.

Daniels, D. & Plomin, R. (1985). Differential experiences of siblings in the same family. *Developmental Psychology, 21,* 747–760.

Dibble, E. & Cohen, D. (1974). Companion instruments for measuring children's competence and parental style. *Archives of General Psychiatry, 30,* 805–815.

Dunn, J. (1988). *The beginnings of social understanding*. Cambridge, MA: Harvard University Press.

Dunn, J. & Kendrick, C. (1982). *Siblings: Love, envy, and understanding*. Cambridge, MA: Harvard University Press.

Dunn, J. & McGuire, S. (1993). Young children's nonshared experiences: A summary of studies in Cambridge and Colorado. In E. M. Hetherington, D. Reiss, & R. Plomin (Eds.), *Separate social worlds of siblings: Importance of nonshared environment on development*. Hillsdale, NJ: Lawrence Erlbaum Associates.

Dunn, J. & Munn, P. (1985). Becoming a family member: Family conflict and the development of social understanding. *Child Development, 56,* 480–492.

Dunn, J. & Plomin, R. (1986). Determinants of maternal behavior toward three-year-old siblings. *British Journal of Developmental Psychology, 4,* 127–137.

Dunn, J. & Plomin, R. (1990). *Separate lives: Why siblings are so different*. New York: Basic Books.

Dunn, J., Plomin, R., & Daniels, D. (1986). Consistency and change in mothers' behavior towards young siblings. *Child Development, 57,* 348–356.

Dunn, J., Plomin, R., & Nettles, M. (1985). Consistency and change in mothers' behavior toward infant siblings. *Developmental Psychology, 21*, 1188–1195.

Dunn, J., Stocker, C., & Plomin, R. (1991). Nonshared experiences within the family: Correlates of behavior problems in middle childhood. *Development and Psychopathology, 2*, 113–126.

Ernst, L. & Angst, J. (1983). *Birth order: Its influence on personality*. Berlin: Springer-Verlag.

Harter, S. (1982). The perceived competence scale for children. *Child Development, 53*, 87–97.

Hetherington, E. M. (1988). Parents, children, and siblings six years after divorce. In R. Hinde & J. Stevenson-Hinde (Eds.), *Relationships within families: Mutual influences* (pp. 311–331). Oxford: Clarendon Press.

Hoffman, L. W. (1991). The influence of the family environment on personality: Accounting for sibling differences. *Psychological Bulletin, 110*, 187–203.

Koch, H. L. (1960). The relation of certain formal attributes of siblings to attitudes held toward each other and toward their parents. *Monographs of the Society for Research in Child Development, 25* (4, Serial No. 78).

Loehlin, J., Horn, J. M., & Willerman, L. (1989). Modeling IQ change: Evidence from the Texas Adoption Project. *Child Development, 60*, 993–1004.

Lytton, H. (1990). Child and parent effects in boys' conduct disorders: A reinterpretation. *Developmental Psychology, 26*, 683–697.

McGuire, S., Dunn, J., & Plomin, R. (1991). *Siblings' nonshared experiences with teachers and friends*. Paper presented at the Biennial Meeting of the Society for Research in Child Development, Seattle, WA.

McGuire, S. & McHale, S. M. (1993). Experiences with siblings during middle childhood: Connections within and between family relationships. Manuscript submitted for publication.

McHale, S. M. & Gamble, W. C. (1989). Sibling relationships of children with disabled and nondisabled brothers and sisters. *Developmental Psychology, 25*, 421–429.

McHale, S. M. & Pawletko, T. M. (1992). Differential treatment in two family contexts. *Child Development, 63*, 68–81.

Patterson, G. R. (1986). The contribution of siblings to training for fighting: A microsocial analysis. In D. Olweus, J. Block, & M. Radke-Yarrow (Eds.), *Development of antisocial and prosocial behavior: Research, theories and issues* (pp. 235–261). New York: Academic Press.

Plomin, R. (1986). *Development, genetics, and psychology*. Hillsdale, NJ: Lawrence Erlbaum Associates.

Plomin, R., Chipuer, H. M., & Neiderhiser, J. M. (1993). Behavioral genetic evidence for the importance of nonshared environment. In E. M. Hetherington, D. Reiss, & R. Plomin (Eds.), *Separate social worlds of siblings: Importance of nonshared environment on development*. Hillsdale, NJ: Lawrence Erlbaum Associates.

Plomin, R., Coon, H., Carey, G., DeFries, J. C., & Fulker, D. W. (1991). Parent–offspring and sibling adoption analyses of parental ratings of temperament in infancy and childhood. *Journal of Personality, 59*, 705–732.

Plomin, R. & Daniels, D. (1987). Why are children in the same family so different from each other? *Behavioral and Brain Sciences, 10*, 1–16.

Plomin, R., DeFries, J. C., & Fulker, D. W. (1988). *Nature and nurture during infancy and early childhood*. Cambridge, England: Cambridge University Press.

Rovine, M. (1993). Estimating non-shared environment using sibling discrepancy scores. In E. M. Hetherington, D. Reiss, & R. Plomin (Eds.), *Separate social worlds of siblings: Importance of nonshared environment on development*. Hillsdale, NJ: Lawrence Erlbaum Associates.

Rowe, D. & Plomin, R. (1981). The importance of nonshared (e_1) environmental influences in behavioral development. *Developmental Psychology, 17*, 517–531.

Scarr, S. (1987). Distinctive environments depend on genotypes. *Behavioral and Brain Sciences, 10*, 38–39.

Scarr, S. & Grajek, S. (1982). Similarities and differences among siblings. In M. E. Lamb & B. Sutton-Smith (Eds.), *Sibling relationships: Their nature and significance across the life span* (pp. 357–382). Hillsdale, NJ: Lawrence Erlbaum Associates.

Scarr, S. & Weinberg, R. A. (1977). Intellectual similarity within families of both adopted and biological children. *Intelligence, 1*, 170–191.

Stocker, C., Dunn, J., & Plomin, R. (1989). Sibling relationships: Links with child temperament, maternal behavior, and family structure. *Child Development, 60*, 715–727.

Stocker, C. & McHale, S. (1992). The nature and family correlates of preadolescents' perceptions of their sibling relationships. *Journal of Social and Personal Relationships, 9*, 179–195.

Chapter 16 Sibling Relationships in Childhood and Adolescence

Bowlby, J. (1969). *Attachment and loss: (Vol. 1). Attachment.* New York: Basic Books.

Bretherton, I. (1985). Attachment theory: Retrospect and Prospect. In I. Bretherton & E. Waters (Eds.), *Growing points of attachment theory and research* (pp. 3–35). Monographs of the Society for Research in Child Development, *50* (1–2, Serial No. 209).

Brody, G., Stoneman, Z., & Burke, M. (1987). Child temperaments, maternal differential behavior, and sibling relationships. *Developmental Psychology, 23*, 354–362.

Bryant, B. K. & Crockenberg, S. B. (1980). Correlates and dimensions of prosocial behavior: A study of female siblings with their mothers. *Child Development, 51*, 529–544.

Buhrmester, D. (1992). The developmental course of sibling and peer relationships. In F. Boer & J. Dunn (Eds.), *Children's sibling relationships: Developmental and clinical Issues* (pp. 19–40). Hillsdale, NJ: Lawrence Erlbaum Associates.

Buhrmester, D. & Furman, W. (1990). Perceptions of sibling relationships in middle childhood and adolescence. *Child Development, 61*, 1387–1398.

Buss, A. H. & Plomin, R. (1975). *A temperament theory of personality development.* New York: Wiley-Interscience.

Coddington, R. D. (1972). The significance of life events as etiologic factors in the diseases of children: II. A study of a normal population. *Journal of Psychosomatic Research, 16*, 205–213.

Dunn, J. (1983). Sibling relationships in early childhood. *Child Development, 54*, 787–811.

Dunn, J. (1988). Relations among relationships. In S. Duck (Ed.), *Handbook of personal relationships* (pp. 193–210). New York: John Wiley & Sons.

Dunn, J. & Kendrick, C. (1982). *Siblings: Love, envy, and understanding.* Cambridge, MA: Harvard University Press.

Dunn, J. & Plomin, R. (1990). *Separate lives: Why siblings are so different.* New York: Basic Books.

Hetherington, E. M. (1988). Parents, children, and siblings six years after divorce. In R. Hinde & J. Stevenson-Hinde (Eds)., *Relationships within families: Mutual inflences* (pp. 311–331). Oxford Clarendon Press.

Howe, N. & Ross, H. S. (1990). Socialization, perspective-taking, and the sibling relationship. *Developmental Psychology, 26*, 160–165.

McGuire, S. & Dunn, J. (1993). *The sibling relationship and life events during middle childhood: Individual and family characteristics as moderating variables.* Manuscript submitted for publication.

McHale, S. M. & Pawletko, T. M. (1992). Differential treatment of siblings in two family contexts. *Child Development, 63*, 68–81.

Patterson, G. R. (1986). The contribution of siblings to training for fighting: A microsocial analysis. In D. Olweus, J. Block, & M. Radke-Yarrow (Eds.), *Development of antisocial and prosocial behavior: Research, theories and issues* (pp. 235–261). New York: Academic Press.

Plomin, R., McClearn, G. E., Pederson, N. L., Nesselroade, J. R., & Bergeman, C. S. (1988). Genetic influence on childhood family environment perceived retrospectively from the last half of the lifespan. *Developmental Psychology, 24*, 738–745.

Rende, R. D., Slomkowski, C. L., Stocker, C., Fulker, D. W., & Plomin, R. (1992). Genetic and environmental influences in maternal and sibling interaction in middle childhood: A sibling adoption study. *Developmental Psychology, 28*, 484–490.

Rowe, D. C. (1983). A biometrical analysis of perceptions of family environment: A study of twin and singleton sibling kinships. *Child Development, 54*, 416–423.

Smith, M. S. (1987). Evolution and developmental psychology: Toward a sociobiology of human development. In C. B. Crawford, M. S. Smith, & D. Krebs (Eds.), *Sociobiology and psychology* (pp. 225–252). Hillsdale, NJ: Lawrence Erlbaum Associates.

Stillwell, R. & Dunn, J. (1985). Continuities in sibling relationships: Patterns of aggression and friendliness. *Journal of Child Psychology and Psychiatry, 26*, 627–637.

Stocker, C., Dunn, J., & Plomin, R. (1989). Sibling relationships: Links with child temperament, maternal behavior, and family structure. *Child Development, 60*, 715–727.

Stocker, C. & McHale, S. (1992). The nature and family correlates of preadolescents' perceptions of their sibling relationships. *Journal of Social and Personal Relationships, 9*, 179–195.

Teti, D. M. & Ablard, K. E. (1989). Security of attachment and infant–sibling relationships: A laboratory study. *Child Development, 60*, 1519–1528.

Vandell, D. L., Minnett, A. M., & Santrock, J. W. (1987). Age differences in sibling relationships during middle childhood. *Journal of Applied Developmental Psychology, 8*, 247–257.

Chapter 17   Genetic Influence on "Environmental" Measures

Ainsworth, M. D. S., Blehar, M., Waters, E., & Wall, S. (1978). *Patterns of attachment.* Hillsdale, NJ: Lawrence Erlbaum Associates.

Baumrind, D. (1971). Current patterns of parental authority. *Developmental Psychology Monographs, 4* (1, Pt. 2).

Bayley, N. (1969). *Manual for the Bayley Scales of Infant Development.* New York: Psychological Corporation.

Bell, R. Q. (1968). A reinterpretation of the direction of effects in socialization. *Psychological Review, 75*, 81–95.

Belsky, J. (1984). The determinants of parenting: A process model. *Child Development, 55*, 83–96.

Bergeman, C. S. & Plomin, R. (1988). Parental mediators of the genetic relationship between home environment and infant mental development. *British Journal of Developmental Psychology, 6*, 11–19.

Braungart, J. M., Fulker, D. W., & Plomin, R. (1992). Genetic mediation of the home environment during infancy: A sibling adoption study of the HOME. *Developmental Psychology, 28*, 1048–1055.

Buss, A. H. & Plomin, R. (1984). *Temperament: Early developing personality traits.* Hillsdale, NJ: Lawrence Erlbaum Associates.

Caldwell, B. M. & Bradley, R. H. (1978). *Home Observation for Measurement of the Environment.* Little Rock: University of Arkansas.

Chipuer, H. M., Merriwether-DeVries, C., & Plomin, R. (1992). *A genetic analysis of perceptions of the family environment: Nonadopted and adopted 7-year-olds and their parents.* Manuscript submitted for publication.

Dibble, E. & Cohen, D. J. (1974). Companion instruments for measuring children's competence and parental style. *Archives of General Psychiatry, 30,* 805–815.

Dunn, J. & Plomin, R. (1986). Determinants of maternal behavior toward infant siblings. *British Journal of Developmental Psychology, 26,* 459–468.

Dunn, J., Plomin, R., & Daniels, D. (1986). Consistency and change in mothers' behavior towards young siblings. *Child Development, 57,* 348–356.

Emery, R. E. (1982). Interparental conflict and the children of discord and divorce. *Psychological Bulletin, 92*(2), 310–330.

Jöreskog, K. G. & Sörbom, D. (1985). *LISREL VI user's guide.* Morrisville, IN: Scientific Software.

Lytton, J. (1977). Do parents create or respond to differences in twins? *Developmental Psychology, 13,* 456–459.

Lytton, J. (1980). *Parent–child interaction: The socialization process observed in twin and singleton families.* New York: Plenum Press.

Moos, R. H. (1976). *The human context: Environmental determinants of behavior.* New York: Wiley.

Moos, R. H. & Moos, B. S. (1981). *Family Environment Scale manual.* Palo Alto, CA: Consulting Psychologists Press.

Plomin, R. (in press). *Genetics and experience.* Newbury Park, CA: Sage Publications.

Plomin, R. & Bergeman, C. (1991). The nature of nurture: Genetic influence on "environmental" measures. *Behavioral and Brain Sciences, 14,* 373–427.

Plomin, R. & DeFries, J. C. (1985). *Origins of individual differences in infancy: The Colorado Adoption Project.* Orlando, FL: Academic Press.

Plomin, R., McClearn, G. E., Pedersen, N. L., Nesselroade, J. R., & Bergeman, C. S. (1988). Genetic influence on childhood family environment perceived retrospectively from the last half of the lifespan. *Developmental Psychology, 24,* 738–745.

Rowe, D. C. (1981). Environmental and genetic influences on dimensions of perceived parenting: A twin study. *Developmental Psychology, 17,* 203–208.

Rowe, D. C. (1983). A biometrical analysis of perceptions of family environment: A study of twins and singleton sibling kinships. *Child Development, 54,* 416–423.

Rowe, D. C. & Plomin, R. (1977). Temperament in early childhood. *Journal of Personality Assessment, 41,* 150–156.

Wachs, T. D. (1992). *The nature of nurture.* Newbury Park, CA: Sage Publications.

Wachs, T. D. & Gruen, G. (1982). *Early experience and human development.* New York: Plenum Press.

Chapter 18   Family Environment in Early Childhood and Outcomes in Middle Childhood: Genetic Mediation

Achenbach, T. M. & Edelbrock, C. S. (1983). *Manual for the Child Behavior Checklist and Revised Child Behavior Profile.* Burlington: Department of Psychiatry, University of Vermont.

Achenbach, T. M. & Edelbrock, C. S. (1986). *Manual for the Teacher's Report Form and Teacher Version of the Child Behavior Profile.* Burlington: Department of Psychiatry, University of Vermont.

Braungart, J. M., Fulker, D. W., & Plomin, R. (1992). Genetic mediation of the home environment during infancy: A sibling adoption study of the HOME. *Developmental Psychology, 28,* 1048–1055.

Braungart, J. M. & Rende, R. D. (1991). *Genetic mediation of the longitudinal associations between family environment and childhood behavior problems.* Paper presented at the 21st annual meeting of the Behavior Genetics Association, June 6–9, 1991, St. Louis, MO.

Caldwell, B. M. & Bradley, R. H. (1978). *Home Observation for Measurement of the Environment.* Little Rock: University of Arkansas.

Coon, H., Fulker, D. W., DeFries, J. C., & Plomin, R. (1990). Home environment and cognitive ability of 7-year-old children in the Colorado Adoption Project: Genetic and environmental etiologies. *Developmental Psychology, 26,* 459–468.

Edelbrock, C., Rende, R., Plomin, R., Thompson, L. A. (in press). Genetic and environmental effects on competence and problem behavior in childhood and early adolescence. *Journal of Child Psychology and Psychiatry.*

Harter, S. (1982). The perceived competence scale for children. *Child Development, 53,* 87–97.

McGee, R., Silva, P. A., & Williams, S. (1984). Perinatal, neurological, environmental and developmental characteristics of seven-year-old children with stable behavior problems. *Journal of Child Psychology and Psychiatry, 25,* 573–586.

Moos, R. H. & Moos, B. S. (1981). *Family Environment Scale manual.* Palo Alto, CA: Consulting Psychologists Press.

Plomin, R. (in press). *Genetics and experience.* Newbury Park, CA: Sage Publications.

Plomin, R. & DeFries, J. C. (1985). *Origins of individual differences in infancy: The Colorado Adoption Project.* Orlando, FL: Academic Press.

Plomin, R., DeFries, J. C., & Fulker, D. W. (1988). *Nature and nurture during infancy and early childhood.* Cambridge, England: Cambridge University Press.

Plomin, R., Loehlin, J. C., & DeFries, J. C. (1985). Genetic and environmental components of "environmental" influences. *Developmental Psychology, 21,* 391–402.

Plomin, R. & Neiderhiser, J. M. (1992). Genetics and experience. *Current Direction in Psychological Science, 1,* 160–163.

Rice, T., Fulker, D. W., DeFries, J. C., & Plomin, R. (1988). Path analysis of IQ during infancy and early childhood and an index of the home environment in the Colorado Adoption Project. *Intelligence, 12,* 27–45.

Thompson, L. A., Fulker, D. W., DeFries, J. C., & Plomin, R. (1986). Multivariate analysis of "environmental" influences on infant cognitive development. *British Journal of Developmental Psychology, 4,* 347–353.

Wachs, T. D. (1992). *The nature of nurture.* Newbury Park, CA: Sage Publications.

Walker, H. M. & McConnell, S. W. (1988). *Walker-McConnell Scale of Social Competence and School Adjustment.* Austin, TX: Pro-Ed.

Chapter 19   Home Environmental Influences on General Cognitive Ability

Caldwell, B. M. & Bradley, R. H. (1978). *Home Observation for Measurement of the Environment.* Little Rock: University of Arkansas.

Coon, H., Fulker, D. W., DeFries, J. C. & Plomin, R. (1990). Home environment and cognitive ability of 7-year-old children in the Colorado Adoption Project: Genetic and environmental etiologies. *Developmental Psychology, 26,* 459–468.

DeFries, J. C., Plomin, R., Vandenberg, S. G., & Kuse, A. R. (1981). Parent–offspring resemblance for cognitive abilities in the Colorado Adoption Project: Biological, adoptive, and control parents and one-year-old children. *Intelligence, 5,* 245–277.

Moos, R. H. & Moos, B. S. (1981). *Family Environment Scale manual*. Palo Alto, CA: Consulting Psychologists Press.

Plomin, R. & DeFries, J. C. (1985). *Origins of individual differences in infancy: The Colorado Adoption Project*. Orlando, FL: Academic Press.

Plomin, R., Loehlin, J. C., & DeFries, J. C. (1985). Genetic and environmental components of "environmental" influences. *Developmental Psychology, 21*, 391–402.

Rice, T., Fulker, D. W., DeFries, J. C., & Plomin, R. (1988). Path analysis of IQ during infancy and early childhood and an index of the home environment in the Colorado Adoption Project. *Intelligence, 12*, 27–45.

Wachs, T. D. (1992). *The nature of nurture*. Newbury Park, CA: Sage Publications.

Chapter 20   Genotype–Environment Interaction and Correlation

Achenbach, T. M. & Edelbrock, C. (1983). *Manual for the Child Behavior Checklist and Revised Child Behavior Profile*. Burlington: Department of Psychiatry, University of Vermont.

Bell, G. (1990). The ecology and genetic fitness in *Chlamydomonas*: I. Genotype-by-environment interaction among pure strains. *Proceedings of the Royal Society of London. B, 240*, 295–321.

Bergeman, C. S., Plomin, R., DeFries, J. C., & Fulker, D. W. (1988). Path analysis of general and specific cognitive abilities in the Colorado Adoption Project: Early childhood. *Personality and Individual Differences, 9*, 391–395.

Buss, A. H. & Plomin, R. (1975). *A temperament theory of personality development*. New York: Wiley-Interscience.

Cattell, R. B., Eber, H., & Tatsuoka, M. M. (1970). *Handbook for the Sixteen Personality Factor Questionnaire (16PF)*. Champaign, IL: Institute for Personality and Ability Testing.

Eaves, L. (1976). The effect of cultural transmission on continuous variation. *Heredity, 36*, 205–214.

Fulker, D. W., DeFries, J. C., & Plomin, R. (1988). Genetic influence on general mental ability increases between infancy and middle childhood. *Nature, 336*, 767–769.

Harter, S. (1985). *Manual for the Self-Perception Scale for Children*. Denver, CO: Department of Psychology, University of Denver.

Lerner, R. M. & Busch-Rossnagel, N. A. (1981). Individuals as producers of their development: Conceptual and empirical bases. In R. M. Lerner & N. A. Busch-Rossnagel (Eds.), *Individuals as producers of their development: A life-span perspective* (pp. 1–36). New York: Academic Press.

Loehlin, J. C. & DeFries, J. C. (1987). Genotype–environment correlation and IQ. *Behavior Genetics, 17*, 263–277.

McCall, R. B. (1991). So many interactions, so little evidence. Why? In T. D. Wachs & R. Plomin (Eds.), *Conceptualization and measurement of organism–environment interaction* (pp. 142–161). Washington, DC: APA Books.

Moos, R. H. & Moos, B. S. (1981). *Family Environment Scale manual*. Palo Alto, CA: Consulting Psychologists Press.

Plomin, R., Coon, H., Carey, G., DeFries, J. C., & Fulker, D. W. (1991). Parent–offspring and sibling adoption analyses of parental ratings of temperament in infancy and childhood. *Journal of Personality, 59*, 705–732.

Plomin, R. & DeFries, J. C. (1985). *Origins of individual differences in infancy: The Colorado Adoption Project*. Orlando, FL: Academic Press.

Plomin, R., DeFries, J. C., & Fulker, D. W. (1988). *Nature and nurture during infancy and early childhood.* Cambridge, England: Cambridge University Press.

Plomin, R., DeFries, J. C., & Loehlin, J. C. (1977). Genotype–environment interaction and correlation in the analysis of human behavior. *Psychological Bulletin, 84,* 309–322.

Plomin, R. & Hershberger, S. (1991). Genotype–environment interaction. In T. D. Wachs & R. Plomin (Eds.), *Conceptualization and measurement of organism–environment interaction* (pp. 29–43). Washington, DC: APA Books.

Rowe, D. C. & Plomin, R. (1977). Temperament in early childhood. *Journal of Personality Assessment, 41,* 150–156.

Rutter, M. (1983). Statistical and personal interactions: Facets and perspectives. In D. Magnusson & V. L. Allen (Eds.), *Human development: An interactional perspective* (pp. 295–319). New York: Academic Press.

Scarr, S. & McCartney, K. (1983). How people make their own environments: A theory of genotype–environment effects. *Child Development, 54,* 424–435.

Wahlsten, D. (1990). Insensitivity of the analysis of variance to heredity–environment interaction. *Behavioral and Brain Sciences, 13,* 109–161.

Wechsler, D. (1974). *Manual for the Wechsler Intelligence Scale for Children–Revised.* New York: Psychological Corporation.

Chapter 21   Applied Issues

Bohman, M. & Sigvardsson, S. (1978). An 18-year prospective, longitudinal study of adopted boys. In E. J. Anthony, C. Koupernik, & C. Chiland (Eds.), *The child in his family: Vol. 4. Vulnerable children.* New York: Wiley.

Bohman, M. & Sigvardsson, S. (1985). A prospective longitudinal study of adoption. In A. R. Nicol (Ed.), *Longitudinal studies in child psychology and psychiatry.* New York: Wiley.

Bohman, M. & Sigvardsson, S. (1990). Outcome in adoption: Lessons from longitudinal studies. In D. M. Brodzinsky & M. D. Schecter (Eds.), *The psychology of adoption* (pp. 93–106). New York: Oxford University Press.

Brodzinsky, D. M. (1987). Adjustment to adoption: A psychosocial perspective. *Clinical Psychology Review, 7,* 25–47.

Brodzinsky, D. M. (1990). A stress and coping model of adoption adjustment. In D. M. Brodzinsky & M. D. Schechter (Eds.), *The psychology of adoption* (pp. 3–24). New York: Oxford University Press.

Brodzinsky, D. M., Schechter, D. E., Braff, A. M., & Singer, L. M. (1984). Psychological and academic adjustment in adopted children. *Journal of Consulting and Clinical Psychology, 52,* 582–590.

Cattell, R. B., Eber, H. W., & Tatsuoka, M. M. (1970). *Handbook for the Sixteen Personality Factor Questionnaire (16PF).* Champaign, IL: Institute for Personality and Ability Testing.

Coddington, R. D. (1972). The significance of life events as etiologic factors in the diseases of children: I. A survey of professional workers. *Journal of Psychosomatic Research, 16,* 7–18.

Coon, H., Carey, G., Corley, R. P., & Fulker, D. W. (1992). Identifying children in the Colorado Adoption Project at risk for Conduct Disorder. *Journal of the American Academy of Child and Adolescent Psychiatry, 31,* 503–511.

Deutsch, C. K., Swanson, J. M., Bruell, J. H., Cantwell, D. P., Weinberg, F., & Baren, M. (1982). Overrepresentation of adoptees in children with the attention deficit disorder. *Behavior Genetics, 12,* 231–238.

Dibble, E. & Cohen, D. J. (1974). Companion instruments for measuring children's competence and parental style. *Archives of General Psychiatry, 30*, 805–815.

DiGiulio, J. F. (1988). Self-acceptance: A factor in the adoption process. *Child Welfare, 67*, 423–429.

Grotevant, H. D. & McRoy, R. G. (1990). Adopted adolescents in residential treatment: The role of the family. In D. M. Brodzinsky & M. D. Schechter (Eds.), *The psychology of adoption* (pp. 167–186). New York: Oxford University Press.

Hardy-Brown, K., Plomin, R., Greenhalgh, J., & Jax, K. (1980). Selective placement of adopted children: Prevalence and effects. *Journal of Child Psychology and Psychiatry, 21*, 143–152.

Harter, S. (1982). The perceived competence scale for children. *Child Development, 53*, 87–97.

Hersov, L. (1985). Adoption and fostering. In M. Rutter & L. Hersov (Eds.), *Child and adolescent psychiatry: Modern approaches* (2nd edition, pp. 101–117). Oxford: Blackwell.

Hersov, L. (1990). The Seventh Jack Tizard Memorial Lecture: Aspects of adoption. *Journal of Child Psychology and Psychiatry, 31*, 493–510.

Ho, H–Z., Plomin, R., & DeFries, J. C. (1979). Selective placement in adoption. *Social Biology, 26*, 1–6.

Holmes, D. L., Reich, J. N. & Pasternak, J. F. (1984) *The development of infants born at risk.* Hillsdale, NJ: Lawrence Erlbaum Associates.

Kirk, H. D. (1964). *Shared fate.* New York: Free Press.

Kotsopoulos, S., Cote, A., Joseph, L., Pentland, N., Stavrakaki, C., Sheahan, P., & Oke, L. (1988). Psychiatric disorders in adopted children: A controlled study. *American Journal of Orthopsychiatry, 58*, 608–612.

Lindholm, B. W. & Touliatos, J. (1980). Psychological adjustment of adopted and non-adopted children. *Psychological Reports, 46*, 307–310.

Moos, R. H. & Moos, B. S. (1981). *Family Environment Scale manual.* Palo Alto, CA: Consulting Psychologists Press.

Plomin, R. & DeFries, J. C. (1985). *Origins of individual differences in infancy: The Colorado Adoption Project.* Orlando, FL: Academic Press.

Schechter, D. M. & Bertocci, D. (1990). The meaning of the search. In D. M. Brodzinsky & M. D. Schechter (Eds.), *The psychology of adoption* (pp. 62–90). New York: Oxford University Press.

Sharma, A. R. & Benson, P. L. (1992). *A comparison of adopted and nonadopted adolescents on psychological at-risk indicators.* Paper presented at the Biennial Meeting of the Society for Research on Adolescents, Washington, DC, March 1992.

Wechsler, D. (1974). *Manual for the Wechsler Intelligence Scale for Children–Revised.* New York: Psychological Corporation.

Wechsler, D. (1981). *Manual for the Wechsler Adult Intelligence Scale–Revised.* New York: Psychological Corporation.

Wierzbicki, M. (1993). *Psychological adjustment of adoptees: A meta-analysis.* Paper presented at the Sixty-fifth Annual Meeting of the Midwestern Psychological Association, April 29, 1993, Chicago.

Chapter 22   Conclusions

Achenbach, T. M. & Edelbrock, C. (1983). *Manual for the Child Behavior Check list and Revised Child Behavior Profile.* Burlington: Department of Psychiatry, University of Vermont.

Aldhous, P. (1992). The promise and pitfalls of molecular genetics. *Science, 257*, 164–165.

Bayley, N. (1969). *Manual for the Bayley Scales of Infant Development.* New York: Psychological Corporation.

Boerwinkle, E., Chakraborty, R., & Sing, C. F. (1986). The use of measured genotype information in the analysis of quantitative phenotypes in man. *Annals of Human Genetics, 50,* 181–194.

Botstein, D., White, R. L., Skolnick, M., & Davis, R. W. (1980). Construction of a genetic map in man using restriction fragment length polymorphisms. *American Journal of Human Genetics, 32,* 314–331.

Cardon, L. R., Fulker, D. W., DeFries, J. C., & Plomin, R. (1992). Continuity and change in general cognitive ability from 1 to 7 years of age. *Developmental Psychology, 28,* 64–73.

Chipuer, H. M., Plomin, R., Reiss, D., & Hetherington, E. M. (in press). A comparison of two behavioral genetic models: Twin/sibling design versus stepfamily design. *Behavior Genetics.*

Chipuer, H. M., Rovine, M., & Plomin, R. (1990). LISREL modelling: Genetic and environmental influences on IQ revisited. *Intelligence, 14,* 11–29.

DeFries, J. C. & Fulker, D. W. (1985). Multiple regression analysis of twin data. *Behavior Genetics, 15,* 467–473.

DeFries, J. C. & Fulker, D. W. (1988). Multiple regression analysis of twin data: Etiology of deviant scores versus individual differences. *Acta Geneticae Medicae et Gemellologiae, 37,* 205–216.

DeFries, J. C. & Hegmann, J. P. (1970). Genetic analysis of open-field behavior. In G. Lindzey & D. C. Thiessen (Eds.), *Contributions to behavior-genetic analysis: The mouse as a prototype* (pp. 23–56). New York: Appleton-Century-Crofts.

DeFries, J. C., Plomin, R., & LaBuda, M. C. (1987). Genetic stability of cognitive development from childhood to adulthood. *Developmental Psychology, 23,* 4–12.

Edwards, J. H. (1991). The formal problems of linkage. In P. McGuffin & R. Murray (Eds.), *The new genetics of mental illness* (pp. 58–70). London: Butterworth-Heinemann.

Fulker, D. W., Cardon, L. R., DeFries, J. C., Kimberling, W. J., Pennington, B. F., & Smith, S. D. (1991). Multiple regression analysis of sib-pair data on reading to detect quantitative trait loci. *Reading and Writing: An Interdisciplinary Journal, 3,* 299–313.

Fulker, D. W., Cherny, S. S., & Cardon, L. R. (1993). Continuity and change in cognitive development. In R. Plomin & G. E. McClearn (Eds.), *Nature, nurture, and psychology,* (pp. 77–97), Washington, DC: APA Books.

Fulker, D. W., DeFries, J. C., & Plomin, R. (1988). Genetic influence on general mental ability increases between infancy and middle childhood. *Nature, 336,* 767–769.

Gusella, J. F., Wexler, N. S., Conneally, P. M., Naylor, S. L., Anderson, M. A., Tanzi, R. E., Watkins, P. C., Ottina, K., Wallace, M. R., Sakaguchi, A. Y., Young, A. B., Shoulson, I., Bonilla, E., & Martin, J. B. (1983). A polymorphic DNA marker genetically linked in Huntington's disease. *Nature, 306,* 234–238.

Hoffman, L. (1991). The influence of the family environment on personality: Accounting for sibling differences. *Psychological Bulletin, 110,* 187–203.

McGuffin, P. & Murray, R. (1992). *The new genetics of mental illness.* London: Butterworth-Heinemann.

Phillips, K. & Fulker, D. W. (1989). Quantitative genetic analysis of longitudinal trends in adoption designs with application to IQ in the Colorado Adoption Project. *Behavior Genetics, 19,* 621–658.

Plomin, R. (1990). The role of inheritance in behavior. *Science, 248,* 183–188.

Plomin, R. (1993). Nature and nurture: Perspective and prospective. In R. Plomin & G. E. McClearn (Eds.), *Nature, nurture, and psychology,* (pp. 457–483). Washington, DC: APA Books.

Plomin, R. (in press). *Genetics and experience*. Newbury Park, CA: Sage Publications.

Plomin, R. & Bergeman, C. S. (1991). The nature of nurture: Genetic influence on "environmental" measures. *Behavior and Brain Sciences, 14*, 373–427.

Plomin, R., Emde, R. N., Braungart, J. M., Campos, J., Corley, R., Fulker, D. W., Kagan, J., Reznick, J. S., Robinson, J., Zahn-Waxler, C., & DeFries, J. C. (1993). Genetic change and continuity from 14 to 20 months: The MacArthur Longitudinal Twin Study. *Child Development, 64*, 1354–1376.

Plomin, R., Loehlin, J. C., & DeFries, J. C. (1985). Genetic and environmental components of "environmental" influences. *Developmental Psychology, 21*, 391–402.

Reiss, D. (1993). Genes and the environment: Siblings and synthesis. In R. Plomin & G. E. McClearn (Eds.), *Nature, nurture, and psychology*, (pp. 415–430). Washington, DC: APA Books.

Rutter, M. (1993). *Discussant for symposium on genetic and environmental influences on development*. Paper presented at the Sixtieth anniversary meeting of the Society for Research in Child Development, New Orleans, LA, March 25–28, 1993.

Sobell, J. L., Heston L. L., & Sommer, S. S. (1992). Delineation of genetic predisposition to multifactorial disease: A general approach on the threshold of feasibility. *Genomics, 12*, 1–6.

Watson, J. D. (1990). The human genome project: Past, present, and future. *Science, 248*, 44–49.

Wilson, R. S. (1983). The Louisville Twin Study: Developmental synchronies in behavior. *Child Development, 54*, 298–316.

Subject Index

Name Index